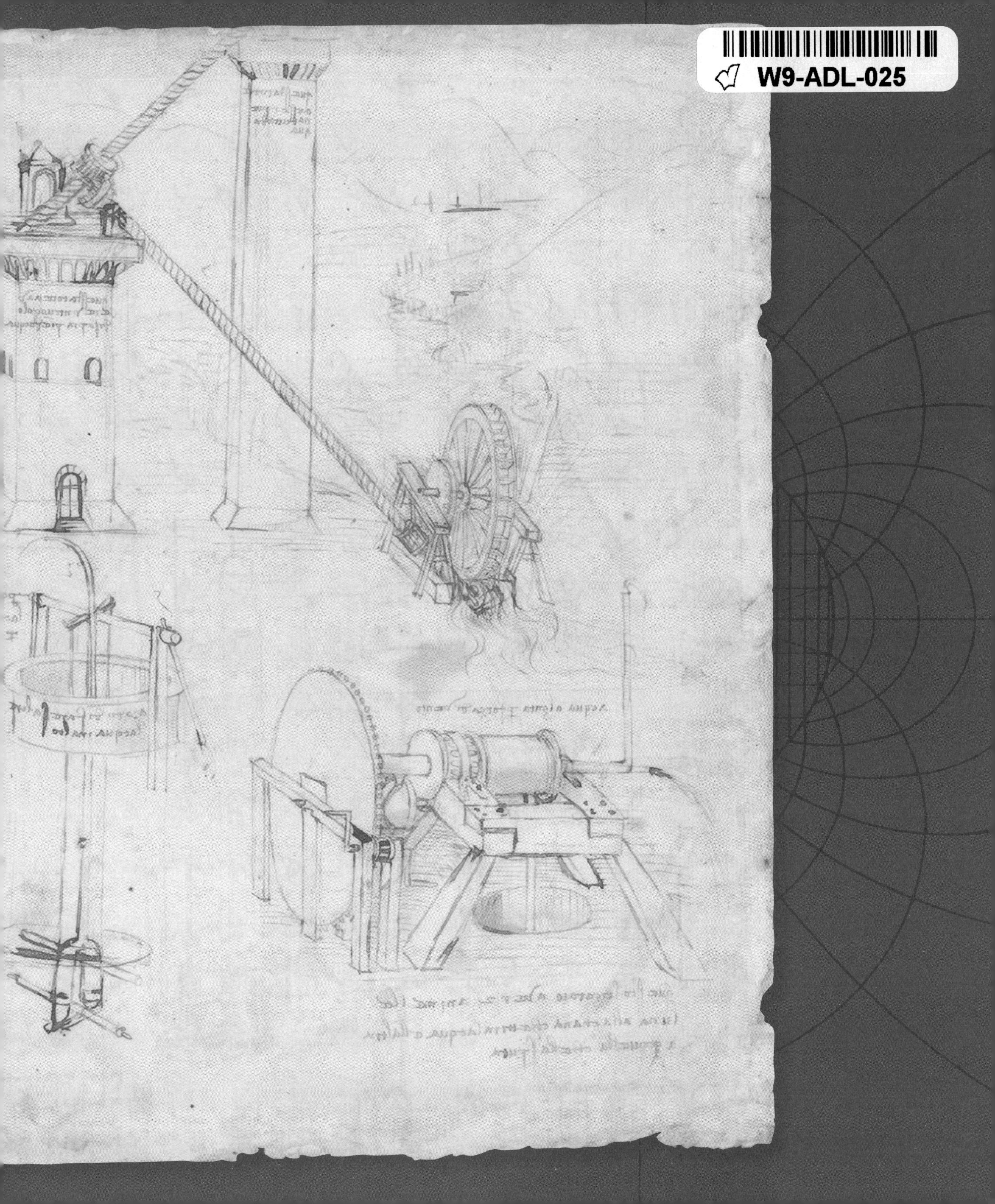
W9-ADL-025

SCIENTIFICA
HISTORICA

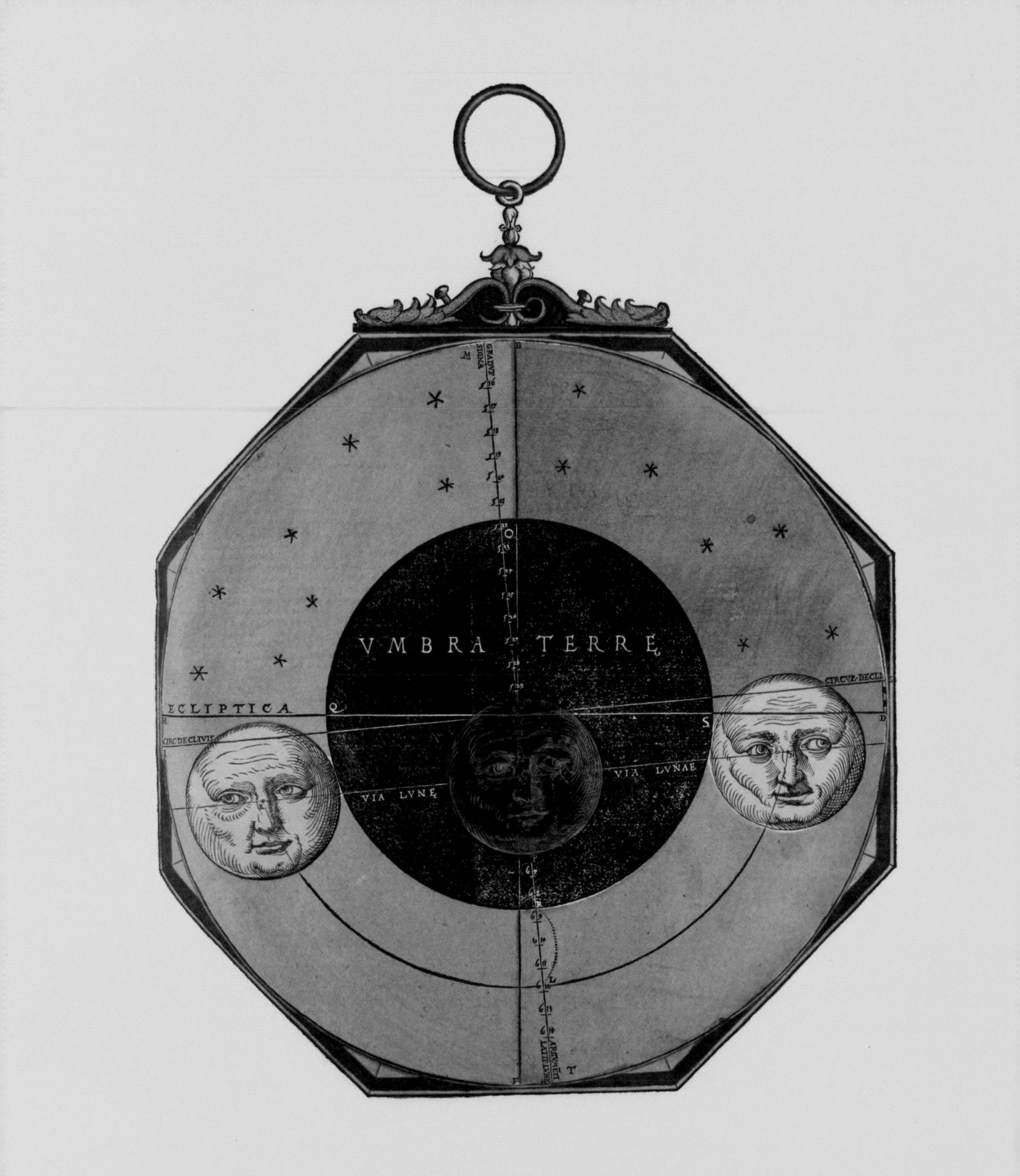

VMBRA TERRE
ECLIPTICA
VIA LVNE
VIA LVNAE

SCIENTIFICA HISTORICA

How the world's great science books chart the history of knowledge

BRIAN CLEGG

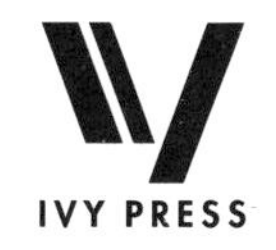

First published in the UK and North America in 2019 by
Ivy Press
An imprint of The Quarto Group
The Old Brewery, 6 Blundell Street
London N7 9BH, United Kingdom
T (0)20 7700 6700
www.QuartoKnows.com

British Library Cataloguing-in-Publication Data
A catalogue record for this book is available from the British Library

ISBN: 978-1-78240-878-9

This book was conceived, designed and produced by
Ivy Press
58 West Street, Brighton BN1 2RA, UK

PUBLISHER David Breuer
EDITORIAL DIRECTOR Tom Kitch
ART DIRECTOR James Lawrence
COMMISSIONING EDITOR Kate Shanahan
PROJECT EDITOR Elizabeth Clinton
DESIGN MANAGER Anna Stevens
DESIGNER Michael Whitehead
COMMISSIONED PHOTOGRAPHY Neal Grundy
EDITORIAL ASSISTANT Niamh Jones

Printed in China

10 9 8 7 6 5 4 3 2 1

Cover image: Science Photo Library/Paul D Stewart

CONTENTS

INTRODUCTION

The Latin word *scientifica* describes something that produces knowledge; from this broad scope, 'science' has come to describe our understanding of the universe and the objects in it. One invention has been central to the development of science. It's not an incredibly complex piece of hardware like the Large Hadron Collider, nor a sophisticated concept like Einstein's general theory of relativity, but something far more familiar. Without this technology, we would be left with little more than folk tales and mysteries, because the invention is writing.

The importance of writing gives us the *historica* of the title, which defines something based on research or producing an account – the fundamental requirement for science to benefit from the written word. In conceptual terms, writing is the technology that frees up communication from the limits of time and space, destroying the shackles of the here and now.

Most animals and even some plants communicate at some level, but usually that communication is immediate and local, after which it is gone forever. Writing transcends that limitation. I can take a book off the shelf and read words that were written thousands of miles away and hundreds, or even thousands, of years ago. There are probably more communications on my bookshelves from dead people than there are from the living – and certainly very few of the books I own were written by authors who live near to me. Writing takes care of time and space. And that is its significance in making science possible.

The power of writing for science is that books act as a storage medium for ideas and discoveries; we don't have to reinvent the wheel every time. Science can only work as it does by building on the discoveries and theories of others. Isaac Newton famously said (probably paraphrasing Robert Burton), 'If I have seen further it is by standing on the shoulders of Giants.' Newton's ability to make use of others' ideas was only possible thanks to the written word. And books have been central to the spread of science in this manner ever since humanity began to look for rational explanations of what they observed around them over 2,500 years ago.

The role of books in transcending time and space is illustrated well in the complex web of written works that ties together the ancient Greek world, Islamic scientists of the latter part of the first millennium and medieval European scientists. The ancient Greeks wrote many books on scientific topics following the revolutionary ideas of Thales of Miletus, who seems to have been amongst the first to make the shift from mythological explanations of the natural world to ones that came closer to a scientific view, from around 600 BCE.

Many of the books of the ancient Greek period were lost as their civilisation fell and their libraries were ransacked. Just one example gives a poignant reminder of this. In a strange little book called *The Sand-Reckoner*, the remarkable third-century BCE

THALES OF MILETUS, FIFTEENTH CENTURY

A representation of the sixth-century BCE ancient Greek philosopher and mathematician Thales of Miletus.

mathematician and engineer Archimedes of Syracuse attempted to work out how many grains of sand it would take to fill the universe. (By 'universe' he had in mind roughly what we would think of as the solar system.) This was not quite as useless a task as it sounds. The Greek number system of the time was very limited. The largest named number was a myriad – 10,000 – which meant that the largest number usually considered was a myriad myriads, or 100 million. But Archimedes wanted to show that it was possible to go far beyond this limitation by devising a new type of number that could easily handle any required value. He demonstrated its flexibility by attempting the remarkable calculation with grains of sand.

The Sand-Reckoner has survived the ravages of time, but in it, Archimedes referred to another volume that otherwise we would not have known existed. To work out the number of sand grains required, Archimedes first used geometry to estimate the size of the universe. He based his calculation on the accepted astronomical model of the time, where the Earth was at the centre of the universe with everything orbiting around it. But he also noted:

Abū Ja'far Muḥammad ibn Mūsā al-Khwārizmī
AL-KITĀB AL-MUKHTAṢAR FĪ ḤISĀB AL-ĞABR WA'L-MUQĀBALA,
COPY, 1342

Covering algebra, calendars, inheritance and more, this was one of the principal mathematics textbooks from its first publication circa 820 CE to the sixteenth century.

Aristarchus of Samos brought out a book consisting of some hypotheses, in which the premises lead to the result that the universe is many times greater than that now so called. His hypotheses are that the fixed stars and the Sun remain unmoved, that the Earth revolves around the Sun in the circumference of a circle, the Sun lying in the middle of the orbit . . .

This lost book by Aristarchus, referenced only by Archimedes, is the first known suggestion of what would become the heliocentric Copernican theory. As is the case for so many other titles from the period, we will never know exactly what Aristarchus wrote.

The books of ancient Greece were largely forgotten in Europe after the fall of the Roman Empire, but as the interest in science grew in the flourishing Islamic world, surviving Greek titles were translated into Arabic and supplemented a growing body of new work, notably in mathematics, physics and medicine. A good example of the new life being brought into the books of the period was *Al-kitāb al-mukhtaṣar fī ḥisāb al-ğabr wa'l-muqābala* (The Compendious Book on Calculation by Completing and Balancing)

GREEK AND ARABIC PHILOSOPHY, FOURTEENTH CENTURY

The left-hand illustration depicts Hippocrates (ca. 460–ca. 377 BCE), Hunayn ibn Ishaq (808–73 CE), and Claudius Galenus (ca. 131–ca. 201 CE) discussing ideas; the illustration on the right shows an Arabic scribe working on a philosophical text.

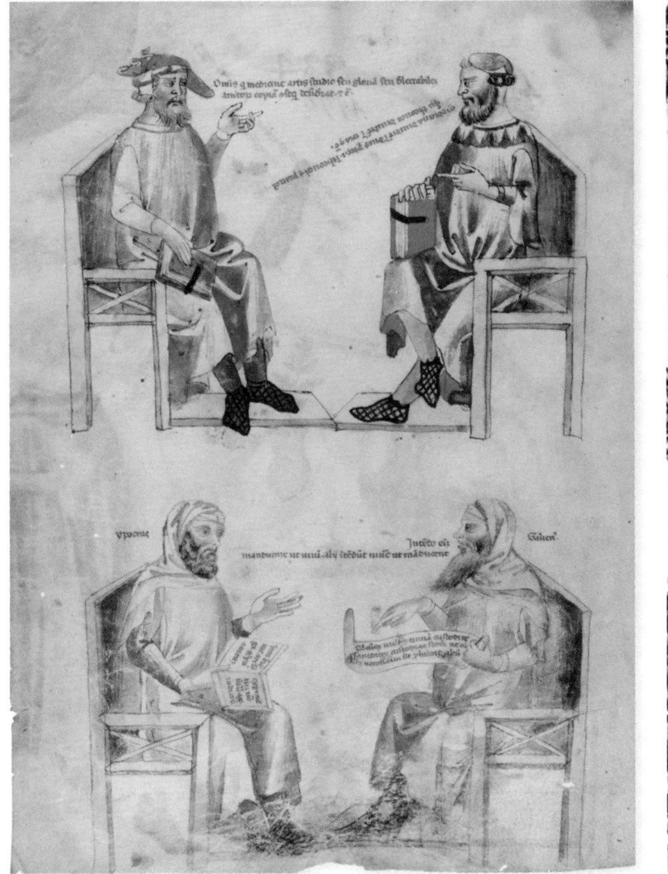

by Abū Ja'far Muḥammad ibn Mūsā al-Khwārizmī, born around 780 CE, possibly in Baghdad in modern-day Iraq. This title was not just influential in the Islamic world. Although some Greek works did start to filter directly back into European awareness in the thirteenth century, Arabic works were first translated a century earlier – both Arabic translations of Greek titles and the original work of scholars such as al-Khwārizmī. *Al-kitāb al-mukhtaṣar* led the way in introducing practical algebra to the West (the word 'algebra' comes from *al-ğabr* in the title). Al-Khwārizmī tells the reader that the book would be useful for 'inheritance, legacies, partition, lawsuits, and trade'.

So, thanks to the medium of the book, ideas from ancient Greece helped inspire the flourishing scientists, medics and mathematicians of the Arabic-speaking world, while translations of the Greek books and new titles by Islamic writers would kick-start a scientific revolution in Europe. These titles linked thinkers who were separated by centuries, languages, distance and culture. It was the books that tied everything together.

The written word

Of course, the physical mechanism used to convey written communication has changed several times from the earliest days of science books. Hippocrates' or Aristotle's notion of a book would be very different from a modern-day ebook on a Kindle. Greek books came in the form of scrolls – continuous sheets of writing material rolled up to form a cylinder. The Greeks inherited the format from ancient Egypt, where papyrus made from reeds would have been the standard medium, though later, parchment (treated animal skins) and paper were also used.

SEATED SCRIBE,
CA. 2500–2350 BCE

An ancient Egyptian scribe working on a papyrus in a statue from the fifth dynasty.

Although scrolls were reasonably practical for relatively small books (which is why ancient books, such as the books that make up the Bible, appear so short to modern eyes), they presented difficulties as a text got longer. A scroll could be a good number of metres long, which made it unwieldy and easy to get into a tangle. With more substantial scrolls there would often be spindles at each end, but managing these presented their own challenges. The reader had to unroll the scroll from one spindle and roll the far side onto another spindle. Depending on the orientation of the text, they would then either read continuously down the scroll like an autocue (which was particularly difficult on the wrists) or across it, with text printed in chunks like pages, in which case, there was a considerable delay in getting onto the next section of text. The format was particularly cumbersome if the aim was to find a particular section rather than read the book from beginning to end.

Although the Romans made surprisingly little contribution to science itself, they gave science books (and literature in general) a huge boost by devising the codex, introduced in the first century CE. This was what we now think of as a traditional book – sheaves of leaves, bound together, which could be flicked through to a particular location and read easily page by page. The codex was also significantly easier than a scroll to copy – a process that was essential to the role of the book in spreading the word of science. A whole industry of book copying sprung up, particularly within religious institutions, which made it possible for books to transmit scientific theories far beyond their initial

TWO SCRIBES, CA. 2400 BCE

Fifth dynasty relief from the Mastaba of Akheteps at Saqqara, the necropolis for the ancient Egyptian capital of Memphis.

WOMAN WITH BOOK, FIRST CENTURY CE

A portrait of a Roman woman known as Sappho holding a book and a stylus, painted on plaster at Pompeii.

sources. This first flowering of copying blossomed wildly with the invention of the printing press, transforming the written word from an extremely expensive vehicle for communication to the few into a mechanism by which science could reach the masses.

Printing presses per se date back as far as the codex, but early presses relied on carving the original words and images in reverse, typically into a wooden block, which would then be used to impress the ink onto the paper. The woodcut technique (and later the process of lithography, based on stones or metal plates marked with ink-resistant materials) would be used for illustrations until modern photographic techniques could be incorporated. However, woodcuts were slow to produce, making it impractical to print many whole books of any length. Nonetheless, from the ninth century, China was producing short scrolls using this method. The earliest known book printed this way was the Dunhuang *Diamond Sutra* from the year 868 CE. The Chinese would continue to print books from wooden blocks well after their development of moveable type.

Like many great ideas, moveable type was a simple one. Rather than trying to carve the whole of a book's page into a single block, a large quantity of individual characters on small blocks were produced, which could be bound together to form a page. The set page could be used until the print run was completed, then dismantled so that the individual blocks could be reused to create another page. Setting up the page (a process called typesetting) took a considerable amount of time – until mechanical typesetting devices were introduced in the nineteenth century – but this was only comparable to the time taken to painstakingly copy a few manuscript pages, after which as many copies as were required could be run off.

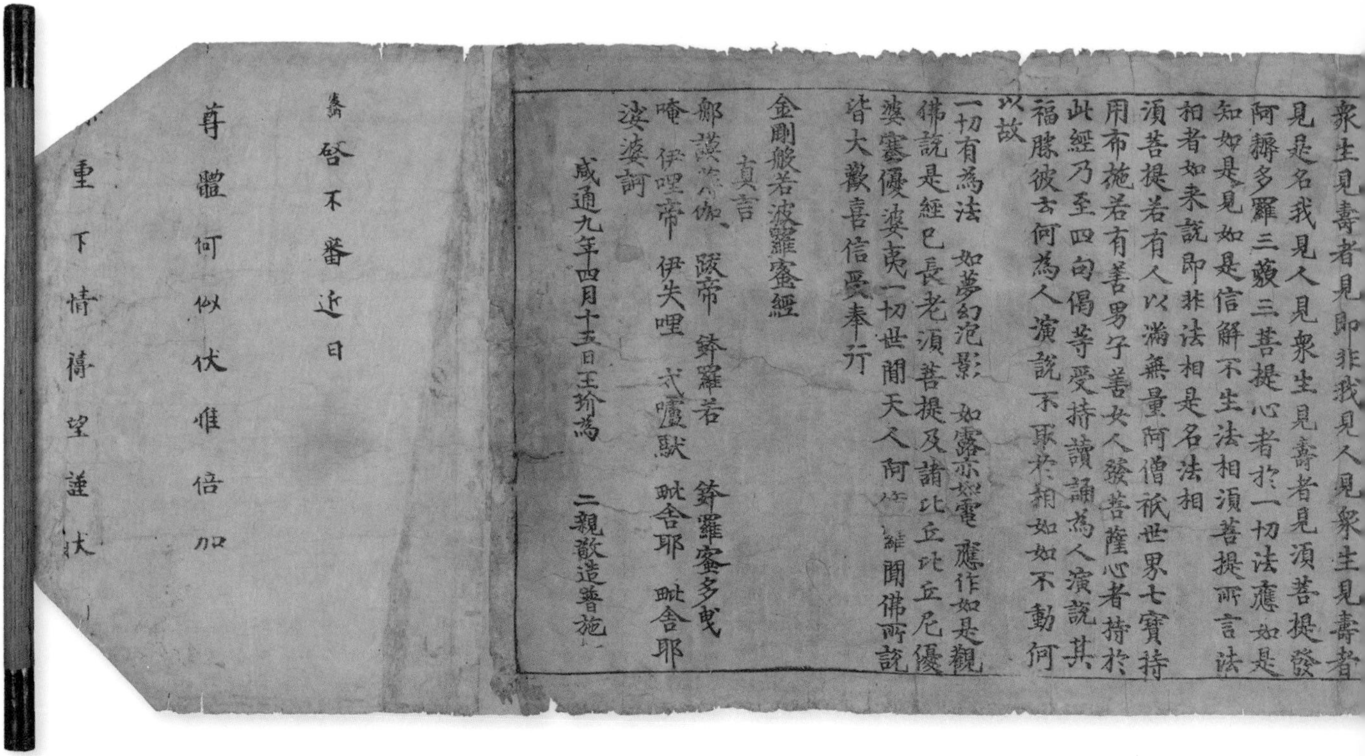

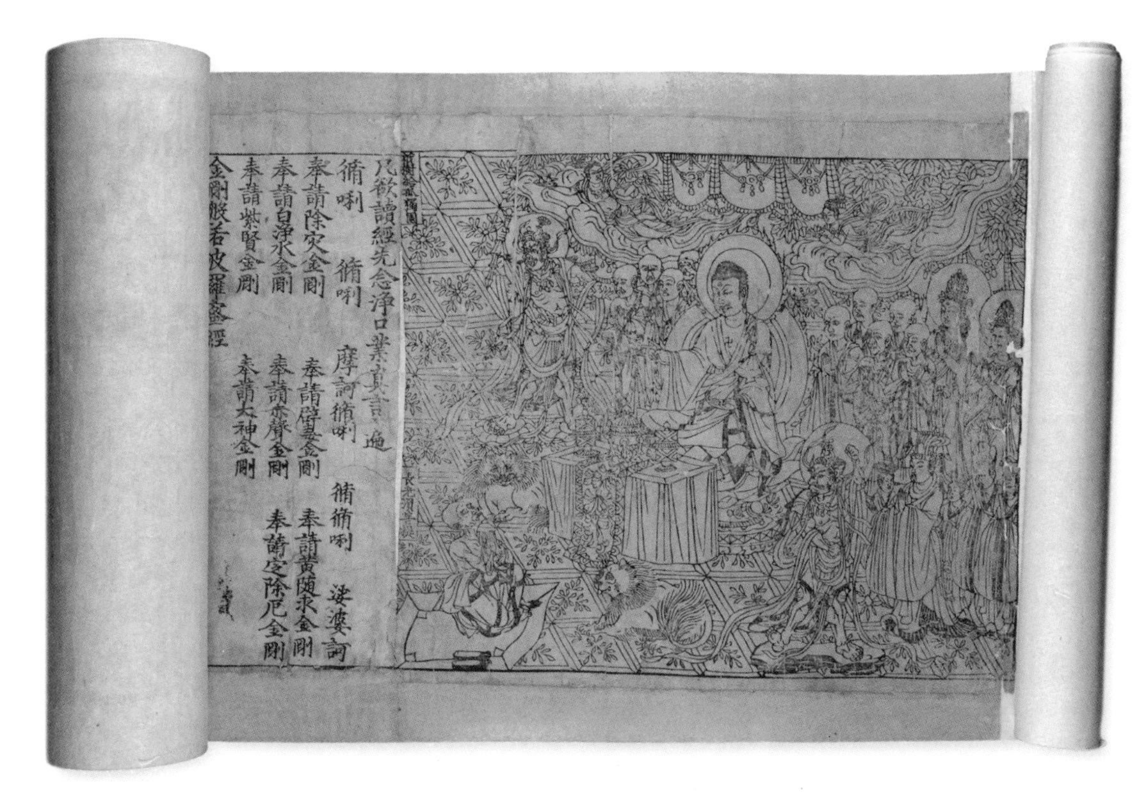

DIAMOND SUTRA, COPY, 868 CE

This Chinese copy of the Indian Buddhist *Diamond Sutra* (below) is the world's earliest dated, printed book. The scroll is made from seven panels and includes a frontispiece (left).

The early Chinese moveable-type blocks were made of ceramic or wood, first developed in the eleventh century, though the earliest known book printed using moveable type is *Notes of the Jade Hall* from 1193, which used fired clay characters. By the fourteenth century, metal, which has the advantage of being more durable, took over. Yet despite this early lead, moveable type would not prove hugely popular in China in the way it would rapidly become so in Europe when it was introduced in the fifteenth century. This seems to have been down to the economy of scale. Those using the Roman alphabet only had to produce around 50 varieties of type block (lower case and upper case), leaving aside special fonts for titles. But for the Chinese market, with characters running into the thousands, there was far less benefit to be gained from moveable type over carving a whole page as a block.

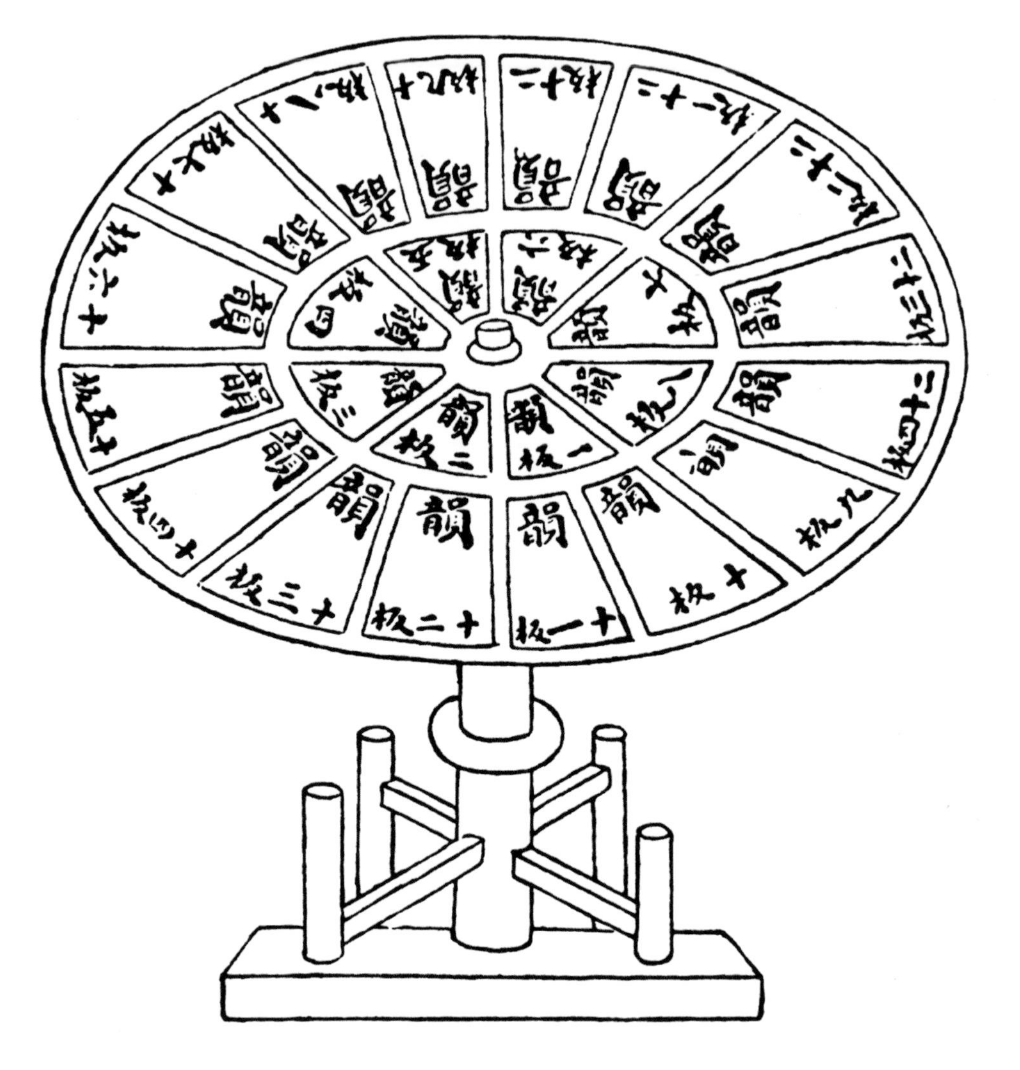

REVOLVING TYPECASE, 1313

A revolving table Chinese typecase with individual moveable type characters arranged primarily by rhyming scheme, from Wang Zhen's *Nong Shu* (Book of Agriculture).

TYPOGRAPHIA HARLEMI PRIMVM INVENTA
Circà Annum. 1440.
Currat penna licet, tantum vix scribitur anno,
Quantum uno reddunt præla Batava die:
Addidit inventis aliquid Germania tantis:
Hollandus cæpit. Theuto peregit opus.
Saenredam invent.
velde sculp.
P. Scriverius.

PRINTING PRESS, 1440

An illustration of Laurens Janszoon Coster's alleged invention of the printing press, the Dutch rival to German printing-press inventor Johannes Gutenberg.

WOODBLOCK OF *AMBROSIA ALTERA*, CA. 1562

Woodblock designed by Giorgio Liberale and cut by Wolfgang Meyerpeck for the illustrated editions of Pietro Andrea Mattioli's *Herbár* (1562), *New Kreüterbuch* (1563) and *Commentarii in sex libros Pedacii Dioscoridis Anarzabei de Medica materia* (1565).

From technical documents to mass communication

As we move through the different periods of scientific writing, the shift in the availability of texts was matched by a transformation in the nature of science books. Initially, books were the means for one natural philosopher (the predecessor to the term 'scientist', which was not introduced until the 1830s) to communicate with his or her peers. The standard language for writing such books in Europe was Latin. This common language was a way to enable information to pass easily from country to country, just as English is used today as the standard language for scientific papers. However, it was also a conscious mechanism to limit access to the information to the cognoscenti. It was the practice of medieval natural philosophers, such as the thirteenth-century English friar Roger Bacon, to keep scientific knowledge from the common herd. Bacon noted (quoting an earlier source) that 'it is stupid to offer lettuces to an ass since he is content with his thistles.'

By the seventeenth century, though, this attitude was changing. Galileo wrote his scientific masterpieces in Italian, rather than Latin, with the aim of reaching out to the public. Isaac Newton had intended the third volume of his crowning glory, *Philosophiae Naturalis Principia Mathematica*, to work for a wider audience, until a falling out with colleagues changed his mind. Other writers would specifically produce books that

Maurice Quentin de la Tour
MADAME DU CHÂTELET-LOMONT, OIL ON CANVAS, EIGHTEENTH CENTURY

Portrait of the French scientist and author Émilie du Châtelet.

simplified the heavyweight works of science for a wider audience. For example, the eighteenth-century French scientist and author Émilie du Châtelet – who wrote her own impressive review of contemporary science in *Institutions de Physique* (Lessons in Physics) – not only translated Newton's masterpiece, the *Principia*, into French, but offered a commentary to make it more approachable to the general reader.

With the establishment of scientific bodies, such as the Royal Society in London, founded in 1660, scientific journals began to be available for the more focused spreading of scientific ideas amongst experts. By the late nineteenth century, scientists were still writing some books for their peers (and textbooks would always be required for students), but these were gradually eclipsed by books intended for the general public. A good example of a title that straddled the two modes of scientific communication was Charles Lyell's *Principles of Geology* from the 1830s. Though relatively technical, the three-volume book with coloured illustrations was sufficiently approachable to bring the latest thinking in geology – with its profound implications that the Earth was much older than had been previously assumed – to a fascinated wider public. Similarly, when the great nineteenth-century Scottish physicist James Clerk Maxwell wrote his heavy-duty *Theory of Heat*, even as lowly a publication as *The Ironmonger* was able to recommend the book to its readers, noting that 'the language throughout is simple and the conclusions striking'.

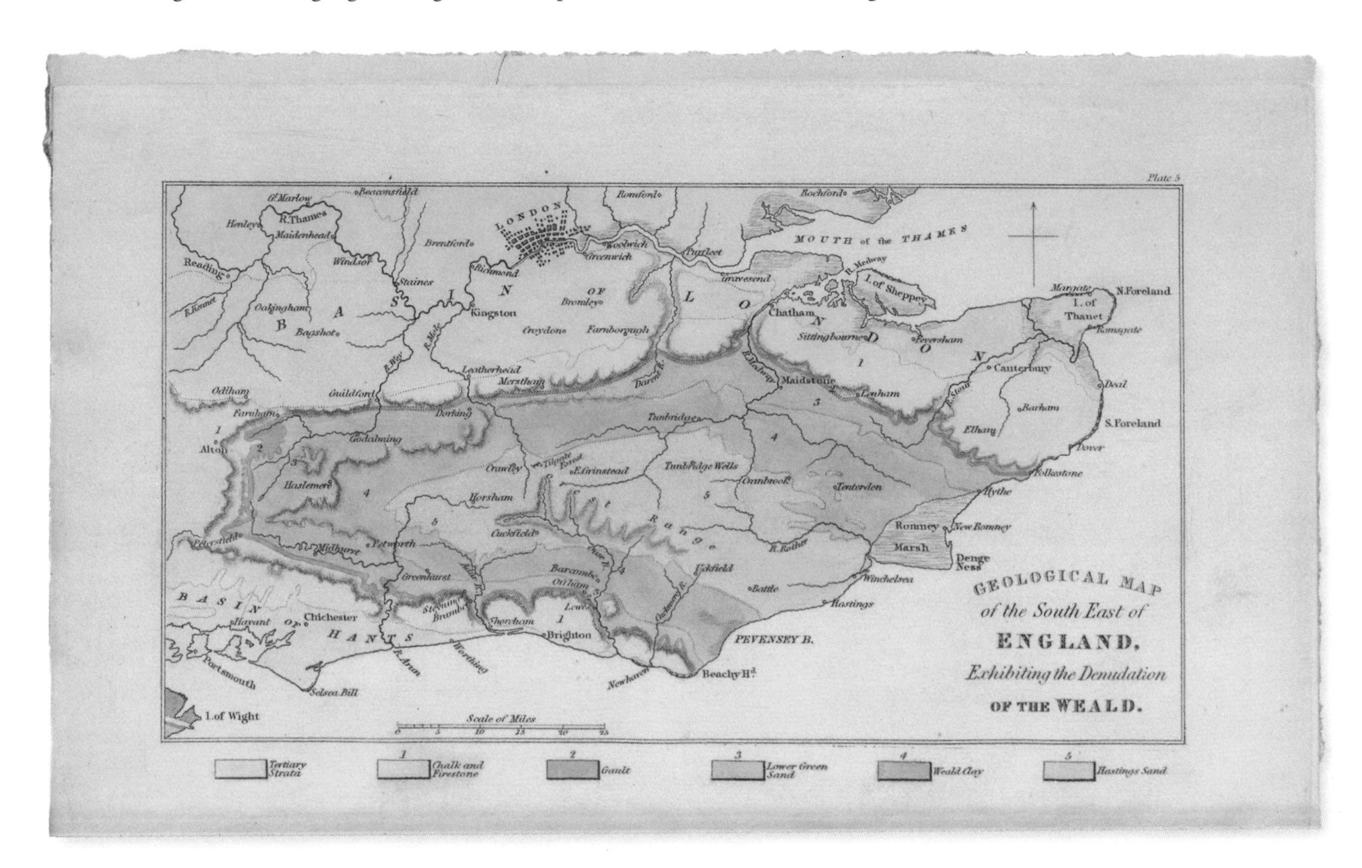

Charles Lyell
PRINCIPLES OF GEOLOGY,
JOHN MURRAY, 3 VOLS.,
1830–3

Geological map of South East England: one of the coloured plates from volume III.

In modern times, while the majority of scientist-to-scientist communication is in the form of emails, papers and press releases, the book remains a significant format for communication of science to the wider world. In that sense, science books have transformed from being a vehicle for insider communication between the cognoscenti to something for all of us to appreciate better just what science is doing and how it affects our lives.

A standard cover

The move to wider science communication has paralleled a change in the nature of book binding. Look at any pre-Victorian title and it is liable to have a dull leather or fabric binding with little more to make it stand out than some decoration on the spine. This is because many books right up to the late nineteenth century would be produced with no covers at all – the publisher would just provide the inner leaves, and these would be taken to a bookbinder, whose job was to produce a uniform cover for the book to match the rest of the owner's library.

As more people began to read, more books were published with cheap paper or board covers, making them instantly accessible. But it is notable that, for example, when the pioneer motion photographer Eadweard Muybridge travelled to America in the 1850s, his first source of income was transporting unbound books from the London Printing and Publishing Company, to sell for binding in the United States, starting in New York and then moving out to the rapidly growing city of San Francisco.

Stephen Hawking
A BRIEF HISTORY OF TIME,
BANTAM PRESS, 1988

Hawking's book, here in first edition, was hugely influential on the development of popular science: the introduction by Carl Sagan was later replaced by Hawking's own.

The idea of a standard, illustrated publisher's cover is a relatively modern addition to the science book. Even as recently as the first half of the twentieth century, most science books for the public were fairly dull in appearance. It was thought that any attempt to be too popularist was simply inappropriate for the material. In fact, many scientists who did write for the public were frowned upon by their peers, who considered this an unworthy role for a true scientist. It was only really in the 1960s that the covers of popular science books would start to match the significance of their contents, and the expectations of their audience.

Though by no means the first to achieve bestseller status, a standout example of a book that was bought by a wide range of the public was *A Brief History of Time* by Stephen Hawking, published in 1988. Famously described as the title most likely to have been bought without ever being read, Hawking's book appeared on many shelves where volumes of science would not normally be seen. Crucially, it made publishers realise that the public had an appetite for popular science titles. Since its publication, the genre has flourished, with hundreds of titles published each year.

The nature of science books has changed throughout the existence of the written word. But they remain an essential marker of the progress of science and of its relevance to our society. Science and the book have gone hand in hand in forging the future.

BOOKBINDERY WORKSHOP, EIGHTEENTH CENTURY

This engraving shows a workshop where existing books are given bindings to fit with the buyer's library.

BIRDSALLS BOOK BINDERS, 1888

Staff at work in the Northampton-based bookbinding company from the late nineteenth century.

About this book

In *Scientifica Historica* we will explore the history of science books, splitting roughly 2,500 years into five periods. The first chapter, *Ancient World*, lays the foundations, from the earliest scientific writing through to around 1200. In the second part of the book, *Renaissance in Print*, running through to the end of the eighteenth century, we see how the move from the hand-copied books of the ancient world to printed titles would transform the nature and availability of science writing. The third chapter, *Modern Classical*, covers the nineteenth century, when we see the role of science writing beginning to transform as science comes of age and journals take over as the primary vehicle for scientist-to-scientist communication, leaving science books to take on a broader audience.

The final two chapters, *Post-Classical* and *The Next Generation*, cover the twentieth and twenty-first centuries, which have seen a huge change in both science and the nature of science writing. *The Post-Classical* chapter reflects the shift in the way that science was undertaken, moving from a mostly amateur activity to a totally professional occupation, where the mathematical component of science grew exponentially, and scientific theories (sometime highly counter-intuitive) became more important than merely collecting information. The *Post-Classical* and *The Next Generation* eras are divided around 1980, when the modern, popular science book began to dominate. Up to this point, many of the key science books were written by leading scientists and tended to take a patronising approach to their audience. But with some notable exceptions, the later popular science titles have been written for a more discerning audience who expect a better quality of writing and accessibility.

Each of the five chapters covers a wide range of titles to illustrate the way that books have been used and how they have changed through the ages. Mostly these have been restricted to original pieces of writing. However, we have also explored the increasingly popular spin-off titles from television series which have indubitably become an important part of science writing in the final period covered.

Timeline of the development of the book

Mesopotamian clay tablet
ca. 4000 BCE

Egyptian scroll
ca. 2600 BCE

Many of these titles are from Europe and North America. This simply reflects the way that the history of modern science communication – and science itself – has developed. For a book to be significant in the history of science, from the Renaissance up to recent times, it will generally have come from these two continents. The reasons that science flourished particularly in Europe and North America from the fourteenth century onwards are disputed, but it seems to have been driven by a combination of growth of trading wealth, luck (particularly in terms of the origins of the Industrial Revolution) and a relative lack of religious suppression. In recent times, countries such as China and India have once more become big players in the sciences, but as yet this has not been reflected in science writing, perhaps in part because of the dominance of English as the universal language of science. This doesn't mean that there have not been very popular titles outside the ones covered in this book, but they have had less influence on our worldwide understanding of scientific matters. The majority of scientific papers are published in English for the same reason – although the great science communicators come from around the world.

Doom-mongers are always announcing the death of books, but science titles have maintained a healthy existence throughout each of our five periods and should do so into the future. Their nature may have changed, but they continue to be an unrivalled mechanism for linking humans to our universe. A television show or YouTube video can only ever skim the surface of a topic. A one-hour programme will typically not be able to cover the contents of a single chapter of a book. A science book enables the reader to take in so many different ways of understanding a topic, to be able to process the information at their own speed and to appreciate it in far more depth than pictures and speech alone can provide.

The science book has been a shining beacon of human progress since the invention of writing – and long may it continue to be so.

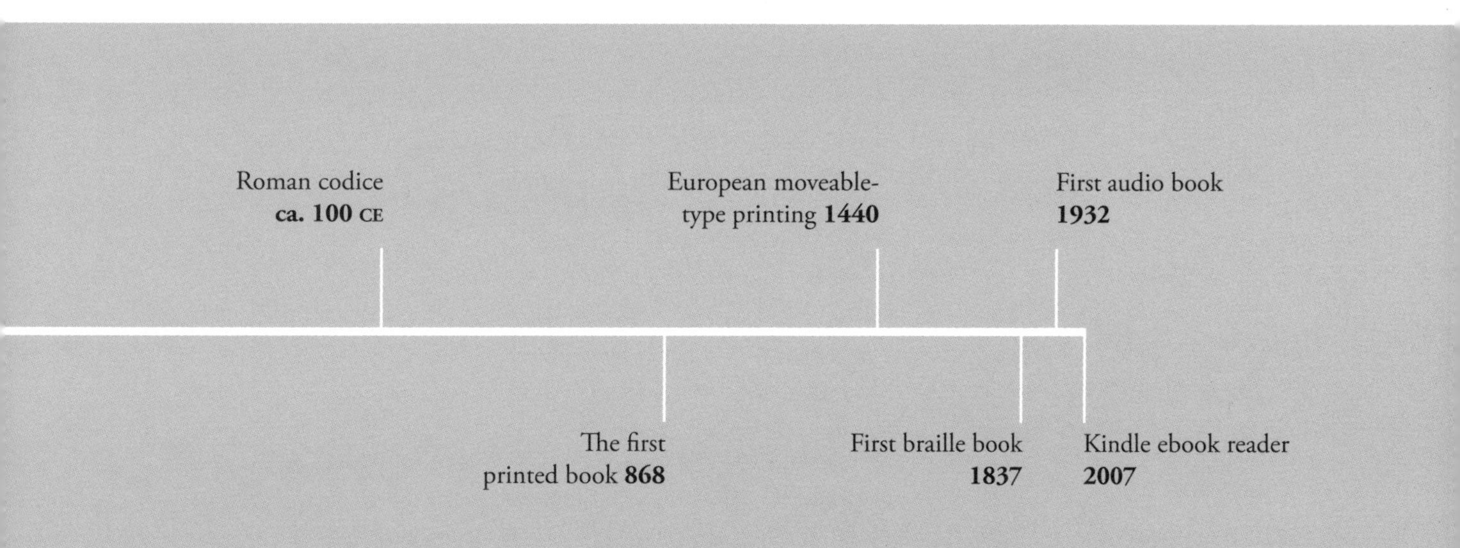

ANCIENT WORLD

LAYING THE FOUNDATIONS

FOR CENTURIES THERE have been debates over what makes humans unique among the animals. Biologists frequently insist that there is nothing special about the species *Homo sapiens*. The term 'exceptionalism' is used in biology circles in a derogatory fashion to describe the attempt to give us a special status. And, certainly, there are few human abilities that aren't duplicated in some fashion by other animals. However, *Homo sapiens* far exceeds other species in its collective capabilities to adapt its environment for life, and the driver for this ability seems to be creativity.

This remarkable trait was present when *Homo sapiens* first evolved, over 200,000 years ago. Creativity means that humans do not simply accept things as they are and live in the present, but can think outside the moment and ask questions such as 'Why does that happen?' or 'What if I did this?' or 'What could I do to make things different?'

When early humans looked beyond scratching an existence to the full might of nature – from the Sun and the stars to the devastating power of lightning and hurricanes – the first responses to the question 'Why does that happen?' involved deities or magic. The assumption was that there had to be supernatural forces, capable of actions that were forever beyond our understanding, even if they perhaps could be placated by human rituals. However, with the establishment of static gatherings of people in the early cities, there was an opportunity to begin to take what we would now consider a more scientific approach.

First came the use of numbers (although arguably a separate discipline to science, mathematics is so tightly tied to the sciences that we will be considering it an integral part of *Scientifica Historica*). More accurately, what seems to have come first was the tally, a mechanism for counting that did not require numbers. Say, for example, a neighbour borrowed some loaves of bread and you wanted to make sure that your loaves were all replaced. Without numbers, you could put a pebble in a safe place for each loaf the neighbour took. When they handed over a replacement loaf, you would throw away a pebble corresponding to it until there were no pebbles left.

We don't know for certain how long such systems were used as they leave no permanent record, but a number of ancient bones have been discovered that appear to have tally marks on them. The Ishango bone, which is over 20,000 years old, is a baboon's leg bone, found on what is now the border of Uganda and the Democratic Republic of the Congo. It has a series of notches carved into it, which are widely interpreted as being a tally. The even older Lebombo bone, dating back over 40,000 years, also has a series of notches, though there is more dispute about their nature.

Tally markers can preserve information remarkably well, as witnessed by the fact that these bones still exist so long after they were first created. Such bones can be considered the earliest ancestor of a written record. Of a similar age to the Lebombo bone are some of the early cave paintings, which provide another form of communication that had the potential to establish traditions across a period of time.

Keeping a long-term written record may not have had significance for the makers of the bone tallies, but as cities and trade grew, the need for accounting meant that records began to be kept. At the time, these may simply have been markers of financial transactions, however the ability to keep information to a later date, and to share it, would be crucial for the development of a scientific view of the world.

FOUR VIEWS OF THE ISHANGO BONE, CA. 20,000–18,000 BCE

The series of notches on this baboon leg bone are thought to be tally marks, housed in the Royal Belgian Institute of Natural Sciences.

From tallies to writing

Over the centuries, straightforward pictures and simple notch-based tallies developed into pictograms. As the name suggests, pictograms were image-based, but unlike cave paintings they were stylised into a standard form to represent individual concepts. Some modern Chinese characters still take this form – the character for 'door' for example, looks a little like a door.

With some thought, pictograms could also be used to convey less concrete notions. For example, a series of pictograms could be used to communicate the process of putting bread into a basket. If we see a loaf of bread, then a hand, then a loaf in a basket, the message is fairly clear. (For a modern example of a message using pictograms, think of an IKEA instruction sheet.) In that basic form there is no separate symbol to display the concept of 'in' or 'into', meaning that we need an awful lot of pictograms. There would need to be, for example, a different symbol for a loaf in a basket and for a dog in a basket. But it is not hard to imagine something like an arrow being used to indicate the relationship of 'in', after which we just need the pictograms for bread (or dog) and basket with that linking arrow image. A symbol such as this arrow is known as an ideogram, as it indicates something significantly more abstract than an object or an action.

It was this kind of gradual abstraction that led to the formation of proto-writing, the precursor to modern scripts, which seems to have developed at least 6,000 years ago. One early example appearing to carry such proto-writing is a set of tablets found in Tărtăria, located in modern-day Romania, in what was once Transylvania. These are clay tablets marked with a mix of pictograms, lines and symbols. As we don't know what they mean, it's possible they were purely decorative, but they are usually assumed to be a precursor to writing, putting across information in a more structured fashion than simple decoration.

TRADITIONAL CHINESE 'DOOR' CHARACTER

Although very stylised, the door character bears a resemblance to a traditional door with a transom.

In some ways similar, Egyptian hieroglyphs also combined pictograms and ideograms, but did so with more distinct structures. The symbols were not restricted to words, but could also form parts of words, making it possible to build compound words from a mix of symbols, requiring a stock of fewer distinct images. We tend to think of hieroglyphics as the standard script of ancient Egypt because it is what we see on ancient tombs and wall paintings, but in fact it was developed as a formal means of writing for special settings and was too complex for everyday use. Another system, hieratic, was developed in parallel and involved far fewer, more stylised symbols – similar to Chinese characters – which could be written more quickly than hieroglyphs.

However, the Egyptians were not to the first to develop a stylised writing system. Another of the ancient powers of the region, the Sumerian civilisation (which later developed into the Babylonians), devised their cuneiform script around 3600 BCE, making it the earliest known writing system. This script originally combined stylus marks representing numbers with a pictogram-based form of writing. A millennium later, it had become more stylised, with all characters made up of combinations of wedge-shaped marks produced on clay tablets using a stylus: this was the origin of the words 'style' and 'stylised'.

The European alphabet has a Greek name (alpha and beta are the first two letters of the Greek alphabet), but a more complex background. It seems to have originally derived from the proto-Canaanite abjad. An abjad is like an alphabet but without vowels, which are implied or shown by accent markers – both Arabic and Hebrew use modern abjads. The proto-Canaanite abjad was in use in parts of the Middle East from around 3,500 years ago. Used by the Phoenicians, it was the source of both Greek and Aramaic letters. Greek, though, appears to have been the first true alphabet, with vowels represented by separate characters, originating about 1000 BCE.

The alphabet used in most Western countries is often called Latin or Roman; our upper-case letters are pretty much the same as those used for carving inscriptions by the Romans – their equivalent of Egyptian hieroglyphs. (The character set is not identical, as the Romans didn't have separate letters J and U, using I and V, which were easier to carve.) Like the Egyptians with hieratic, the Romans also had an everyday set of characters, known as Roman cursive, which morphed into our lower-case letters. For the Romans these were two totally separate styles which would not be mixed, but after the fall of the Roman Empire various options of combining them were tried, such as using capitals to emphasise new sections of writing, or to pick out nouns (as is still the case in modern German).

When first introduced, though, these letters would not have been called upper case and lower case. This terminology dates from the moveable type printing era, when pages of type were set using individual metal letters, bound together to form a page (see page 14). The two kinds of character were kept in separate boxes, with the basic letters (technically referred to as minuscule) in a lower case and the fancier capital versions in a higher 'upper case'.

Why is the development of writing so important? Because without writing, it is hard to see how a scientific tradition could be built. Stories of the gods at work in the heavens or throwing lightning do not need precision. They benefit, if anything, from the

FRESCO FROM THE TOMB OF NEFERTARI, TWELFTH CENTURY BCE

The burial chamber of Nefertari in the Valley of the Queens, Luxor, features excerpts and scenes from several chapters of the *Book of the Dead*. Here, three genii guard the Second Gate of the Kingdom of Osiris.

GREEK STELE, 163/4 CE

Stelae were stone slabs with inscriptions often used as grave or boundary markers. This stele records the eponymous archon (chief magistrate), Philisteides, and the kosmetes (military trainer of young men), Claudius, at the top, beneath is a list of trainers and trainees, as well as a variety of festivals and events.

THE LYON TABLET, 48 CE

The Latin text reproduces emperor Caludius's speech in favour of some leading citizens from Gaul being allowed to sit on the Senate.

embellishment and modification that inevitably accompanies an oral tradition. As verbal stories are passed from person to person, less and less of the original remains. But for scientific ideas to be tested and built on, nothing else could match the unchanging foundation provided by the written word.

The permanence of clay

As we have seen, the earliest written records were not books, but pieces of clay. Working on a far greater scale than the Tărtăria tablets, the Sumerians and the later Babylonians of Mesopotamia produced vast quantities of clay tablets, originally for accounting purposes. These blocks of clay could be easily marked using the stylus-end 'cuneiform' markers which were first used to represent numbers, but soon also used in combination to form the stylised characters derived from earlier pictograms.

If the markings were just a temporary note, the clay could be moistened, wiped and reused – but by baking the clay tablet in an oven it became a permanent store of the information recorded on it. It would be an exaggeration to describe these tablets as scientific, but some did give guidance on, for example, practical mathematics. They did not contain mathematical proofs, but there were examples of Pythagorean triples – collections of numbers such as 3, 4, 5 and 8, 15, 17, which reflect the relationship of lengths of the sides of a right-angled triangle that would later be proved in Pythagoras's theorem.

ASSYRIAN CLAY TABLETS

An accounting tablet (left) from Anatolia, ca. twentieth – nineteenth century BCE, and an astronomical tablet (right) from Nineveh, whose date and origin is debated.

Unknown
EDWIN SMITH PAPYRUS, CA. 1600 BCE

The 4.7 metre (15 foot)-long scroll may have originated in Thebes and was bought by American archeologist Edwin Smith in 1862.

Remarkably, these numerical records date back around 3,800 years. Such tablets also began to be used to record what we would now think of as scientific data, specifically astronomical observations. This information provided the basis both for calendars and for astrological use – there is no evidence at this stage of the application of scientific theories – yet like the invention of writing itself, such collections of data were necessary precursors to the scientific approach.

Similar practical examples (rather than work that had a detailed theoretical basis) began to crop up in the Egyptian civilisation. Practical geometry was an essential for both the measurement of fields and the construction of buildings, again bringing in the guidance of Pythagorean triples. And medicine took the first steps in its long journey from magic to science. The oldest-known example of a written document giving medical guidance with some resemblance of a scientific approach – although not long enough to be considered a true book – is the Egyptian Edwin Smith papyrus, which is around 3,600 years old. It takes the form of a papyrus scroll around 4.7 metres (15 feet) in length, and deals primarily with injuries and surgical techniques, though it does also include a number of magic spells intended for medical purposes.

China was the next of the great civilisations to venture into proto-scientific fields, with mathematical documents dating back at least 3,000 years. It would be relatively late coming to physical or biological sciences, however, as there were philosophical barriers in the way of accepting a purely mechanistic view of the world. India, too, would produce impressive mathematical and later astronomical works from around 500 BCE, which would feed into the development of modern science.

However, the foundations of the approach that has come to dominate science worldwide were primarily developed in ancient Greece. The Greeks built on mathematical ideas from Babylonia and Egypt, but they would take the lead in attempting to build a rational explanation for nature that would eventually become science. They were also the earliest to produce what is close to the modern concept of a science book, though many of the early examples no longer survive.

The early Greeks

The Greek philosopher Thales of Miletus was one of the first to move away from ascribing the forces and structures of nature to the mythical actions of the gods, constructing instead a philosophy that built on theories of the interaction of natural objects. Thales was alive at the same time as the now better-known Pythagoras, who was born around 570 BCE and whose school put numbers at the centre of the explanation of the universe.

With many of the early Greek philosophers it is difficult to know exactly which ideas belonged to the big names that get remembered and which were produced by their followers. Using a famous name added weight to an argument (rather like having a celebrity endorse a product today), and it was common to deploy the big names in a piece of writing even if they weren't directly involved. We know that Pythagoras did not come up with the mathematical theorem named after him. As we have seen, Pythagorean triples predated him by 1,000 years, and proofs of the theorem were developed well before he was born. It is possible, though, that he was responsible for the first scientific theorising on the nature of music, showing how specific ratios of lengths of vibrating objects (strings or organ pipes, for example) produced notes that sounded harmonic and pleasant.

With the output of Thales and Pythagoras we have the problem that not a single piece of their writing has survived. Everything we know is hearsay. Among the earliest extant examples of what could be considered scientific books is the *Hippocratic Corpus*, a collection of disparate works on the subject of medicine, which includes the famous Hippocratic Oath requiring a physician to behave ethically with patients. Again, we don't know if the fifth-century BCE Greek physician Hippocrates of Kos wrote any of the 60 or so titles in this collection. Certainly, the majority of the volumes date back to his period and a little later, though the last was added as much as nine centuries afterwards.

Because of having multiple authors over a period of time, the *Corpus* is a mix of ideas with no consistency of viewpoint: some of its texts are aimed at other physicians, others at lay readers. If these books can be considered amongst the earliest of scientific titles, they very much take the form of a compendium presenting competing theories, rather than providing the reader with the scientific consensus of the time. There was no 'standard text' here. However, some ideas were better supported than others, notably that of the 'four humours' – blood, yellow

François Langlois (after Claude Vignon),
PYTHAGORAS,
SEVENTEENTH CENTURY

An etching of an imagined Pythagoras (ca. 570–495 BCE).

HIPPOCRATIS
COI MEDICORVM OMNIVM
longe Principis, octoginta Volumina, quibus
maxima ex parte, annorum circiter duo mil
lia Latina caruit lingua, Græci uero, Arabes,
& Priſci noſtri Medici, plurimis tamen utilibus
prætermiſſis, ſcripta ſua illuſtrarunt, nunc
tandem per. M. Fabium Caluum Rhauenna
tem uirum undecunq; doctiſſimum latinita-
te donata, CLEMENTI.VII. Pont.Max.
dicata, ac nunc primum in lucem ædita, quo
nihil humano generi ſalubrius fieri potuit.

Hippocrates
HIPPOCRATIC CORPUS,
FRANCISCUS MINUTIUS CALVUS, 1525

The title page of a 1525 edition of the *Corpus*, translated from Greek into Latin by Marcus Fabius Calvus.

Unknown
THE GUILD BOOK OF THE BARBER SURGEONS OF YORK, FIFTEENTH CENTURY

This text for surgeons shows the four medical humours.

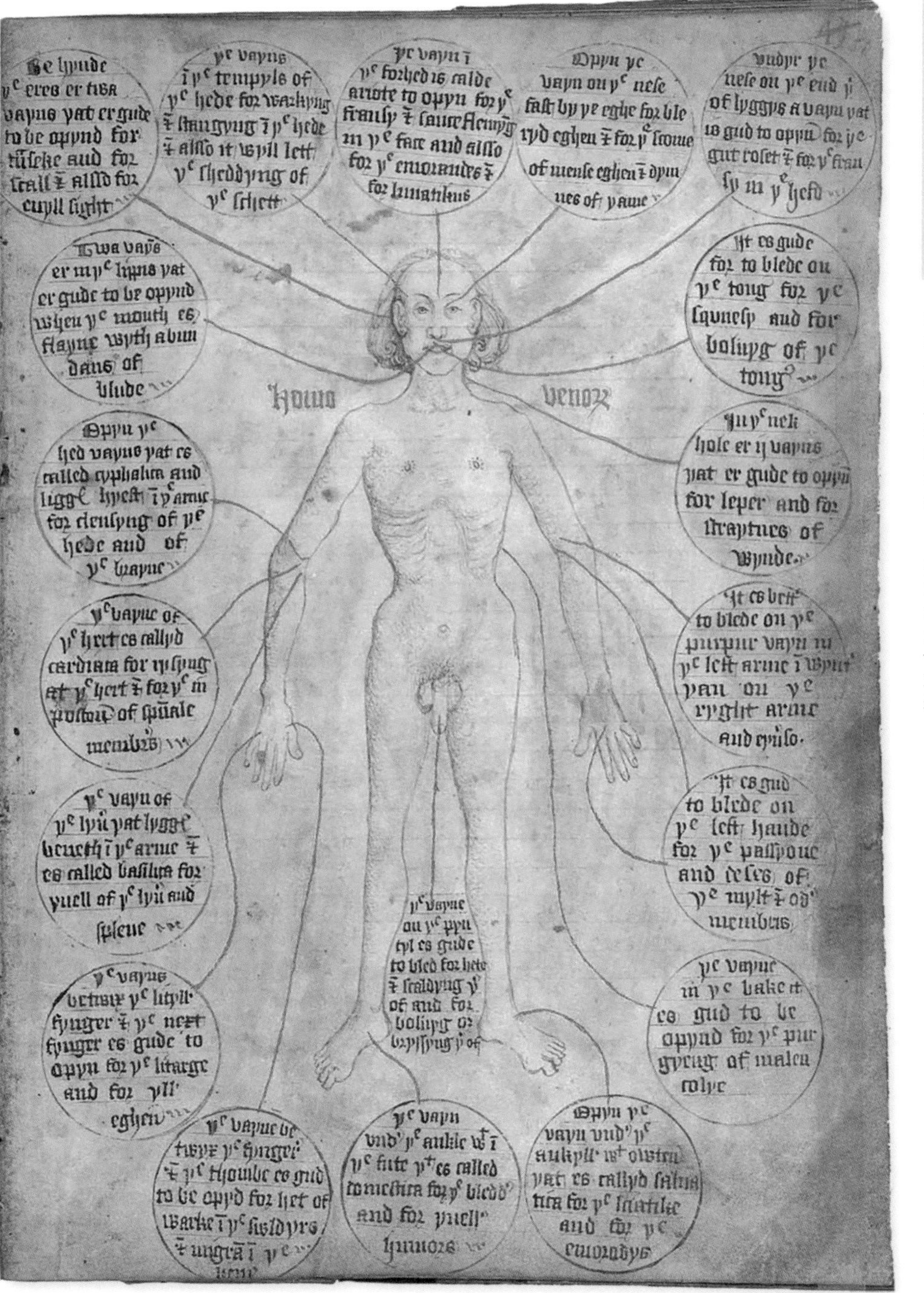

Leonhard Thurneisser
THE BOOK OF ALCHEMY, 1574

Above, an illustration of the four humours (phlegm, blood, black bile and yellow bile) within a half female and half male figure.

Unknown
THE GUILD BOOK OF THE BARBER SURGEONS OF YORK, FIFTEENTH CENTURY

Left, a 'phlebotomy chart' for bleeding.

bile, black bile and phlegm – fluids in the body which it was believed, incorrectly, had to be kept in balance for health. This led to such treatments as bloodletting to reduce an 'excess' of blood – a life-threatening and useless practice that would remain central to medical work all the way through to the nineteenth century. Like the earlier Egyptian medical documents, these books were originally produced in the form of scrolls, though later editions would see them copied into the familiar codex book form, where eventually all the books of the *Corpus* would be made available as a single volume.

The survival of books from this period is very much hit and miss (and far more miss than hit). In modern times, many of the books that are conventionally published will be produced in the thousands. However, prior to the printing press, each volume had to be painstakingly copied by hand. It's entirely possible that initially only a handful of copies of a title may have existed, though if a title became successful there would be a branching out of copies, copies of copies, and so on.

Though this copying process helped preserve some text it also presented a distinct danger to the accuracy of the contents. Copyists regularly introduced variations in the text, either accidentally or intentionally if they disagreed with the message. Examples of deliberate later additions and 'improvements' are often found in much-copied ancient works, where modern analysis can show how the original message was modified to match the cultural requirements of a later period. This presented a particular danger for scientific books where preserving the detail was essential. However, copying did at least mean that there were fall-backs if an original book was lost. A much greater danger than either errors or deliberate changes introduced in copying was the instability of ancient societies – and no better example of this can be found than the fate of the Library of Alexandria.

The eighth wonder of the ancient world

We're used to being told of the seven wonders of the ancient world: the Great Pyramid of Giza, the Hanging Gardens of Babylon, the Temple of Artemis at Ephesus, the Statue of Zeus at Olympia, the Mausoleum at Halicarnassus, the Colossus of Rhodes and the Pharos at Alexandria. Indubitably many of these structures were impressive. This was particularly true of by far the most ancient of the seven, the Great Pyramid, which is not just the only one to survive to the present day, but would also continue to be the tallest man-made structure in the world from its construction around 4,500 years ago all the way up to 1311, when Lincoln Cathedral became the first of many churches and subsequent towers to top it. However, in terms of lasting value to civilisation, surely by far the greatest wonder of the ancient world was the Library of Alexandria.

Alexandria, said to have been founded by the Macedonian king and conqueror Alexander the Great in 332 BCE, was the cultural centre of the European, Middle Eastern and North African world in this period. Located on the northern coast of Egypt, the location cemented Alexander's integration of Egypt into the Greek civilisation. It's not clear exactly when the library was first opened – it is believed to be sometime between 300 and 250 BCE – but it was the ancient equivalent of a national library such as the British Library or the US Library of Congress, intended as much as possible to collect all the books of its culture. In fact, the Library of Alexandria had a wider remit still, no longer conceivable in the modern world: the intention was to collect all the world's wisdom. Any books that arrived in the city – on a ship, for example – would be copied, with the original retained by the library and the copy returned to the (presumably disgruntled) owners.

The library's books were largely held in the form of papyrus scrolls. The number of books it contained will never be known, as the catalogue has been lost. Claims from relatively near to the time of the library's destruction put the number in the high 400,000s, though modern estimates, inevitably little more than educated guesswork, place the number at anywhere between 40,000 and 400,000. Whichever end of the scale is accurate, it was a huge number of books for the time, which inevitably would have included all the major scientific literature of the classical world.

Today we only have a small fraction of those early texts, partly due to the library's destruction, when a significant part of the collection was lost. This involved a number of unrelated attacks on Alexandria when the library suffered collateral damage, rather than the single burning that has often been depicted. Thankfully, some books did survive to be translated and appreciated by the Arabic-speaking scholars between the eighth and fourteenth centuries. These texts travelled to the West, preserving works of the most influential scientific ancient Greek philosopher: Aristotle.

Unknown
BURNING BOOKS IN THE LIBRARY OF ALEXANDRIA, ENGRAVING, SIXTEENTH CENTURY

A representation of the books in the library being burned – in reality this was less of a single organised burning, more the result of repeated damage.

Otto von Corven
A HALL IN THE LIBRARY OF ALEXANDRIA, NINETEENTH CENTURY

An imagined scene in the library, where scholars converse over scrolls, which can be seen stacked end-on in the shelves on the far wall.

Aristotle's universe

Born in 384 BCE in Stagira in Greece and educated at Plato's Academy in Athens, Aristotle is said to have been tutor to Alexander the Great. In recent years, Aristotle has had something of a bad press amongst science writers, and it has become popular to deride his lack of a modern scientific approach. It's certainly true that some of Aristotle's theories on nature were far more driven by the ancient Greek tradition of deciding what was correct by intellectual debate than they were by observation and experiment. Infamously, Aristotle is said to have pronounced that women have fewer teeth than men, based purely on his philosophical arguments, rather than actually bothering to check and discover that, in fact, women and men have exactly the same number of teeth.

Nonetheless, though Aristotle's ideas on science have been almost universally shown to be incorrect, it would be unfair to ignore his books, as he was a hugely influential figure. The concepts that he developed (often based on older ideas, but refined in Aristotle's approach), would continue to be supported for nearly 2,000 years. These notably included the model of the universe that had the Earth at the centre with the Sun, planets and stars travelling around it on crystal spheres, and the five-element theory, which considered everything on Earth to be made from earth, water, air and fire, with a fifth element (quintessence) limited to the heavenly bodies.

Of the volumes of Aristotle's scientific work that have survived, one of the most influential was the *Physics*. Given the modern usage of the word, the title is misleading. The book deals with the nature of change and motion (something of an obsession in Greek scientific philosophy, particularly after the Eleatic school had decided that

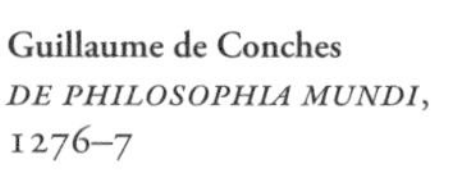

Guillaume de Conches
DE PHILOSOPHIA MUNDI, 1276–7

Below left, the world (universe) according to Aristotle from this French medieval scholastic's book.

Raphael
THE SCHOOL OF ATHENS, 1509–11

This fresco (below right) from the Apostolic Palace in the Vatican includes over 20 philosophers, with Plato and Aristotle (shown here) central.

movement was an illusion and didn't exist). Where now the mechanics of motion is a subset of physics, Aristotle had in mind not just the physical mechanisms of the motion of bodies but of all things involved in movement and change – so his 'physics' took in aspects of what now would be regarded as biology (and, inevitably, philosophy too, as Aristotle's *Physics* was, in the end, philosophy applied to scientific topics rather than science in the modern sense).

In the eight books of the *Physics*, Aristotle introduced the notion of matter and explored the nature of motion. Though his concepts have all been proved wrong, they tie together into a rational whole. In fact, this strong interlinking of his concepts is one of the reasons that his model of the universe (the solar system in modern terms) with the Earth at its centre rather than the Sun, persisted so long.

If we only consider the Sun and the Earth, it seems logical that the Sun travels around the Earth: that's certainly what it appears to do. Of course, we now know that the Sun's motion in the sky is due to the Earth turning, but we still say for convenience that 'the Sun rises' not, 'the Sun becomes visible over the horizon due to the turning of the Earth'. However, when using Aristotle's model where all heavenly bodies turn around the Earth, the motion of the planets with orbits outside of the Earth's – Mars, for instance – seems bizarre. These planets appear to occasionally reverse their motion through space. The whole business is greatly simplified if we assume that the Sun is at the centre of the solar system, where the odd reversals of planets like Mars is explained by the relative motion around the Sun at different rates of the Earth and Mars.

However, if the Earth were not at the centre of things, Aristotle's grand plan built up in the *Physics* would have fallen apart. He argued that matter has natural tendencies. Of the four earthly elements, he believed that earth and water have an in-built tendency to head for the centre of the universe, while air and fire tend to move away from it. So, he argued, heavy objects, containing more earth-like matter, would naturally fall towards the Earth. The more matter, the stronger the tendency, and the faster they would fall. As Galileo demonstrated nearly 2,000 years later, this simply isn't true, but it wasn't such a bad assumption. Feathers do fall slower than rocks – but unfortunately not for the reason that Aristotle devised.

In the second volume of *Physics*, Aristotle examined causes, separating, for example, the cause of something existing, both in terms of its material and its form, from what we would normally think of as cause – making it happen – and finally coming to the type of cause that has caused many problems in the development of science, the teleological cause – the purpose behind its existence. We now only see a teleological cause (in science, at least) when there is intervention in nature by a thinking entity. A computer has a teleological cause – it was made for a purpose – but earthquakes, say, or evolution, aren't created with a goal in mind. They don't have a purpose. But assumptions of a teleological cause have proved a significant problem in the development of some areas of science, and can continue to be a problem when religious beliefs are not separated from scientific theories.

The *Physics* goes on to consider everything from infinity to the nature of motion. Another example of the interlocking nature of Aristotle's grand vision is the way that he used his (incorrect) understanding of motion to argue against the existence of atoms. Although we tend to think of atoms as being a modern concept, the fifth-century BCE

Aristotle
PHYSICS, COPY, THIRTEENTH CENTURY

A thirteenth-century Latin version of Aristotle's fourth century BCE *Physics*, translated from Arabic in the twelfth century by Gerard of Cremona.

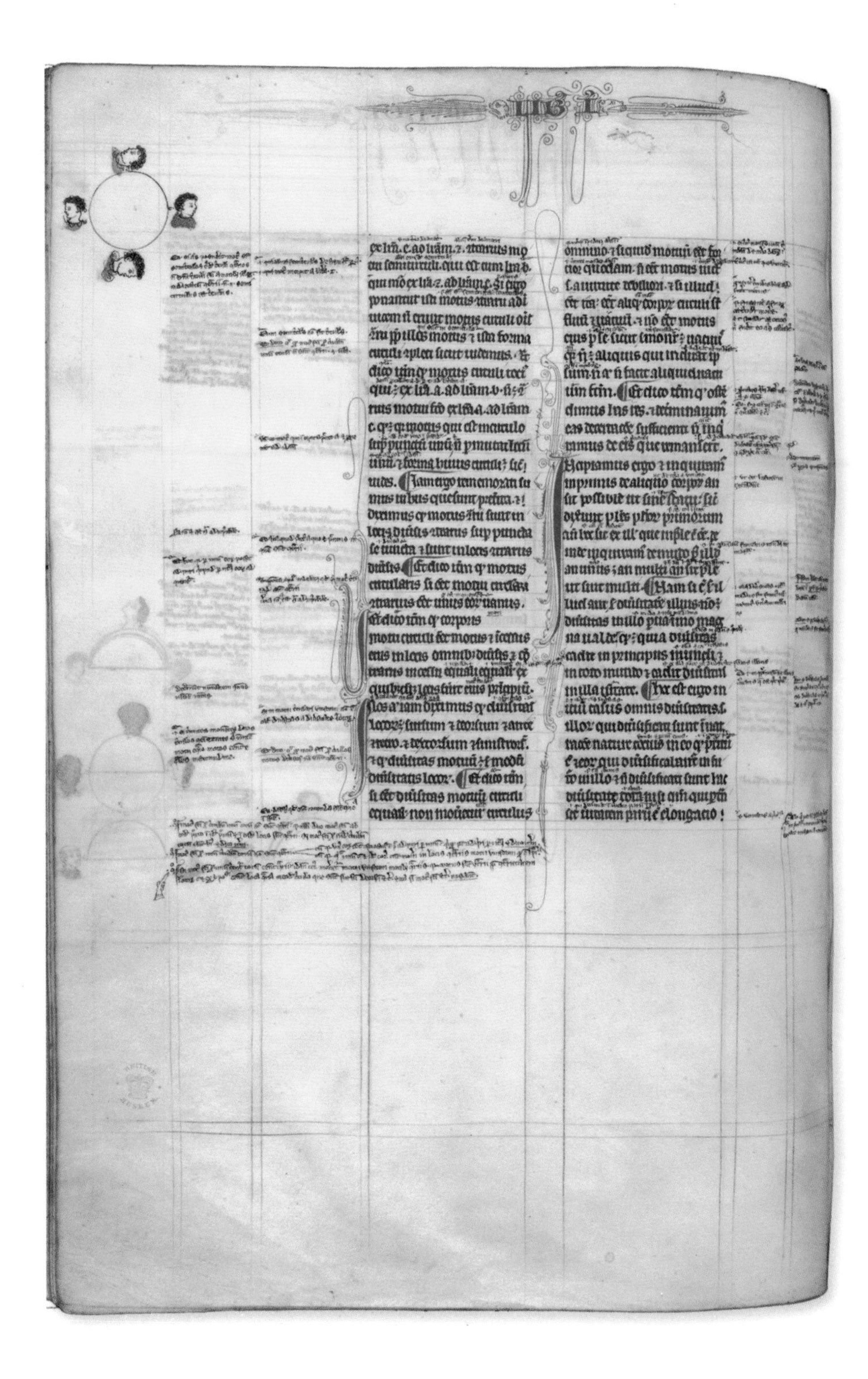

ancient Greek philosophers Leucippus and his pupil Democritus had argued that matter was made of tiny fragments, which were so small that they were uncuttable – *atomos* in Greek. If there were atoms, then there had to be a void – a space containing nothing where there were gaps between atoms. (The early atomic theory assumed that each atom was a different shape, and there are very few shapes that can be packed together to fill space without there being any gaps.) But Aristotle argued that such a void could not exist. Fascinatingly, in doing so, he pretty much came up with Newton's first law of motion – albeit for the purposes of showing that the idea was ludicrous.

Aristotle said that if there were a void, 'no one could say why something moved will come to rest somewhere; why should it do so here rather than there? Hence it will either remain at rest or must move on to infinity unless something stronger hinders it.' His version of physics assumed that for something to keep moving, it must be pushed. When we stop pushing it, it naturally comes to a stop. This was, for example, what happened to a cart. But what about an arrow in flight? Why did it keep moving after it left the bowstring? To explain this, Aristotle decided that the air must continue to push the arrow after it left the bow. But if there were a total void, it seemed to him that there was nothing to influence the moving object in any way, which seemed so counter-intuitive that he dismissed the idea of a vacuum. Aristotle's physics matched what was observed in everyday life, but he couldn't accept that the underlying reality could be different from this, so the void – and therefore the existence of atoms – had to go.

Physics was by no means the limit of Aristotle's many books on scientific subjects. He wrote significantly more titles on biology and zoology than physics and cosmology. Yet the *Physics* stands out because Aristotle's views on cosmology, motion and mechanics sat at the heart of the Western understanding of the universe right through to the sixteenth and seventeenth centuries. Other titles were, in their way, also influential. For example, Aristotle's *History of Animals* gave us the approach of grouping animals by similar characteristics, and by contrast with the *Physics* it was less philosophical and more focused on observation, recording a considerable amount of accurate data on a range of species. However, *History of Animals* is still less significant as a scientific work than the *Physics*, despite having far more in it that is correct. In the twentieth century, the physicist Ernest Rutherford is said to have remarked, 'All science is either physics or stamp collecting.' The implication of this distinctly snippy remark is that real science has to include explanations and theories; it should not just be the collection of data. Books like *History of Animals* were important, but certainly fall more into the stamp-collecting camp than anything else.

The elements of mathematics

For a book with lasting influence, it is hard to beat Euclid's *Elements*, written around 290 BCE. This multi-volume masterpiece would still be used as a textbook at the start of the twentieth century – and even now has a strong influence on the underlying foundation of mathematics in axioms and proofs. The topic of the book is one that centuries of children have found painful at school, yet it was one of the earliest aspects of mathematics to be studied as it was so practically useful: it is geometry.

Euclid
THE ELEMENTS OF GEOMETRY, COPY, 888

A page of the oldest surviving manuscript of the *Elements*, the D'Orville Euclid, written by 'Stephanos the clerk', showing a detail of Pythagoras's theorem.

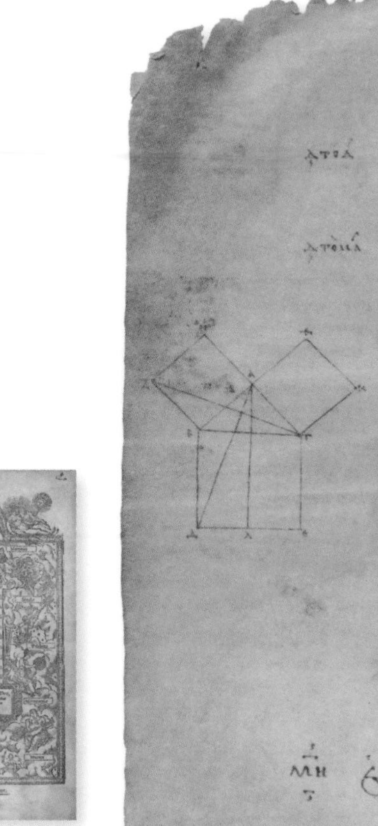

Euclid
THE ELEMENTS OF GEOMETRY, JOHN DAYE, 1570

Frontispiece of first English translation by Henry Billingsley of Euclid's work from ca. 300 BCE, printed in London, with a preface by Elizabethan mathematician and occultist John Dee.

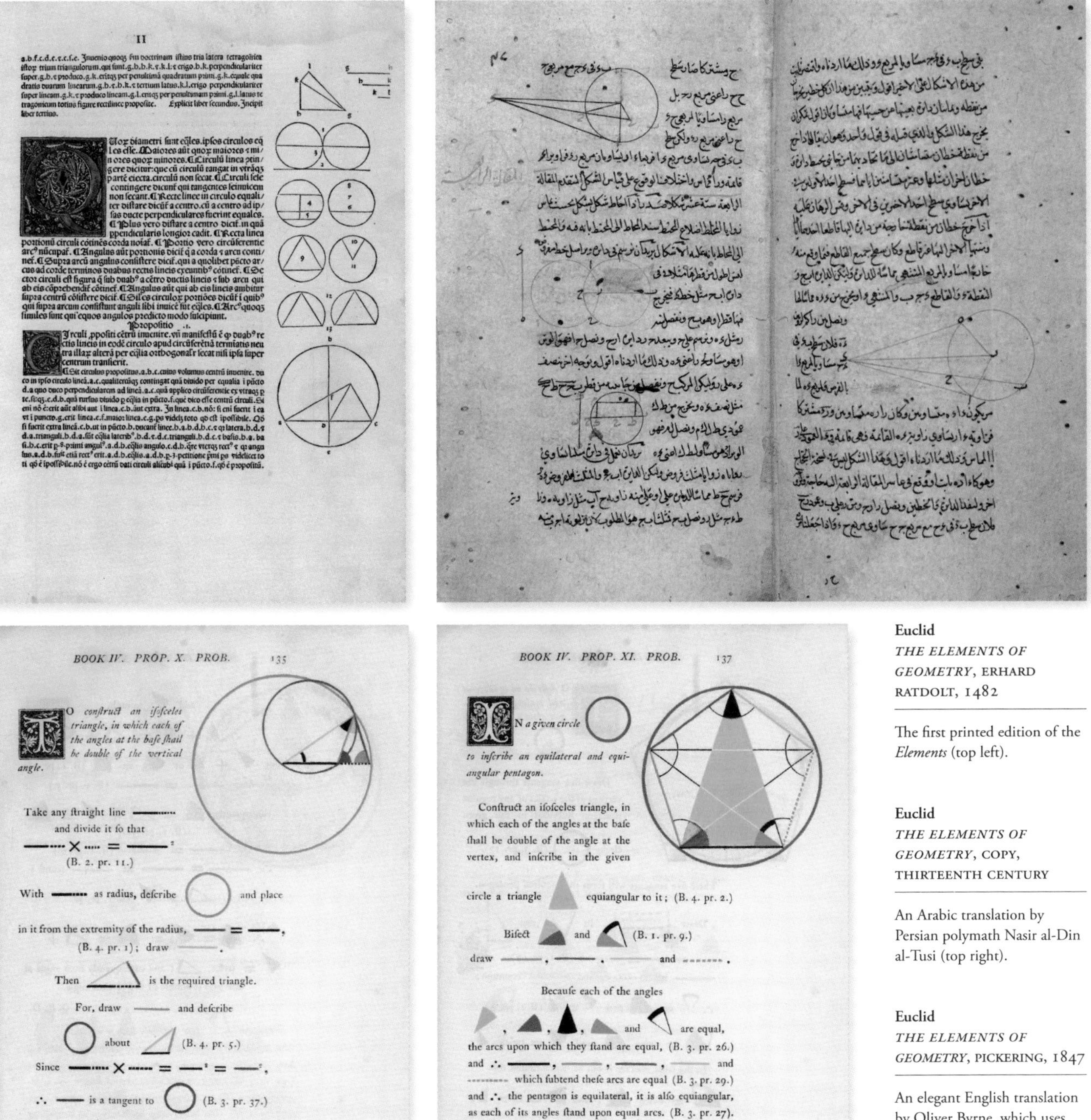

BOOK IV. PROP. X. PROB. 135

TO construct an isosceles triangle, in which each of the angles at the base shall be double of the vertical angle.

Take any straight line and divide it so that

(B. 2. pr. 11.)

With as radius, describe and place in it from the extremity of the radius, (B. 4. pr. 1); draw

Then is the required triangle.

For, draw and describe about (B. 4. pr. 5.)

Since

∴ is a tangent to (B. 3. pr. 37.)

∴ (B. 3. pr. 32),

BOOK IV. PROP. XI. PROB. 137

IN a given circle to inscribe an equilateral and equiangular pentagon.

Construct an isosceles triangle, in which each of the angles at the base shall be double of the angle at the vertex, and inscribe in the given circle a triangle equiangular to it; (B. 4. pr. 2.)

Bisect and (B. 1. pr. 9.)

draw , , and .

Because each of the angles , , , and are equal, the arcs upon which they stand are equal, (B. 3. pr. 26.) and ∴ , , , and which subtend these arcs are equal (B. 3. pr. 29.) and ∴ the pentagon is equilateral, it is also equiangular, as each of its angles stand upon equal arcs. (B. 3. pr. 27).

Q. E. D.

T

Euclid
THE ELEMENTS OF GEOMETRY, ERHARD RATDOLT, 1482

The first printed edition of the *Elements* (top left).

Euclid
THE ELEMENTS OF GEOMETRY, COPY, THIRTEENTH CENTURY

An Arabic translation by Persian polymath Nasir al-Din al-Tusi (top right).

Euclid
THE ELEMENTS OF GEOMETRY, PICKERING, 1847

An elegant English translation by Oliver Byrne, which uses coloured graphics to illustrate the proofs (left).

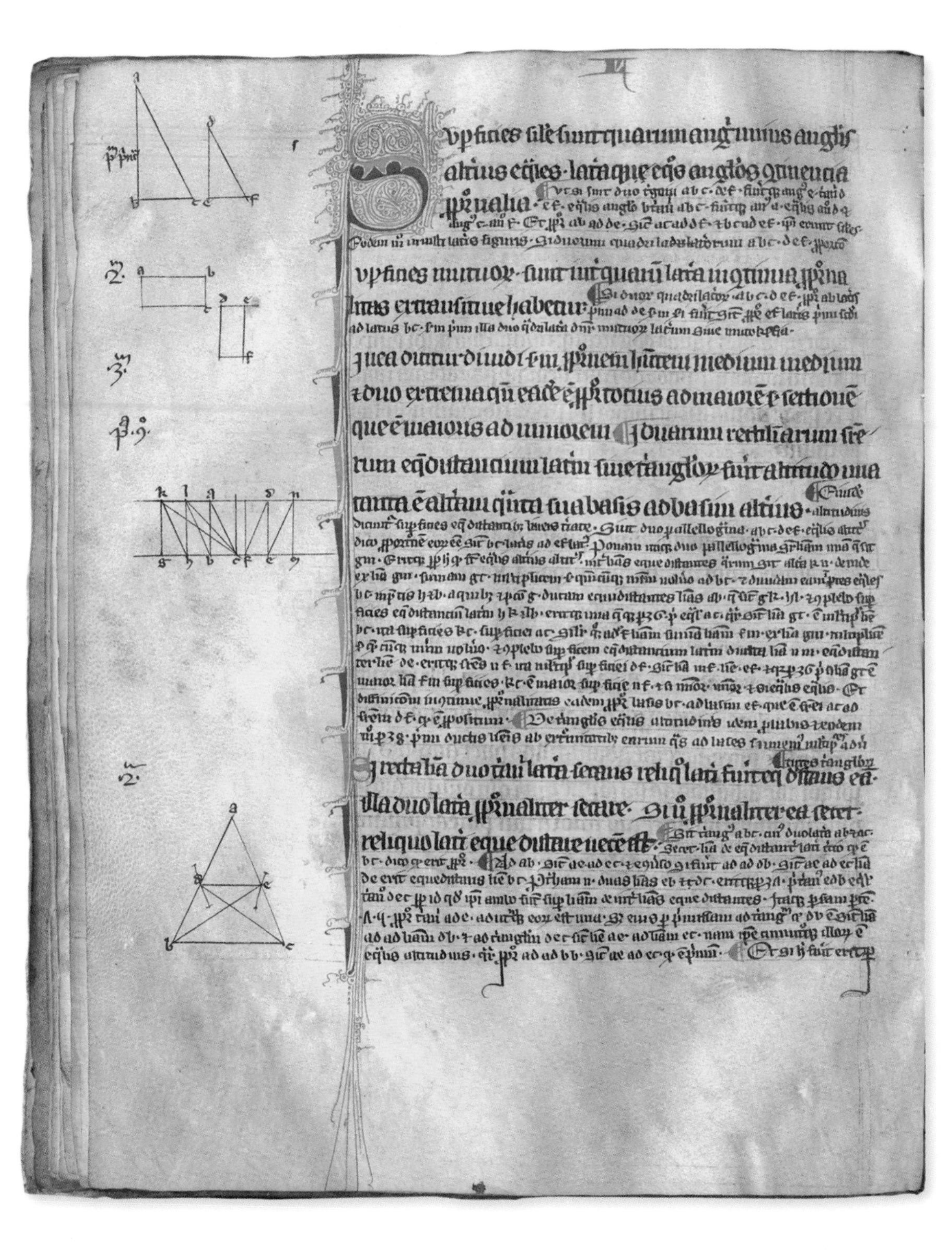

Euclid
THE ELEMENTS OF GEOMETRY, CA. 1294

A handsome Latin manuscript version of the *Elements*.

Geometry literally means measuring the Earth. It was devised to help make the measurements required to divide up land and construct buildings, bringing together the concepts of linear dimensions and of measuring angles. The ancient Egyptians made extensive use of geometry, but their approach was purely pragmatic. They used what worked without worrying about why it did so. What the Greeks brought to the table, and what was crystallised in the *Elements*, was a step-by-step process of logical construction and proof. It was a move from rule-of-thumb practice to the more precise requirements of science. Euclid did not originate the concept of geometric proofs, but what he did in the tour-de-force that is the *Elements* was bring together a whole range of constructions and proofs, building the whole in incremental stages, starting with the most basic of assumptions.

We refer to Euclid as the author of the *Elements*, but there is a degree of uncertainty over whether or not Euclid existed at all as a person. The biographical details we have for him all seem to have been later imaginings. It has been suggested that the *Elements* might have been the work of a group, rather than an individual, who picked the name Euclid as a tribute to the (definitely real) earlier philosopher Euclid of Megara. Whether or not the Euclid of the *Elements* truly existed, though, this was a breakthrough book.

The *Elements* consists of a total of 13 volumes. It starts with 'postulates and common notions' – what mathematicians would now call axioms, the essential assumptions on which the constructions and proofs (or 'propositions') that follow would be based. Axioms are apparently simple statements such as, for example, the idea that a straight line can be drawn between any two points. The book then goes on to give constructions using a straight edge and a pair of compasses of, for example, a circle, and begins to build its geometric theorems, from the famous Pythagoras's theorem to those giving the relationship between pairs of triangles with similar proportions. Each of the proofs ends with the Greek letters OEΔ, which in Latin translation would become the familiar QED for *quod erat demonstrandum*, roughly 'what was to be shown'.

The *Elements* not only provided the tools necessary for much practical geometry, but laid out the pattern for constructing mathematical proofs, building on simpler proofs to reach more complex conclusions. In its later volumes, the book also contains other mathematical principles, including some basic aspects of prime numbers, lowest common denominators and highest common factors. It explores irrational numbers such as the square root of 2, which can't be made from a ratio of whole numbers, and covers a small amount of 3D geometry, such as the volumes of simple 3D shapes and the construction of the so-called Platonic solids, where each side is made up of an identical flat shape.

The *Elements* was one of the many ancient Greek books that were lost after the fall of Greek civilisation, but came back to the West via Arabic translations in the early Middle Ages. Because of its continued importance – the *Elements*, for example, was central to the mathematics part of the European university syllabuses for centuries – the *Elements* had many translations and was one of the earliest scientific books to be mass-produced in printed form, rather than hand-copied.

Moving the world

We also shouldn't overlook the impact of the books of Archimedes, the mathematician and engineer who lived from around 287 to 212 BCE, when it is said he was killed by a Roman invader on Syracuse. Archimedes was without doubt a leading mathematician, who worked on geometry, spirals and the value of pi, and was responsible for an ancestor of integral calculus used for calculating areas and volumes of shapes.

We tend to remember Archimedes for the legend of him leaping from his bath shouting 'Eureka!', and for his claim that with a fulcrum and a long-enough lever, he could move the Earth. Best known of his engineering feats are his screw-based pump and the 'Archimedes principle' – his method to measure the volume of an irregularly shaped object (that Eureka moment), which he famously used to test the gold content of a crown. The crown had been supplied to King Hiero, who wanted to check if the goldsmith had dishonestly replaced some of the gold with another metal. By combining the crown's weight with its volume, Archimedes was able to work out its density and show that it

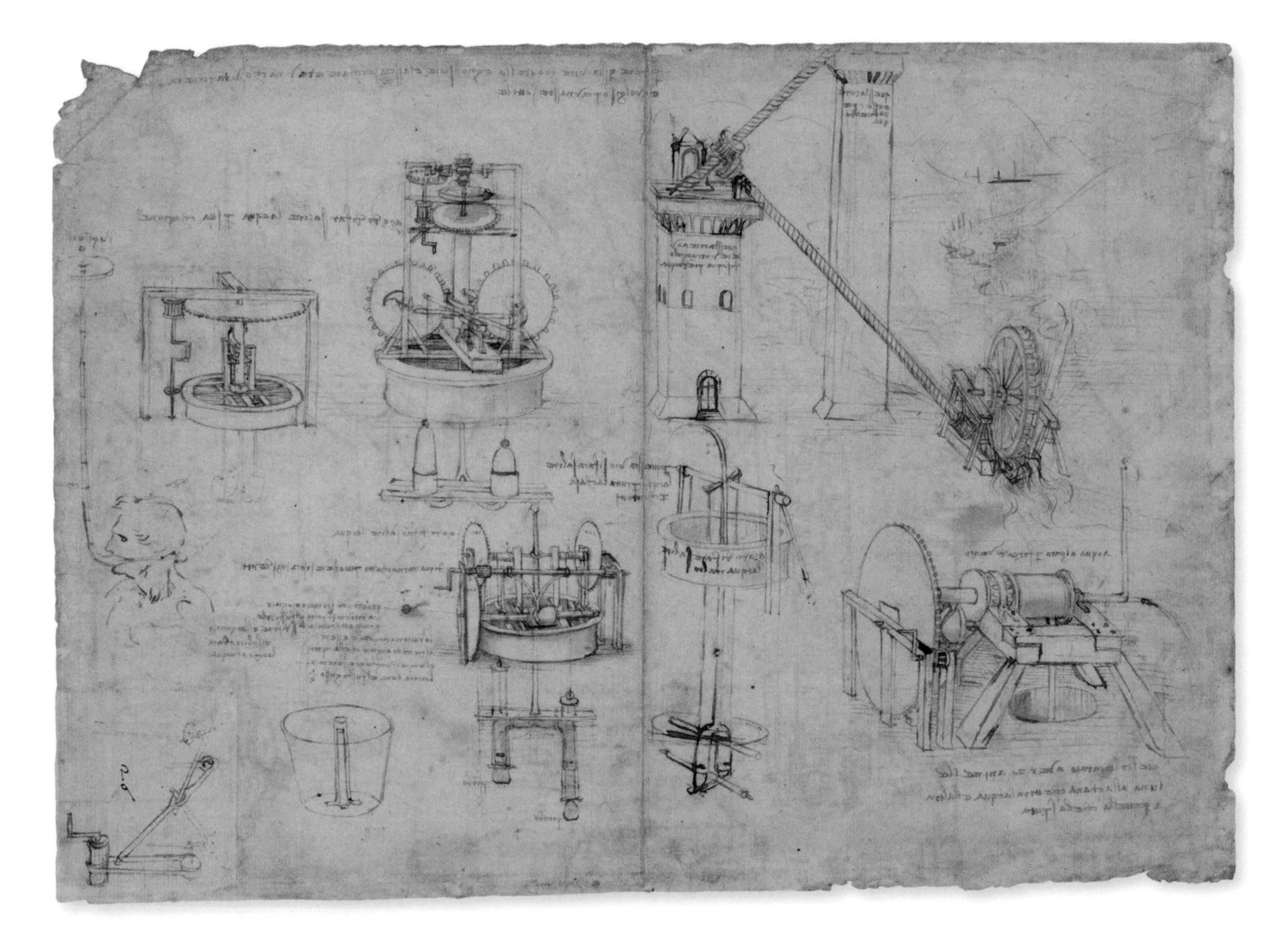

Leonardo da Vinci
CODEX ATLANTICUS,
1478–1519

A page from one of da Vinci's notebooks showing machines for lifting and pumping water. Leonardo drew on ideas from Archimedes and others.

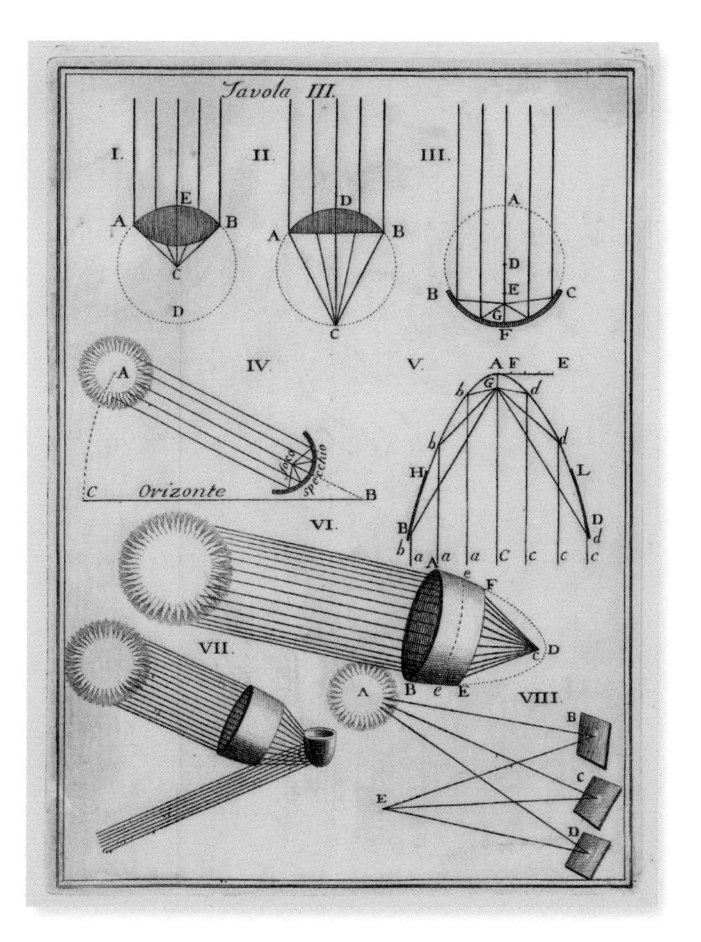

Giammaria Mazzuchelli
NOTIZIE ISTORICHE DI ARCHIMEDE, RIZZARDI, 1737

An illustration from Mazzuchelli's book *Historical and critical information about the life, inventions and writings of Archimedes of Syracuse.*

was not pure gold. This was a limited example of a wider principle he determined, that an object dipped into a fluid is pushed up by a force equal to the weight of the water it displaces. Archimedes also devised a number of weapons of war, including a death ray in the form of mirrors used to focus the rays of the Sun to start a fire on a ship.

A relatively high number of Archimedes' books have survived, notably *The Sand-Reckoner* (see page 8), *On Floating Bodies*, *On the Sphere* and the *Cylinder*, *On Spirals* and *On the Equilibrium of Planes*. They cover the physics of floating bodies, a range of geometric techniques for calculating sizes of objects, and, notably, in *The Sand-Reckoner*, the calculation of the number of grains of sand that would fill the universe as a way to demonstrate extending the Greek number system.

Although it may not have included anything original, or directly had an influence on the development of science, one remarkable Roman work from this period also deserves a mention. Roman civilisation was too focused on practical and militaristic goals to deliver any new science. Rome did introduce the codex, making books far more practical to read, but interesting scientific examples from Roman authors are few and far between.

Archimedes
ON THE SPHERE AND THE CYLINDER, CA. 1450

Renaissance painter Piero della Francesca was also a mathematician; he translated into Latin a number of works by Archimedes and added illustrations to help explain the mathematical theorems.

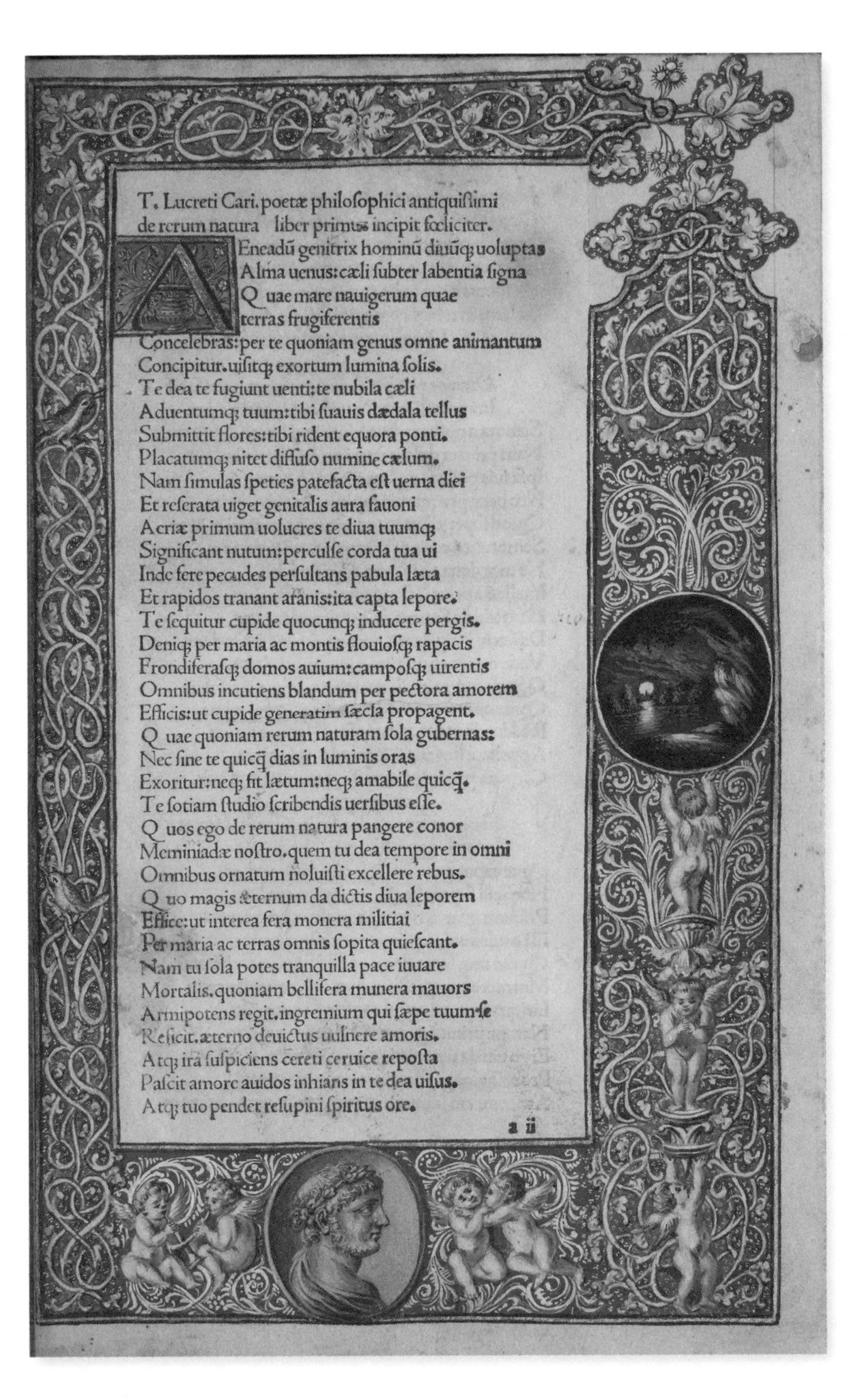
T. Lucreti Cari. poetæ philosophici antiquissimi
de rerum natura liber primus incipit fœliciter.
AEneadū genitrix hominū diuūq; uoluptas
Alma uenus: cæli subter labentia signa
Q uae mare nauigerum quae
terras frugiferentis
Concelebras: per te quoniam genus omne animantum
Concipitur. uisitq; exortum lumina solis.
Te dea te fugiunt uenti: te nubila cæli
Aduentumq; tuum: tibi suauis dædala tellus
Submittit flores: tibi rident equora ponti.
Placatumq; nitet diffuso numine cælum.
Nam simulas speties patefacta est uerna diei
Et reserata uiget genitalis aura fauoni
Aeriæ primum uolucres te diua tuumq;
Significant nutum: perculse corda tua ui
Inde fere pecudes persultans pabula læta
Et rapidos tranant amnis: ita capta lepore.
Te sequitur cupide quocunq; inducere pergis.
Deniq; per maria ac montis flouiosq; rapacis
Frondiferasq; domos auium: camposq; uirentis
Omnibus incutiens blandum per pectora amorem
Efficis: ut cupide generatim sæcla propagent.
Q uae quoniam rerum naturam sola gubernas:
Nec sine te quicq̄ dias in luminis oras
Exoritur: neq; fit lætum: neq; amabile quicq̄.
Te sotiam studio scribendis uersibus esse.
Q uos ego de rerum natura pangere conor
Meminiadæ nostro. quem tu dea tempore in omni
Omnibus ornatum noluisti excellere rebus.
Q uo magis æternum da dictis diua leporem
Effice: ut interea fera monera militiai
Per maria ac terras omnis sopita quiescant.
Nam tu sola potes tranquilla pace iuuare
Mortalis. quoniam bellifera munera mauors
Armipotens regit. ingremium qui sæpe tuum se
Reiicit. æterno deuictus uulnere amoris.
Atq; ira suspiciens tereti ceruice reposta
Pascit amore auidos inhians in te dea uisus.
Atq; tuo pendet resupini spiritus ore.
a ii

Titus Lucretius Carus
'DE RERUM NATURA',
PAULUS FRIDENPERGER,
1486

A page from an early printed version of Lucretius's first century BCE poem.

The exception is 'De Rerum Natura' (On the Nature of Things) by Titus Lucretius Carus, written in the first century BCE. This takes the form of a long poem made up of 7,400 lines, based on the natural philosophy of Epicurus, a Greek philosopher from the third century BCE, whose ideas included atomism. Lucretius's poem takes in everything from the nature of space and matter to agriculture and disease. Other books of the period, most notably Pliny's massive *Naturalis Historia*, collected together existing scientific knowledge, but added little or nothing new.

Wheels within wheels

By the second century CE, Greece was part of the Roman Empire. The most prolific medical author of the ancient world, Galen, was one of the empire's most famous physicians. Born in the Greek city of Pergamon in 129 CE, Galen stuck largely to the theories of Hippocrates and his followers, though he did have significantly more experience of anatomy than his predecessors – albeit his knowledge was largely based on monkeys and pigs, so suffered a little when applied to humans. There was a slightly more scientific basis to some elements of his medicine, notably in his suggestion that

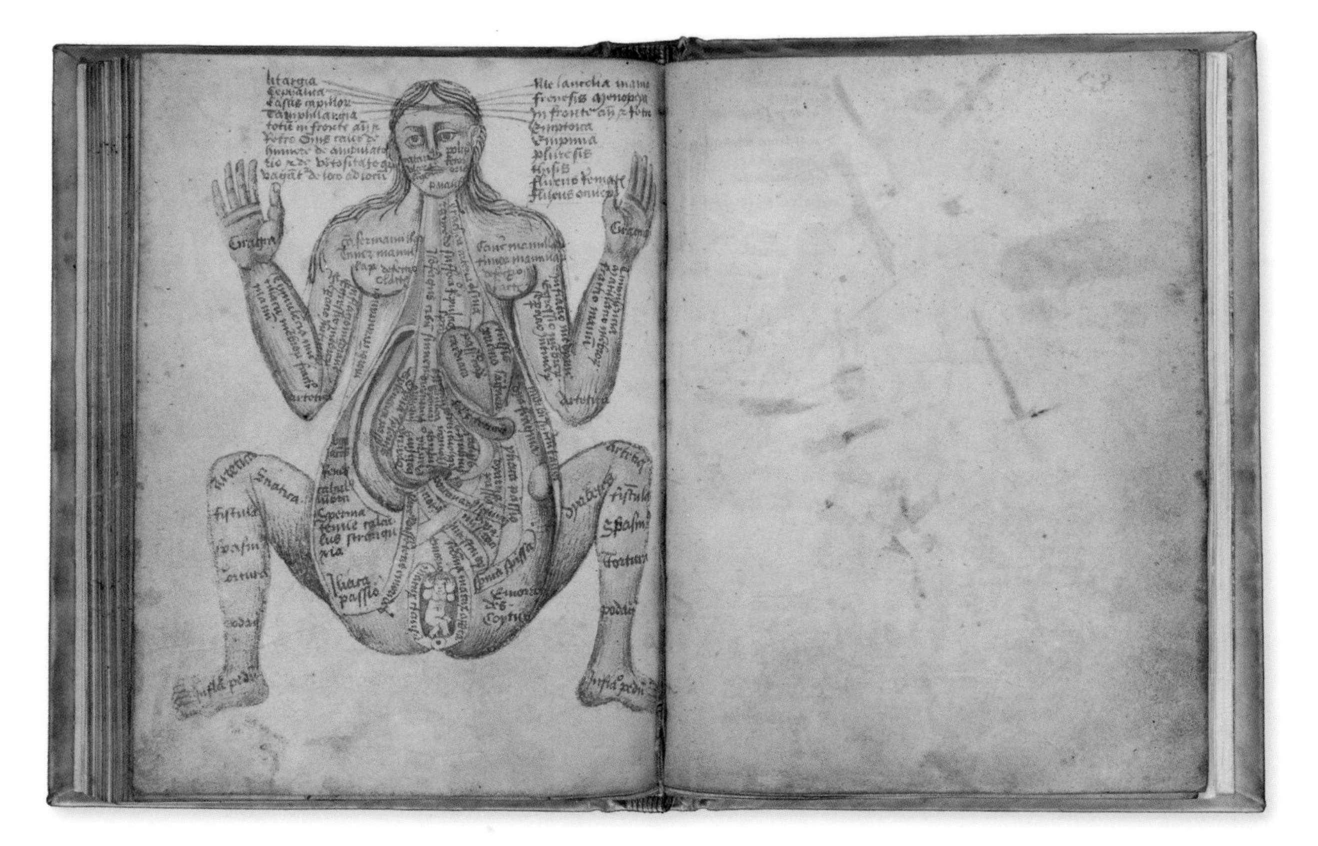

Pseudo-Galen
ANATHOMIA,
FIFTEENTH CENTURY

Galen, like many classical authors, had a number of texts attributed to him. This page depicts a pregnant woman.

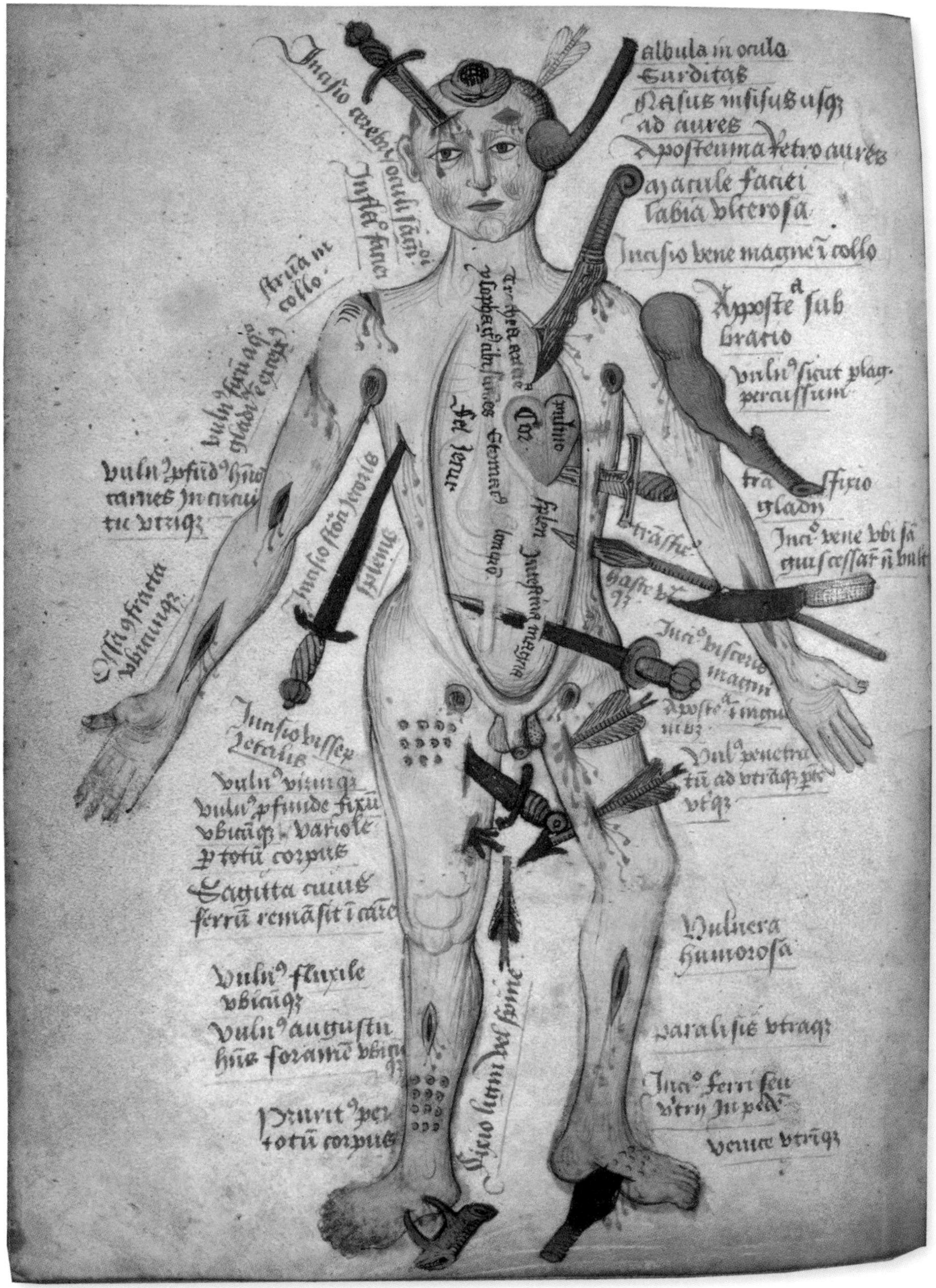

Guy of Pavia

LIBER NOTABILIUM PHILIPPI SEPTIMI, 1345

Above, a dissection from the abdomen of a cadaver from a book claiming to be taken from Galen's work, housed in the Musée Condé, Chantilly.

Pseudo-Galen

ANATHOMIA, FIFTEENTH CENTURY

A 'wound man' from the same volume as the image opposite.

Bartolomeu Velho
COSMOGRAPHIA, 1568

Ptolemy's geocentric universe (without the epicycles) drawn by this Portuguese cosmographer and cartographer.

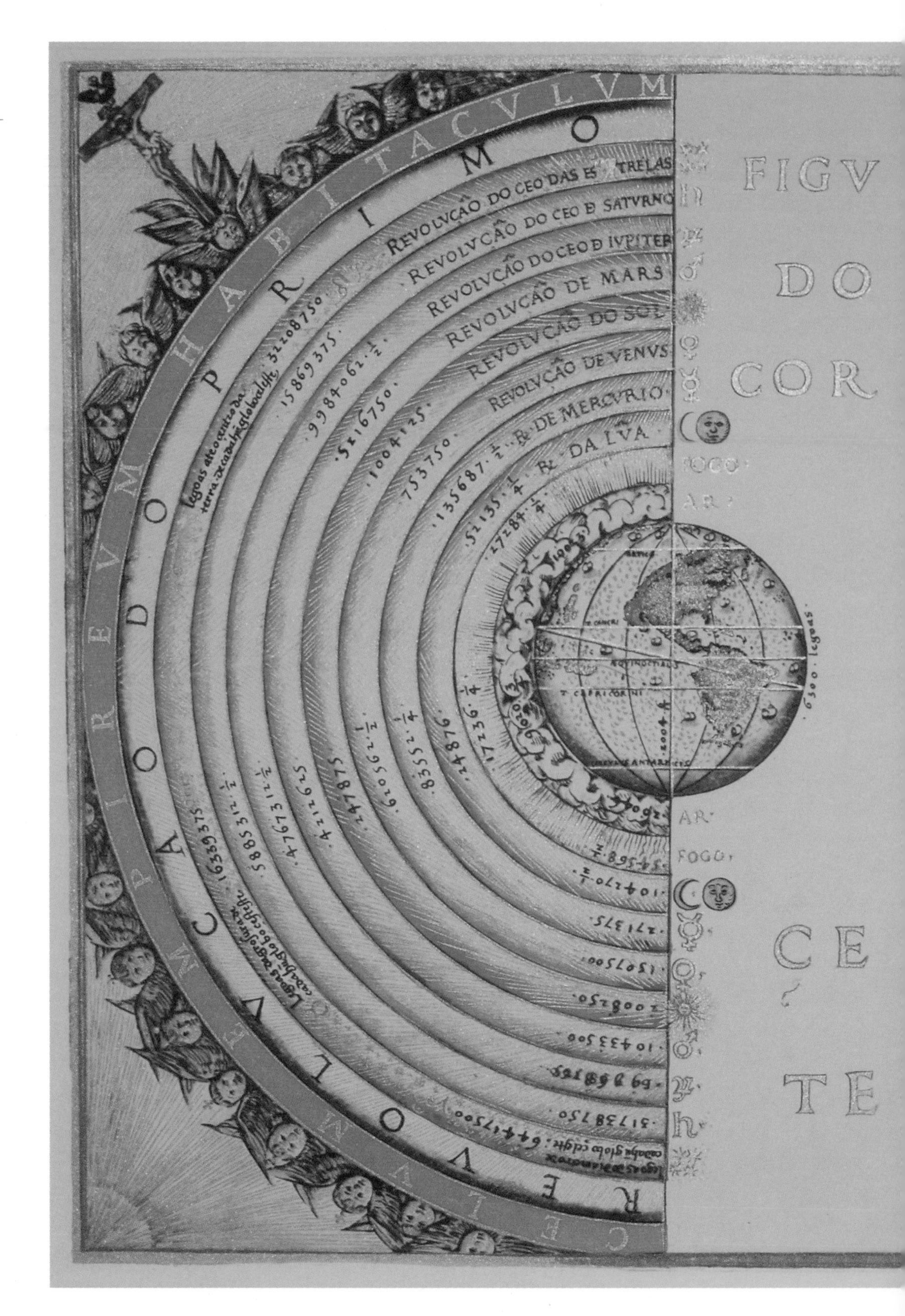

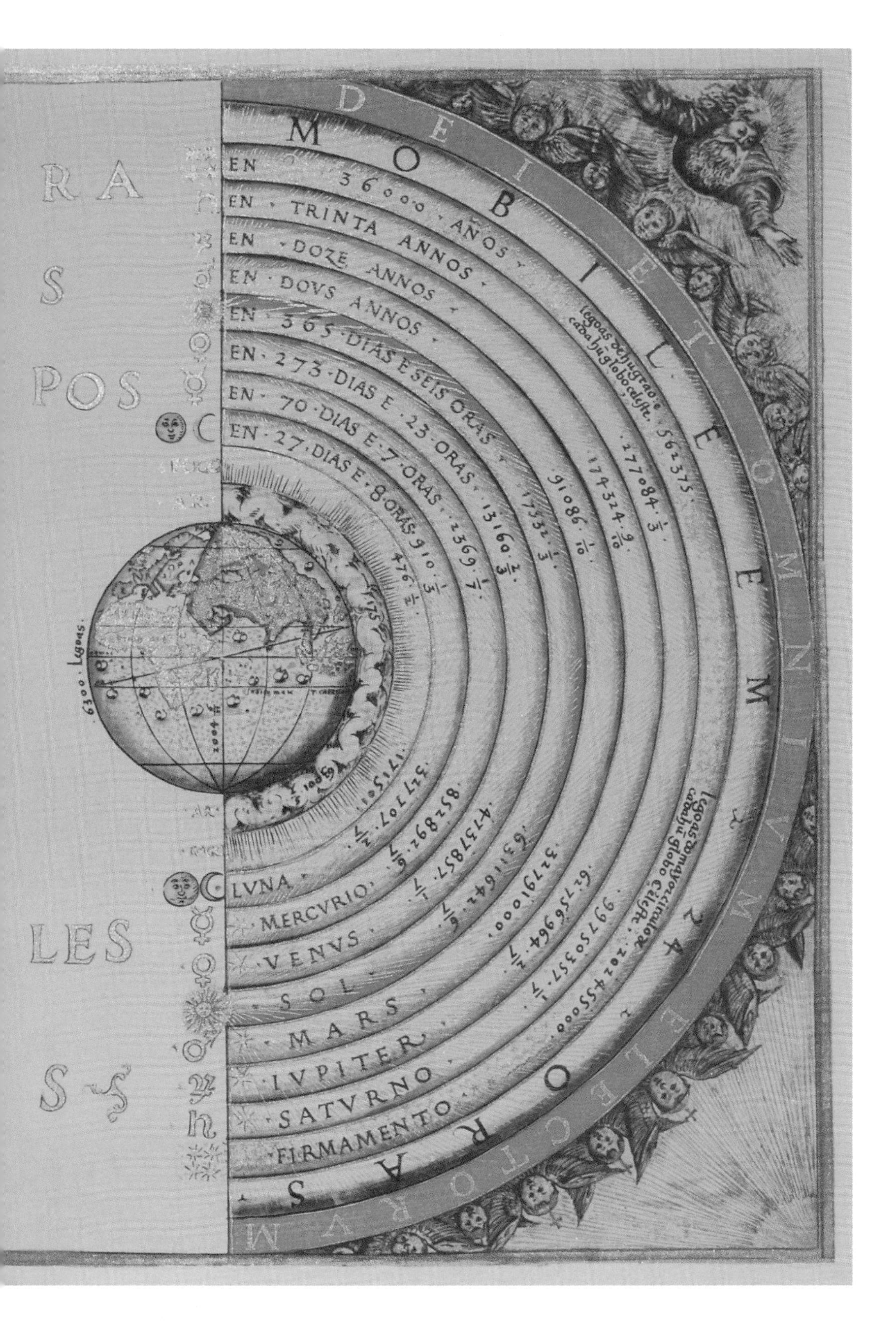

EN 36000 AÑOS
EN TRINTA ANNOS
EN DOZE ANNOS
EN DOUS ANNOS
EN 365 DIAS E SEIS ORAS
EN 273 DIAS E 23 ORAS
EN 70 DIAS E 7 ORAS
EN 27 DIAS E 8 ORAS
LUNA
MERCURIO
VENUS
SOL
MARS
IUPITER
SATURNO
FIRMAMENTO

the arteries carry blood. His vast corpus of books would go on to influence the practice of medicine through to the seventeenth century.

Another book from the same time would also have a long impact: based on Aristotle's physics, it succeeded in keeping alive an incorrect view of the structure of the solar system for 1,500 years. This was the book now known as the *Almagest*, written by the Egyptian-Greek astronomer Ptolemy. The name *Almagest* – translating as 'The Greatest' – is not Greek (or Egyptian), but Arabic, as the book was introduced to the West, like so many Greek texts, via an Arabic translation.

Dating to around 150 CE, the *Almagest* was actually titled *Mathēmatikē Syntaxis* in the original Greek (roughly, 'Mathematical Treatise'). The title might seem odd given it dealt with astronomy, but we need to remember that until the nineteenth century, astronomy and cosmology were considered part of mathematics, not the physical sciences where they have now been more sensibly reassigned.

This hugely influential book in 13 volumes starts by describing the Earth-centred view of the universe that Aristotle had made the standard model, and goes on to cover the motions of the Sun and planets – including the related concepts as seen from the Earth of, for example, eclipses, the equinoxes and the solstices. Ptolemy also described the constellations and put together a catalogue of the fixed stars. This catalogue was not all Ptolemy's own work, but was largely based on an earlier catalogue by the ancient Greek astronomer Hipparchus that was already around 280 years old.

As we have seen, there was a problem arising from the requirement of Aristotelian cosmology that everything orbited around the Earth: the way that planets outside the Earth's orbit would suddenly reverse their direction in so-called retrograde motion. To offer a scientific explanation of how this was possible, Ptolemy was forced to introduce a series of fixes to the older, simple Aristotelian model that had each orbiting planet embedded in a nested crystal sphere.

Andreas Cellarius
HARMONIA MACROCOSMICA,
JOHANNES JANSSONIUS,
1660

Cellarius's *The Celestial Atlas, or the Harmony of the Universe* depicted the world systems of Ptolemy, Tycho Brahe and Nicolaus Copernicus.

Andreas Cellarius
HARMONIA MACROCOSMICA, JOHANNES JANSSONIUS, 1660

A further engraving by Cellarius setting Ptolemy's structure for the universe against the constellations of the zodiac.

PTOLEMY, FIFTEENTH CENTURY

In this illustration from Gautier de Metz's *Image du Monde* (Image of the World), Ptolemy is mistakenly depicted as a king as he observes the heavens.

Ingeniously (if painfully), Ptolemy managed to match observation, and to keep the (also incorrect) assumption that all orbits had to be circular as this was the 'perfect' shape and the heavens required perfection. He did this using a structure known as epicycles. The idea was that instead of Mars, say, simply travelling around the Earth in a circular motion – which it clearly didn't – there was instead an empty point in space that orbited the Earth, called a 'deferent'. Mars then orbited that moving empty point in space in another circular orbit (an epicycle). It was the original case of 'wheels within wheels'. To make things even more complex, because this model still didn't quite match what was observed, Ptolemy stated that instead of travelling on circles around the Earth, deferents orbited a point a little way from the Earth called the 'eccentric'.

If, frankly, eccentric sounds the ideal word for such a contrived structure, the complexity of Ptolemy's model was necessary to make actual observation fit with the Aristotelian model of the universe. Even today scientists can get very attached to their theories, and in the light of contradictory evidence will repeatedly modify aspects of the theory to keep it working. This has happened, for example, several times with the Big Bang theory, which is still our best idea of the way that the universe came into being, but has had to be patched up repeatedly to match observations.

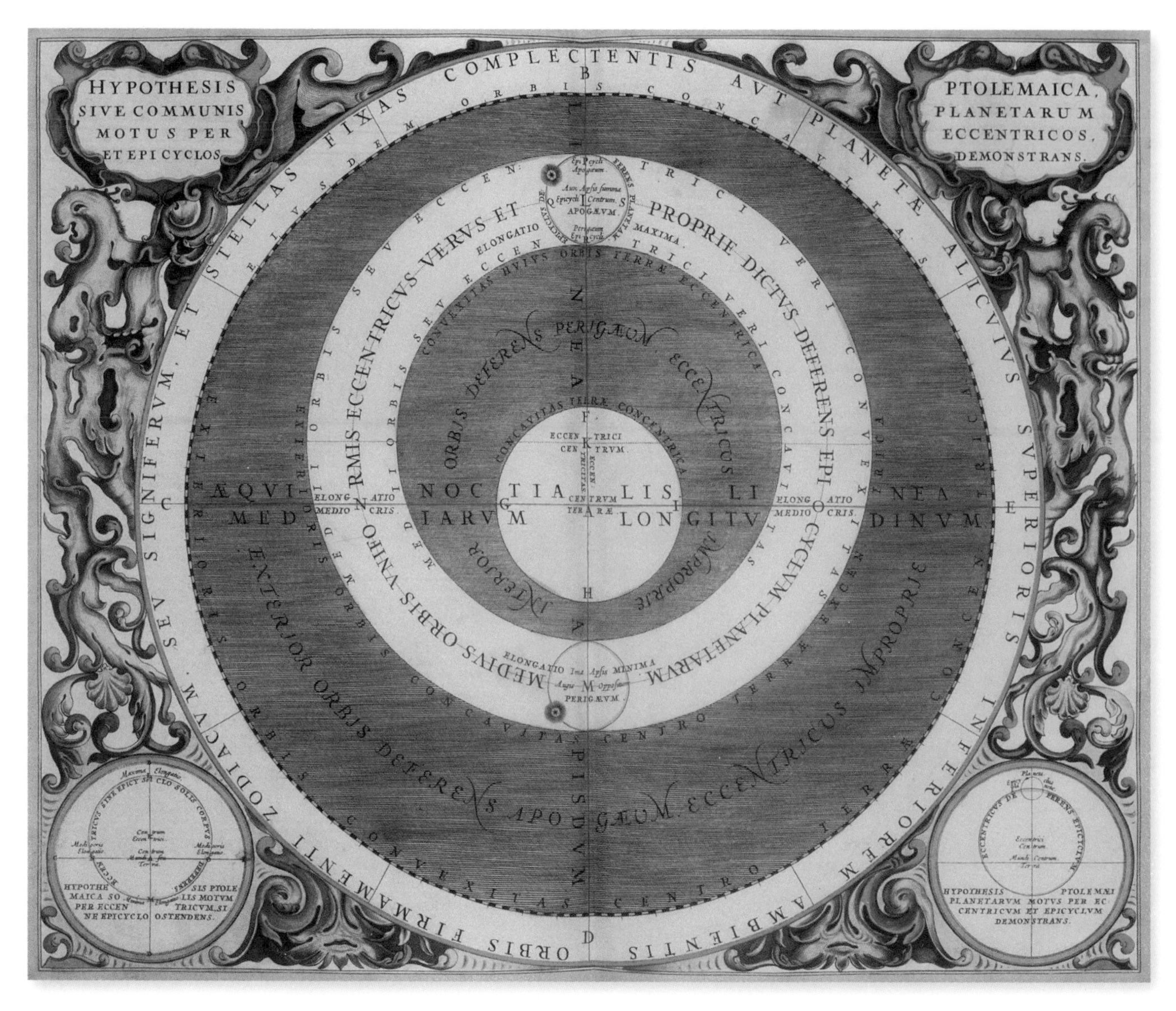

Andreas Cellarius
HARMONIA MACROCOSMICA,
JOHANNES JANSSONIUS,
1660

The engraving above illustrates the planetary motions in eccentric and epicyclical orbits, while the engraving opposite is of the early Christian view of the structure of the universe.

COELI
CHRISTI
SPHÆRIUM
STELLATI
ANI HÆMI
POSTERIUS.
Circulus Polaris Sept.
Circulus Polaris Merid.

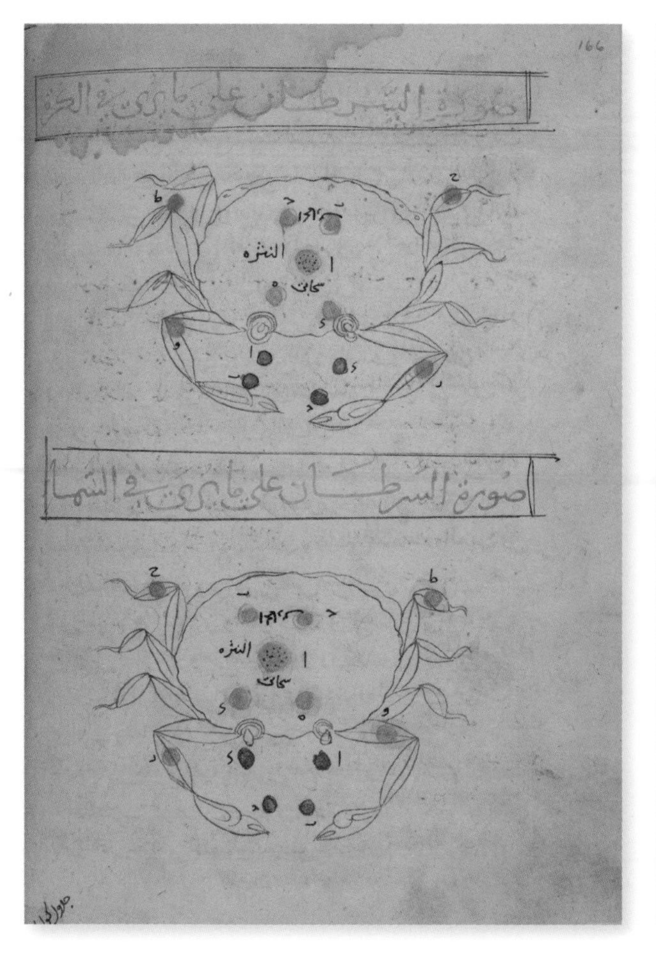

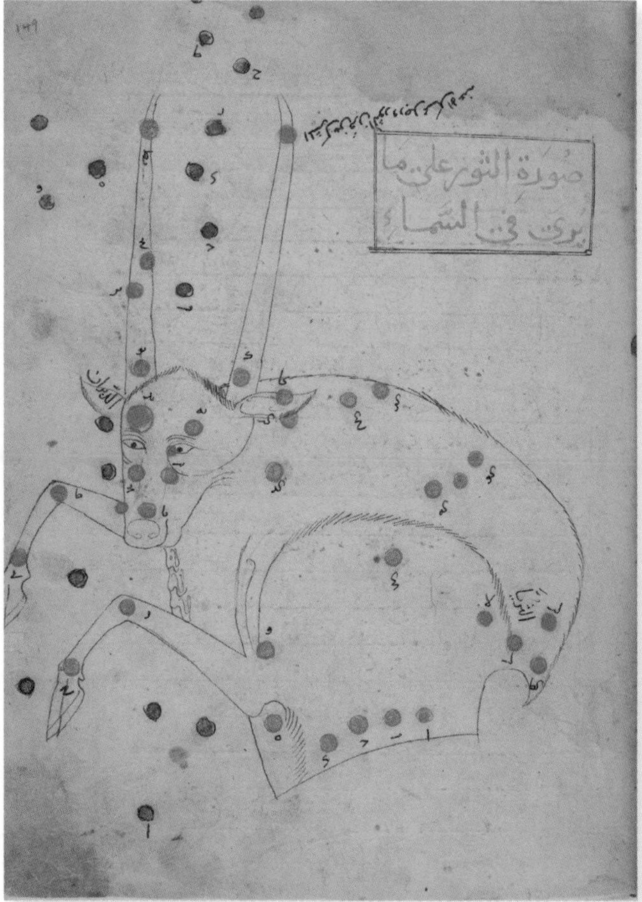

Abd al-Rahman al-Sufi
KITAB SUWAR AL-KAWAKIB AL-ATHABITA,
FIFTEENTH CENTURY

Based on Ptolemy's *Almagest*, this late fifteenth-century Iranian text illustrates 48 constellations including Cancer and Taurus and opposite Aquila.

Mathematics in the early years of the first millennium was not limited to Ptolemy's epicycles or the geometrical wonders of the *Elements* and was being studied around the world. Several other important mathematical volumes would emerge during this period. The earliest of these, developed over a long period starting several centuries earlier but coming to final fruition around 200 CE, was the anonymously authored *Jiuzhang Suanshu* (The Nine Chapters on the Mathematical Art), a summary of state-of-the-art mathematics in China. The approach taken in the book was one of problem-solving – a more pragmatic one than that of the *Elements* with its formal proofs, but using significantly more mathematical logic than was present in the simple use of Pythagorean triples in Sumerian clay tablets. Given this practical approach, it's not surprising that *Jiuzhang Suanshu* gives us such day-to-day requirements of an early civilisation as measuring the areas of shapes (and, yes, calculating the size of Pythagorean triangles), computations required for trade and taxation, and basic equations. It also includes some slightly more abstract concepts, such as square and cube roots and the volumes of solid objects.

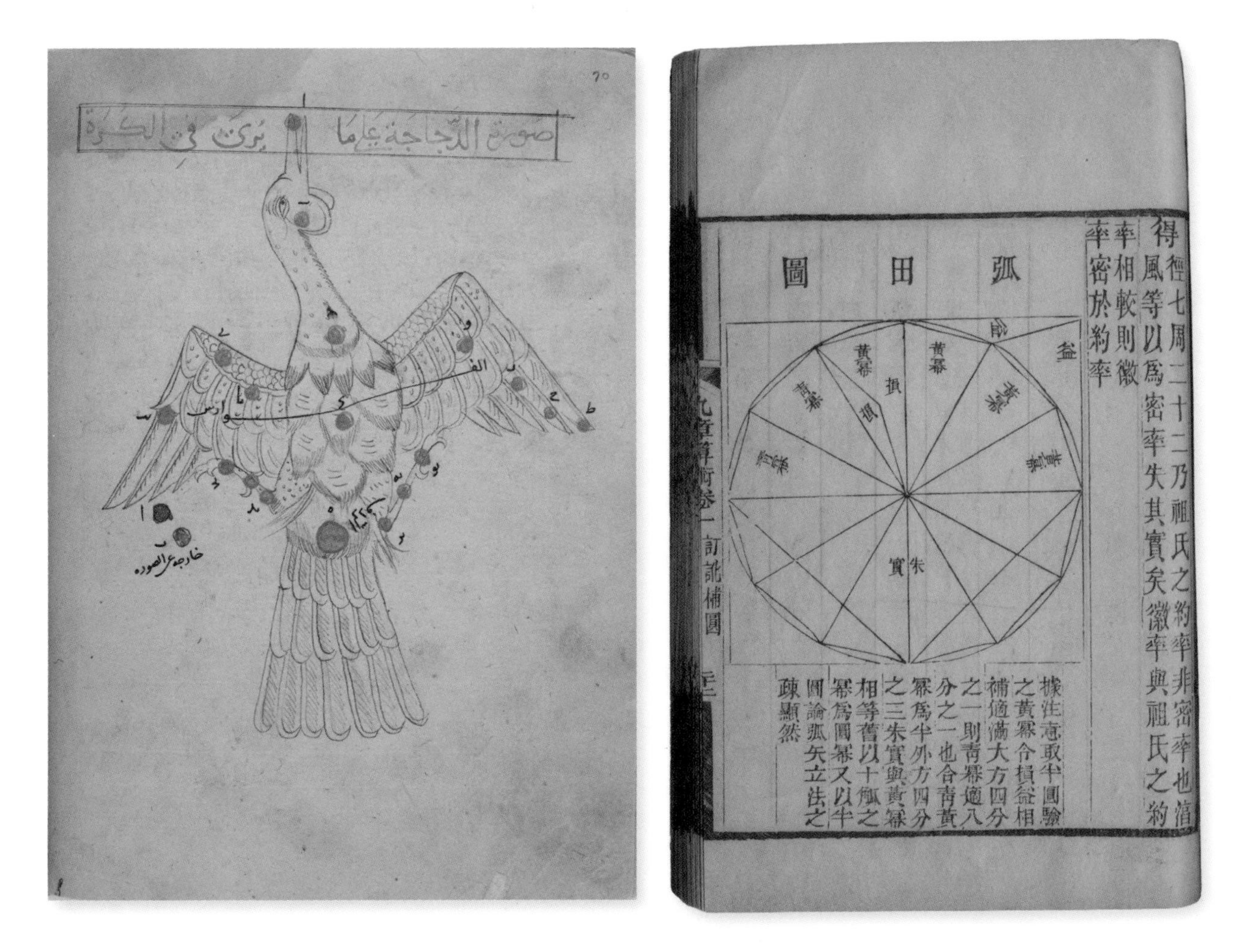

There is no surviving book to attribute to her, but it's important that we also note the late-Greek philosopher and mathematician Hypatia, who lived in Alexandria from the middle of the fourth century. Hypatia was known to have written a number of commentaries and is thought to have edited the *Almagest*. Historically, women's contribution to science, and science writing, has only relatively recently become commonplace, as it is only recently that women have had the same opportunities as men. Hypatia provides our first certain example. This doesn't mean that Hypatia was the first woman to write science books; all the way up to the nineteenth century it was not uncommon for books by women to be published anonymously or under a pseudonym (think, for instance, of the Brontë sisters, who initially wrote as Currer, Ellis and Acton Bell). It is only in recent years that women writers have produced influential science books with anything like the same frequency as men. This move towards gender equality has come significantly later than it did for fiction authors, though thankfully considerable progress has now been made. For the moment, though, female writers will be thin on the ground.

Unknown
JIUZHANG SUANSHU,
SIXTEENTH CENTURY

Late edition, showing the method of estimating pi by drawing a polygon with more and more sides, making an increasingly accurate approximation to a circle.

Mathematical transformations

Just as the *Elements* would have a huge influence on Western mathematics, *Jiuzhang Suanshu* was central to China's mainstream development of maths. Like China, India also had a flourishing ancient mathematical tradition, but there, individual mathematicians were given more credit, and so, unlike the anonymous *Jiuzhang Suanshu*, we know quite a lot about the author of a key Indian mathematical work, *Brāhmasphuṭasiddhānta* (roughly, 'The established (or improved) treatise of Brahma'), dating to 628. Its author was Brahmagupta, born around 598, who like many mathematicians of the time studied astronomy as well as mathematics – the book often uses astronomical examples in its content.

As was the case with the Chinese book, this was mathematics presented as statement of fact, rather than as a result of logical proof, and it was provided in an unusually complex fashion, as *Brāhmasphuṭasiddhānta* was written in the form of poetry. The book was important not only for its various geometric results, but also for its developments in algebra, including one of the two solutions to the quadratic equation familiar to high-school students. Probably the book's most important innovation was in dealing with non-positive integers. It covered negative numbers, a concept not then widely in use, and treated zero as a number, rather than a simple placeholder for numbers with no value in a particular column. Brahmagupta didn't get this entirely right – he thought that $0 \div 0 = 0$. However, this was still a major development in mathematical thought. The use of zero would be essential for the development of modern mathematics.

The concept of zero as a number (as opposed to a placeholder) originated in India, as did the useful Hindu numerals, but they would come to the West through the thriving new academic centres of the Islamic world. A text that would have significant influence was *Al-kitāb al-mukhtaṣar fī ḥisāb al-ğabr wa'l-muqābala* (The Compendious Book on Calculation by Completing and Balancing), written around 820 in Arabic by Abu Ja'far Muḥammad ibn Mūsā al-Khwārizmī and translated into Latin by Robert of Chester in 1145. We have few reliable biographical details about al-Khwārizmī. He was born in

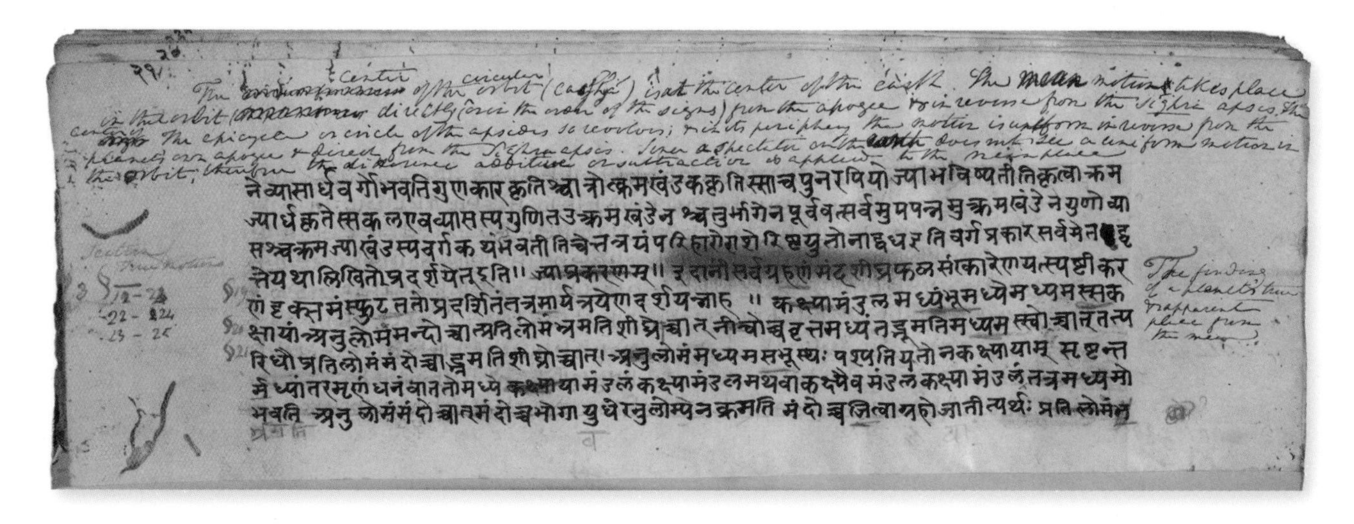

Pṛthūdhaka
COMMENTARY ON BRĀHMASPHUṬASIDDHĀNTA, TENTH CENTURY

A commentary on Brahmagaputa's celebrated work of 628, with annotation by its translator into English, Henry Thomas Colebrooke.

Persia around 780, possibly in Baghdad, and definitely worked in Caliph al-Mamun's Baghdad House of Wisdom, as the *Al-kitāb al-mukhtaṣar* was dedicated to the caliph.

The author and his book give us two technical mathematical terms: 'algorithm', from the Latinised version of al-Khwārizmī's name, Algorithmi; and 'algebra', from the 'al-ǧabr' of the title. It is in the exploration of algebra that this book proved such a success. Al-Khwārizmī was not the first to work on algebraic problems outside India. The third-century Greek philosopher Diophantus had dealt with algebraic equations with powers of a variable value in his book *Arithmetica*, but he did not attempt to produce generalised solutions that would work for any equation of the same form. Al-Khwārizmī's version of algebra was in some ways more different from our current approach than that of Diophantus – al-Khwārizmī only works in terms of words, where Diophantus used something closer to a modern equation – but, crucially, al-Khwārizmī dealt with general solutions, so his approach could be applied to a wide range of possibilities.

Another important book by the same author, preserved only in a Latin translation, *Algoritmi de numero Indorum* (Al-Khwārizmī Concerning the Hindu Art of Reckoning), was a description of the Indian number system (which as we have seen would become known as Arabic numerals in the West, because of the route they took). However, this content proved less significant in terms of the wider development of mathematics, as Indian/Arabic numerals were not widely adopted in the West until Fibonacci's *Liber Abaci* (see page 66–7) reintroduced the numbering system at the start of the thirteenth century.

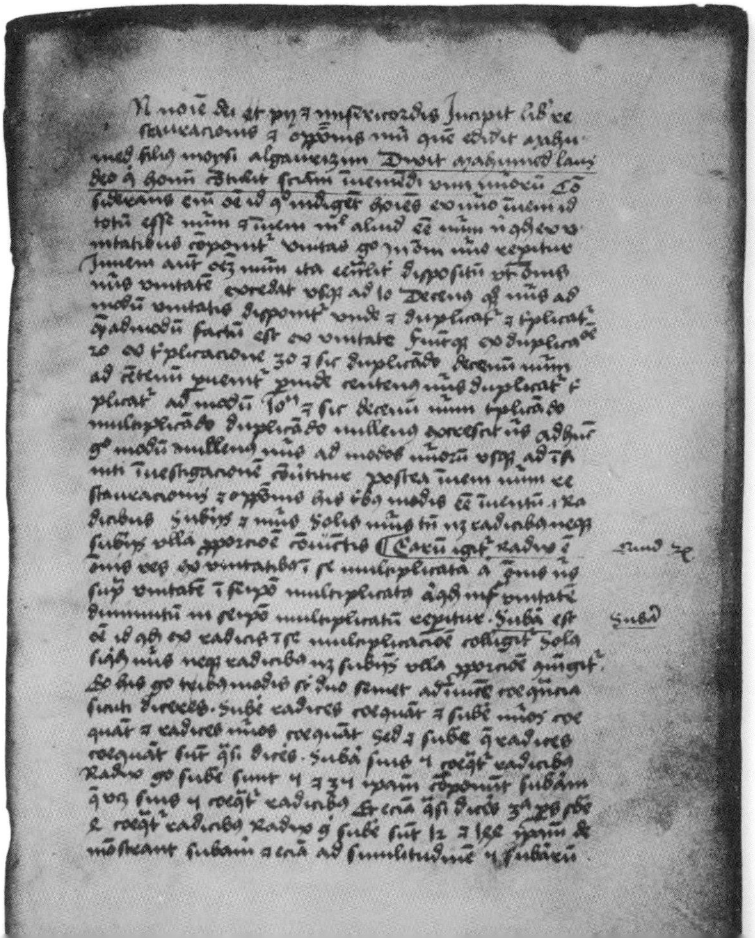

Abū Ja'far Muḥammad ibn Mūsā al-Khwārizmī
AL-KITĀB AL-MUKHTAṢARFĪ ḤISĀB AL-ĠABR WA'L-MUQĀBALA, CA. 1145

A page from Robert of Chester's translation, which he produced when living in Segovia in Spain.

Arabic optics and medicine

A major figure in the translation of ancient Greek texts into Arabic was Ḥunayn ibn Isḥāq, born in Al-Hira, in what is now Iraq, in 809. Ibn Isḥāq was a physician, so had a particular interest in medical sources, though he translated a wider range of scientific documents. However, he was also responsible for his own contribution to science writing, notably in *Al-Ashr Makalat Fi'l'ayn* (The Book of the Ten Treatises on the Eye), which has been described as the earliest extant systematic textbook on ophthalmology. There is no doubt that ibn Isḥāq was heavily influenced by ancient Greek texts, particularly Galen, who he widely referenced, but there are also elements of *Al-Ashr Makalat Fi'l'ayn* which seem to be original observations from ibn Isḥāq's experimental work, including some impressive illustrations.

Mathematics and medicine were not all that was being produced in the thriving Islamic culture and being passed on to the West. There were a number of titles that covered optics and light, notably those written by Abū 'Alī al-Ḥasan ibn al-Haytham, whose name was Latinised to Alhazen. Born in what is now Basra in Iraq in 965, al-Haytham was said to have had a similar character to Leonardo da Vinci, the Italian Renaissance polymath who took on roles designing anything from bridges to war machines despite having no previous experience.

According to legend, al-Haytham bit off more than he could chew by promising the Caliph al-Hakim that he could divert the Nile to control flooding and irrigation. Failing to do so, al-Haytham is said to have pretended to be insane, a pretence he had to maintain for years until the caliph's death. What we know more certainly is that al-Haytham wrote a number of books, including *Mizan al-Hikmah* (Scale of Wisdom) and particularly notably *Kitāb al-Manāẓir* (Book of Optics), which contain a wide range of experimentally-based observations on light, both in the way it reflects from mirrors – flat and curved – and is refracted (bent) when passing from one material to another. Al-Haytham even used the refraction of the atmosphere that allows light to continue in the sky after the Sun is below the horizon to estimate the thickness of the atmosphere.

Al-Haytham's work would be highly influential on medieval Western optics, but even more impact was felt from another Arabic text, ibn Sīnā's *al-Qānūn fī al-Ṭibb*, known in the West as Avicenna's *Canon of Medicine*. Ibn Sīnā – whose full name was Abū 'Alī al-Ḥusayn ibn 'Abd Allāh ibn al-Ḥasan ibn 'Alī ibn Sīnā – was born around 980 in what is now Uzbekistan, though his work as a physician was primarily in Persia. The five volumes of the *Canon of Medicine* would become central to medical practice both in the Islamic world and Europe.

Like many of the originators of great science books of the period, ibn Sīnā used information from the best of ancient Greek writing on the topic – taking in the work of Hippocrates and Galen, including the four humours, and mixing in Aristotle's natural philosophy. He then added in his own ideas to make a heady mixture that would importantly include both a guide to natural substances with medical properties (the *Materia Medica*) and a formulary describing the manufacture of compound drugs. Although a wide range of these would no longer now be considered effective (or even safe), some indubitably had a genuine medical benefit.

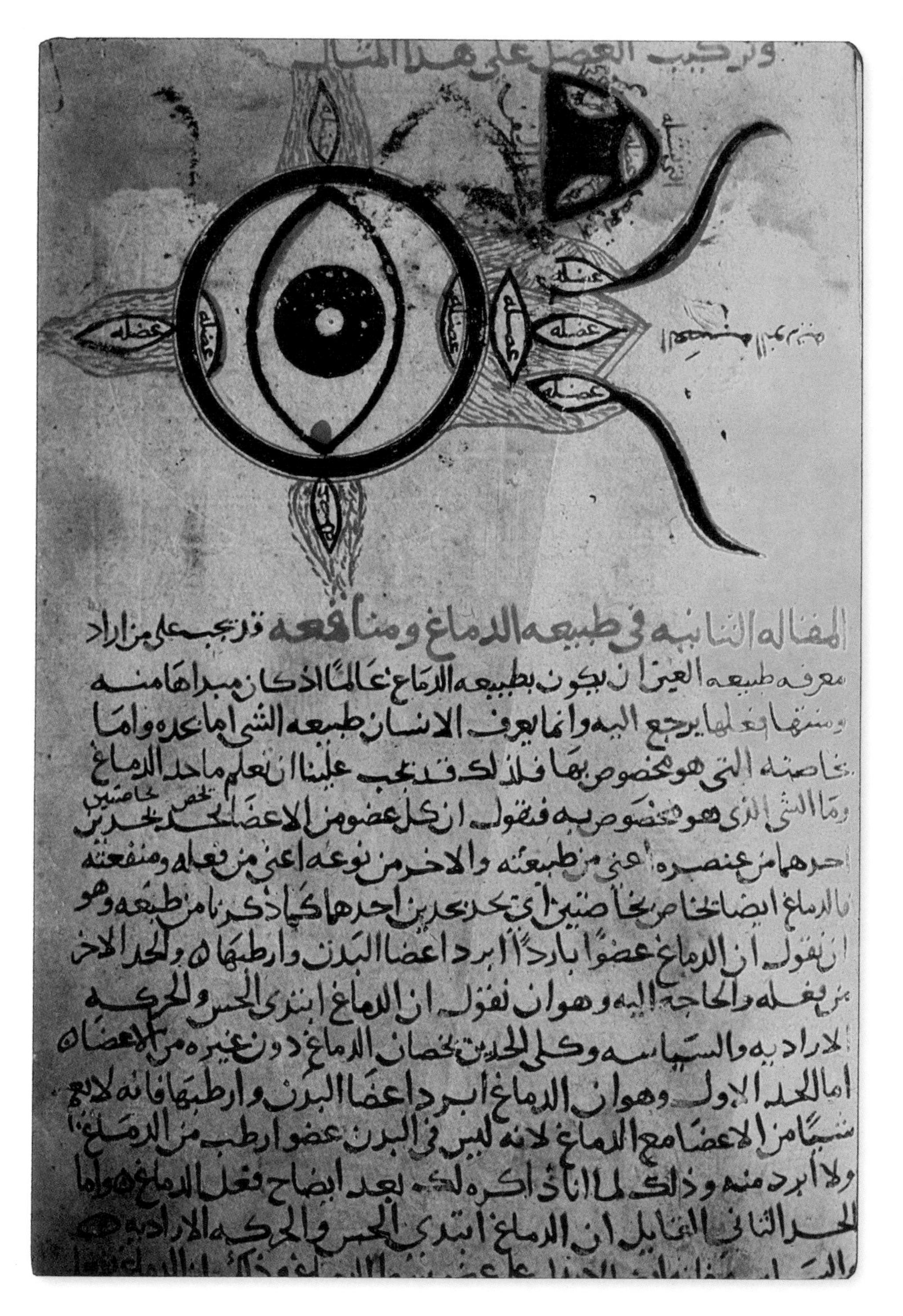

Ḥunayn Ibn-Isḥāq
AL-ASHR MAKALAT FI'L'AYN,
TWELFTH CENTURY

A twelfth-century copy of the ninth-century work *Al-Ashr Makalat Fi'l'ayn* on the structure, diseases and treatment of the eye.

Ḥunayn Ibn-Isḥāq
AL-ASHR MAKALAT FI'L'AYN,
TWELFTH CENTURY

A page from one of two known manuscript copies of the ninth-century work *Al-Ashr Makalat Fi'l'ayn* showing a schema of the eye.

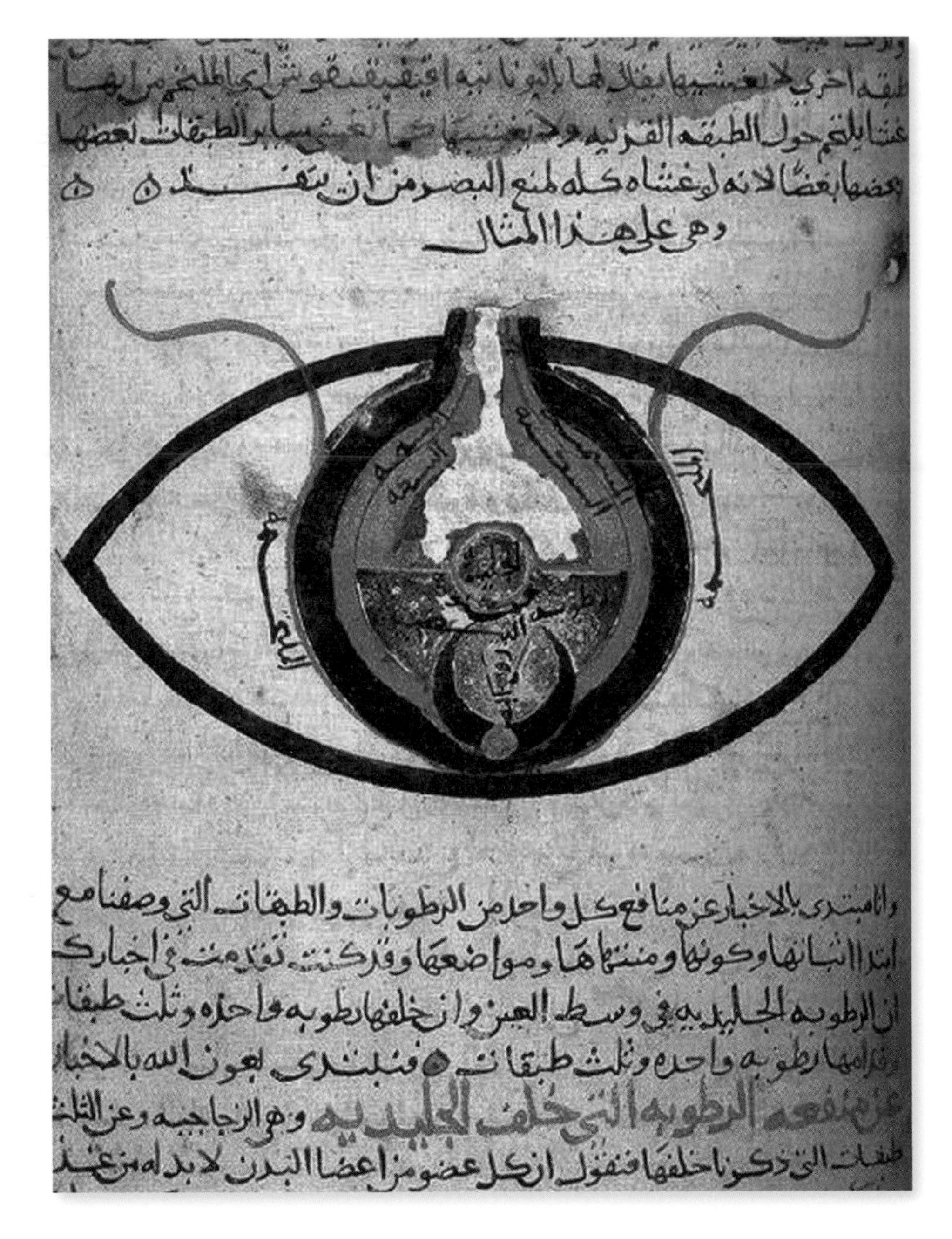

Ibn Al-Haytham
KITĀB AL-MANĀẒIR,
THIRTEENTH CENTURY

A thirteenth-century copy of the tenth/eleventh-century work *Kitāb al-Manāẓir* on light and optics, revised by Kamal al-Din al-Farisi, here showing the optical workings of the eye.

Ibn Sīnā
AL-QĀNŪN FĪ AL-ṬIBB,
1632

A copy of the eleventh-century work *al-Qānūn fī al-Ṭibb*. The cover shows a doctor taking a woman's pulse; inner pages show internal organs and the nervous system.

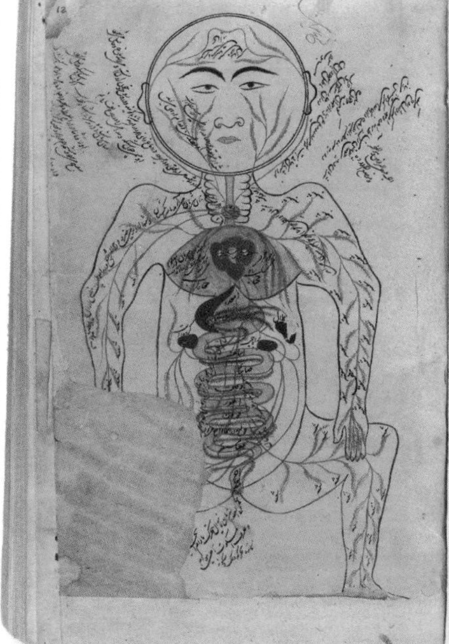

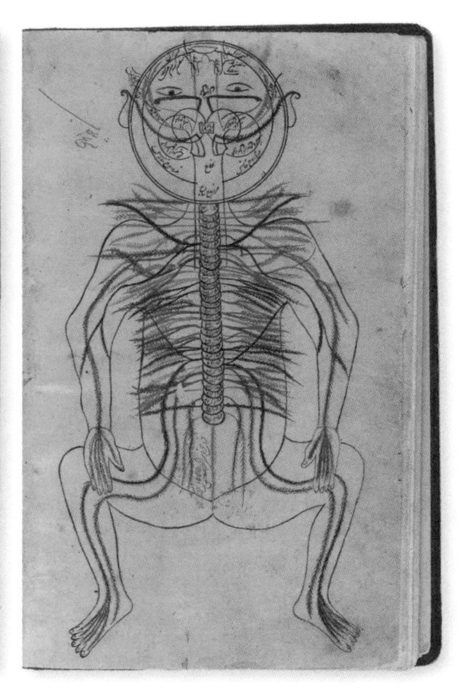

Mathematics moves on

While the Arabic-speaking world was spreading the word for mathematics that at least in part originated in India, Indian mathematicians were not standing still. By the twelfth century, another mathematical genius was rivalling Brahmagupta. His name was Bhāskara, often known as Bhāskara II to avoid confusion with a seventh-century mathematician. Born in 1114, probably in the modern state of Karnataka, Bhāskara is known for a single important work, the *Siddhānta Śiromaṇī*, a name not dissimilar to Ptolemy's *Almagest* in meaning, translating as 'Crown of Treatises'.

The four parts of the book cover arithmetic and measurement, algebra, the movement of the planets and the rotation of the heavens. The first volume has many practical applications, such as the calculation of interest, but also includes more sophisticated number theory concepts such as zero and negative numbers. (Like Brahmagupta, Bhāskara had problems with $0 \div 0$, in his case declaring the result to be infinite.) However, more impressive was the algebra in the second volume, which covers a far wider range of equations than earlier texts and develops some early ideas on what would become calculus (though there is no evidence that the subsequent seventeenth-century development of calculus as we now know it was influenced by this book). Much of the astronomical work in the book was based on existing models, both from ancient Greece and earlier Indian philosophers, but Bhāskara seems to have improved on their calculations to give more accurate values.

Mathematical skill was also central to a highly influential work from this period – Fibonacci's *Liber Abaci* (Book of Calculation), written in 1202. Properly Leonardo of Pisa, Fibonacci (a nickname based on 'son of Bonacci') was a master Italian

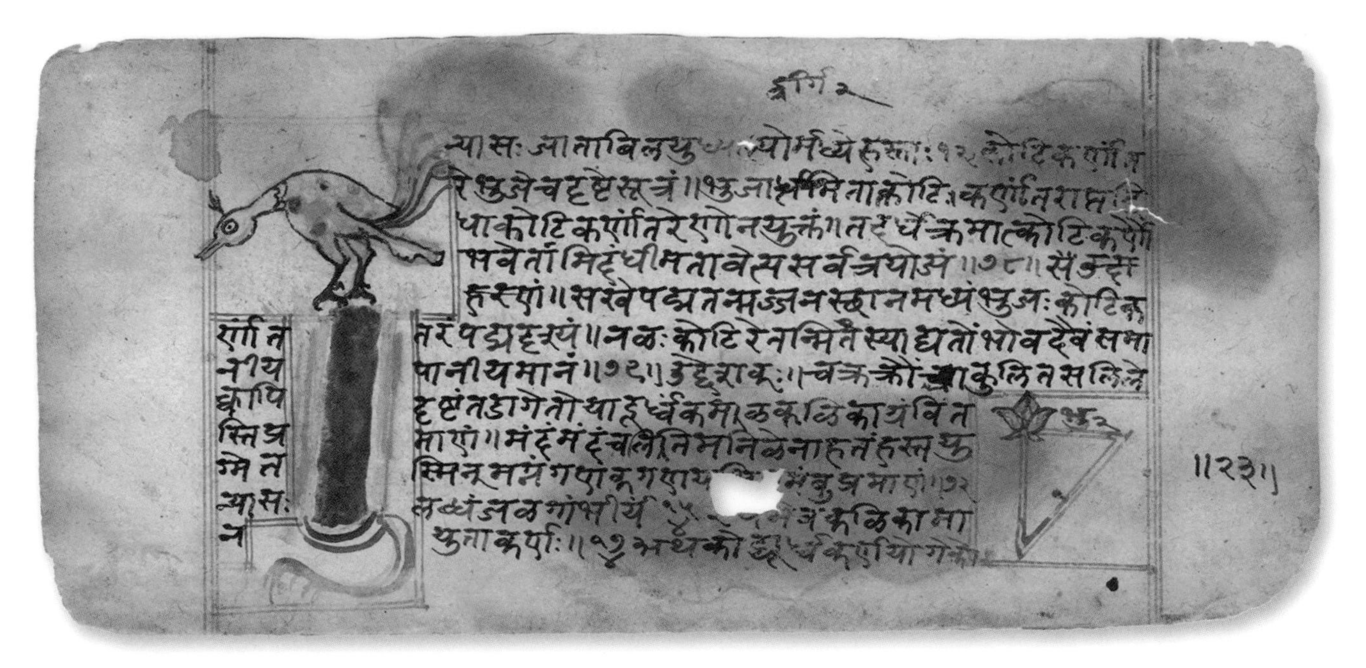

Bhāskara II
LILAVATI, 1650

A page from Lilavati, the first volume of *Siddhānta Śiromaṇī*. It uses the Pythagorean theorem to work out a problem where a snake is heading from a pillar to its hole and a peacock has to fly down along the hypotenuse to catch it.

mathematician who was born in Pisa around 1175. Like most mathematical books of the period, the *Liber Abaci* contains practical mathematical tips on, for example, the calculation of interest. It also came closer to our modern representation of fractions. Before Fibonacci, these were written out as the Greeks had as combinations of 1/x rather than having larger numbers above the dividing line. So, for instance, what we would denote ¾, prior to Fibonacci could only be represented as ½ + ¼. *Liber Abaci* also introduced the Fibonacci series – the series produced by adding together the previous two numbers that begins 1, 1, 2, 3, 5, 8, 13, 21… which Fibonacci illustrated with the growth in population from a breeding pair of rabbits.

However, the book's main claim to fame was in succeeding where translations of *Al-kitāb al-mukhtaṣar* (see page 60) hadn't in spreading the use of Arabic numerals into Europe, which Fibonacci accurately described as 'Indian style'. At the time, numbers were either written out as words or represented using the clumsy Roman numerals, making the more modern system so much more practical. As we have seen, *Liber Abaci* was not the first foreign book to praise the benefits of the Indian system – in fact, they had already been highlighted as far back as 662 by a Syrian bishop – but it seems to have been the one that really lit the spark for the use of these representations.

Along with numerals, *Liber Abaci* introduced the use of the zero to Europe, an essential for the development of modern mathematics. The zero proved more popular initially with mathematicians than it did with accountants. It was easy to change a zero into a 6, 8 or 9 in a manuscript, so it was treated with suspicion. In 1299, for example, the council of the Italian city of Florence banned the use of Indian numerals in accounts for this reason. Even as late as the sixteenth century, a Belgian priest informed his suppliers that they should only use words as numerical values in their contracts.

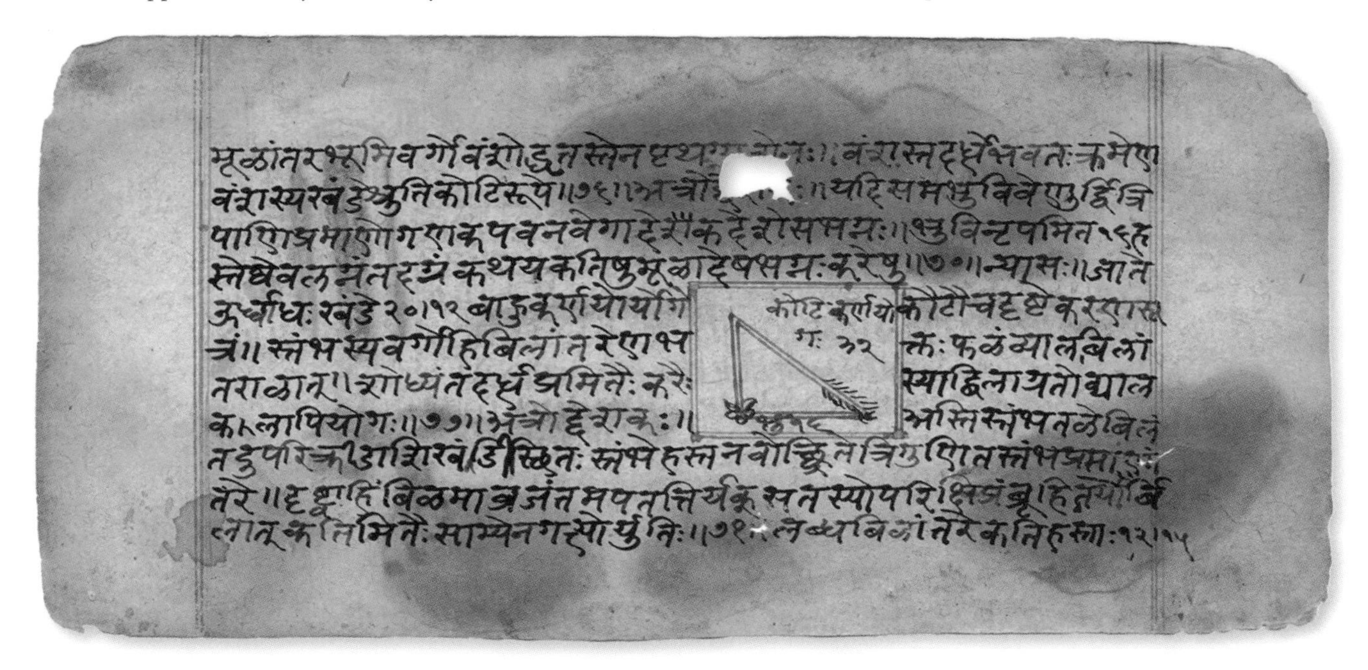

Bhāskara II
LILAVATI, 1650

This page features another example of Pythagoras's theorem.

Leonardo of Pisa (Fibonacci)
LIBER ABACI, 1227

Pages from the book that introduced the Fibonacci series and popularized Arabic/Indian numerals, showing fractions, in the collection of the Biblioteca Nazionale Centrale, Florence.

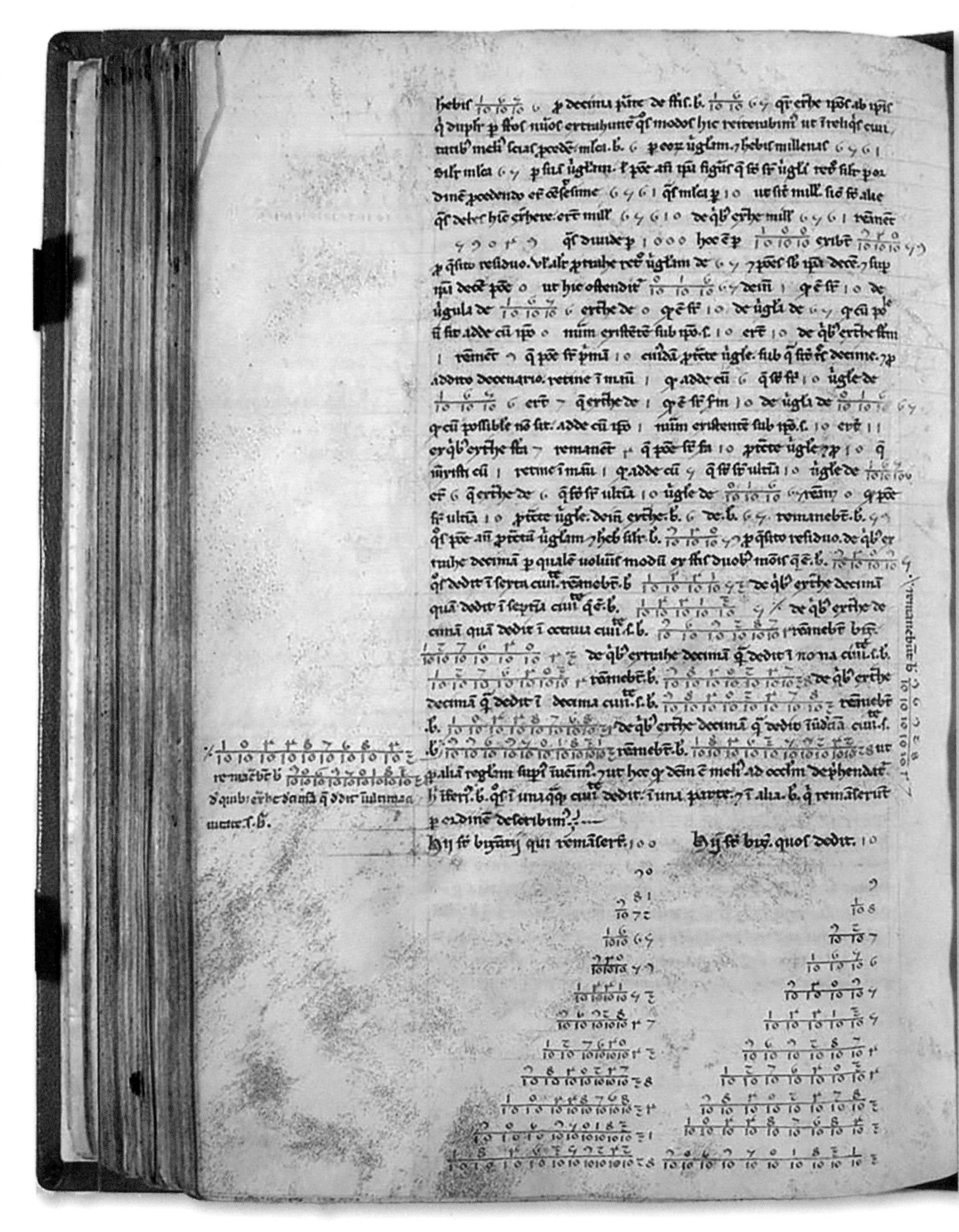

1[illegible]0

[illegible] qđ est de bucee i qua sunt barriles 100 uini ex quib' p singlos menses extra
hit' decimu residui. et qrit' qt barriles i fine anni remanserit. s. p' 1[illegible] menses.
Illa si cit' ꝓpositu eet. q quidā hns bizātios precit p 1[illegible] ciuitates et dedit i unaquaq
ciuitate decimā residui suor bizātiorum et remanserūt ei. b. [illegible] [illegible]
et qrit qta sit suma bizātior ipsi. descibe i ordine duodecies [illegible] sub quadā u
gula sta ree ut hic ostendit [illegible] et mlca
bizātios [illegible] cu suis fractionib' p o [illegible] 10 q sunt sub duoceci nouenarijs p[illegible]
ugle. et diuide sumā ipsi multiplicationis p illos duodecim nouenarios. cu positi
fuerint sub quadā ugla. et sic hebis. b. 100 p suma illi.
[illegible]uidā hns bizātios uoluit exire de quadā ciuitate hnte portas 10 et opo
tuit eu dare prime porte [illegible] suor bizātior. et insup [illegible] uni biz. s. dimi
diu biz. qs ibi attulit et isti [illegible] uni biz. tercie tertiu et [illegible] uni biz. et [illegible]
qrtā et [illegible] uni biz. et ita p ordinē usq q i decima porta dedit deci
mā biz. qs ibi attulit. et [illegible] uni biz. et remansit ei. b. 1 qrit qt
biz. hūit ille. Potest itaq h qstio dupli solui. pmū qdē ut de porta i
porta ab ultima icipiendo [illegible] sic. Illā i fine remansit ei. b. 1 et de
dit ultime porte decimā uni biz. g' cu ipse dedit ipam decimā. hebat
biz. [illegible] et q' [illegible] dederat decimā biz. qs ibi attulit. et remansit ei ipsi
biz. [illegible] iuueniend' e nus ex q extracta [illegible] remaneat. b. [illegible] que num po
ne ut sit 10 [illegible] regulas arbor. de q extracto [illegible] remanet [illegible] q [illegible]
uellet ee [illegible] qre mlca 10 p [illegible] et diuide p [illegible] exibt. b. [illegible] et tm re
mansit ei p' nonā portā. cu q adde [illegible] uni. b. q dedit none porte. erit
b. [illegible] [illegible] inuenies num de q extracta [illegible] ipsi remaneat. b. [illegible] [illegible]
g' q num ille sit [illegible] de q extracto [illegible] remanet [illegible] q uellet ee [illegible]
qre mlca [illegible] p [illegible] et diuide p [illegible] exib. b. [illegible] et tm remansit ei p' octa
ua portā. cu q adde [illegible] uni. b. q dedit i ipsa porta. erit. b. [illegible] que de
[illegible]strata [illegible] mlca p [illegible] et diuide p 7 exib. b. [illegible] cu q adde [illegible] uni
b. q dedit septie porte erit. b. [illegible] qs mlca p 7 et diuide p 6 exibt
b. [illegible] cu qb' adde [illegible] uni. b. que dedit sexte porte. erit. b. [illegible]
qs mlca p 6 et diuide p [illegible] exibt. b. [illegible] cu qb' adde [illegible] uni. b. q
dedit quite porte erit. b. [illegible] qs mlca p [illegible] et diuide p [illegible] exibt. b. [illegible]
cu qb' adde [illegible] uni. b. q dedit qrte porte. erit. b. [illegible] qs mlca p [illegible]
et diuide p [illegible] exibt. b. [illegible] cu quib' adde [illegible] uni. b. q dedit tercie
porte. erit. b. [illegible] qs mlca p [illegible] et diuide p [illegible] exibt. b. [illegible] et tot rema
serūt ei p' scdā portā. cu qb' adde [illegible] uni. b. q dedit i scda porta. erit. b. [illegible]
p qb' inuenies num ex q extracto dimidio ipsi remaneat. b. [illegible] p q nūo
pones [illegible] ex quo extracto dimidio remanet [illegible] q uellet ee [illegible] qre mlca
[illegible] p [illegible] et diuide p 1 exibt. b. [illegible] et tot remanserūt ei p' primā por
tā. cu qb' adde [illegible] uni. b. qs dedit i prima porta. erit. b. [illegible] p qb'

Suma
100

Against all odds

By the thirteenth century, when Fibonacci was writing, an increasing number of Europeans were producing books that primarily commented on ancient Greek and Islamic science, but also added new ideas of their own. Few of these books have had a lasting impact, but the *Opus Majus* of Roger Bacon is worth picking out, both as one of the most impressive examples of its kind and also because of Bacon's remarkable story.

A Franciscan friar based mostly in the English city of Oxford, Bacon was born in either 1214 or 1220 (there are two interpretations of his only autobiographical comment). Bacon seems to have had a driven personality. Despite a ruling from the head of his order that friars should not write books, Bacon was determined to produce a scientific encyclopaedia and looked for political support in the church to enable him to do so. The French cardinal Guy de Foulques had shown an interest in Bacon's work, so Bacon asked de Foulques to get him an exception from the rule and some funding. Unfortunately, the request seems to have become garbled along the way; after two years, de Foulques finally replied giving his support to Bacon and asking that the (non-existent) book be sent to him immediately. To make matters worse, de Foulques told Bacon to do this secretly, so gave him no defence against the Franciscan prohibition of writing books and provided no cash, which was a disaster as Bacon had already used up his inheritance.

ROGER BACON, 1617

No contemporary images are known to exist of this thirteenth-century English friar, proto-scientist and author; this illustration is from *Symbola Aureae Mensae* by Michael Maier (1568–1622).

Roger Bacon
OPUS MAJUS,
FIFTEENTH CENTURY

This copy of Bacon's *Opus Majus* was donated to the Bodleian Library, Oxford, in 1634, where it has remained ever since.

With things looking dire, Bacon had a stroke of luck. To wide surprise, de Foulques was elected Pope, as Clement IV. Bacon obtained formal papal blessing on his venture and decided to write a proposal for his great encyclopaedia, intending to provide a covering letter and a short synopsis of the book. To say Bacon got carried away would be an understatement: his proposal ended up 500,000 words long. While this was being copied, he started on the covering letter – which also became a major manuscript. And this happened once again – all in two years from 1266 to 1267. These three volumes became known as the *Opus Majus* (Great Work), *Opus Minus* (Lesser Work) and *Opus Tertius* (Third Work). Between them they included mathematics and astronomy, optics, geography, philosophy and much more – notably including a section describing the importance of experience and experiment to understanding nature, rather than relying purely on philosophical musing.

Bacon sent off the first two volumes of his vast proposal to the Pope – it's likely the third was still being copied at this point. Before he could get a response, however, the news broke that the Pope had died. In all likelihood, Clement never saw Bacon's remarkable books. Clement's successor as Pope had no interest in science, and, according to legend, by his order Bacon was imprisoned for his actions for as much as 13 years. At the time, Bacon's books were suppressed, but remarkably they survived and provide an exceptional picture of the science of the period. While Bacon contributed some original ideas, particularly on calendar reform and the nature of light, and even dared to question some of Aristotle's thinking, the importance of the *Opus Majus* is in the scale of Bacon's vision.

The books covered in this chapter, important though they were, were restricted in the number of people they could reach, both by limited literacy and the need to manually copy each volume. However, in the next chapter, with the advent of the printing press and more widespread literacy, we will see books that increasingly bore fruit.

Roger Bacon
OPUS MAJUS,
THIRTEENTH CENTURY

The earliest known copy of some of Bacon's work, housed in the British Library, London. The page shown is from part five of the *Opus Majus*, titled 'Tractatus perspective'.

2

RENAISSANCE IN PRINT

THE REVOLUTION IN BOOKS

It is fashionable to talk of revolutions in science, but the period covered in this chapter – from around 1200 to the end of the eighteenth century – featured two parallel revolutions. In the world of books, the introduction of moveable-type printing technology made it possible for the science book to reach a much wider audience. And in science itself, the work of Copernicus, Newton and others transformed our view of the universe, and of the way that science was undertaken. It was during this period that natural philosophy evolved into science.

The historian of science David Wooton points out in his excellent 2015 book *The Invention of Science* how this period, for example, saw the literal invention of the concept of discovery. When Columbus attempted to sail west to China in 1492 and instead hit on the New World, he did not have an appropriate word to describe what he had done; amongst European languages, the word 'discovery' or its equivalent only existed in Portuguese at the time (and even there had only been introduced a few years before). The idea of looking outwards and making discoveries was the hallmark of this new era in science. Up until this period, the tendency was not to look outwards but inwards to philosophical musings, and backwards to try to use and interpret ancient wisdom. The Renaissance brought the urge to discover and think anew.

It might seem odd to make use of what seems like simply a voyage of exploration to illustrate the changing nature of science – yet Columbus's voyages were amongst the first observations to clearly counter Aristotle's model of the universe. The four earthly elements in Aristotle's world might be combined in various ways, but his primary

COLUMBUS'S LETTER, 1494

A woodcut Latin version, printed in Basel, of Columbus's letter announcing the discovery of the New World, written to King Ferdinand of Spain on the caravel ship Nina on 15 February 1493.

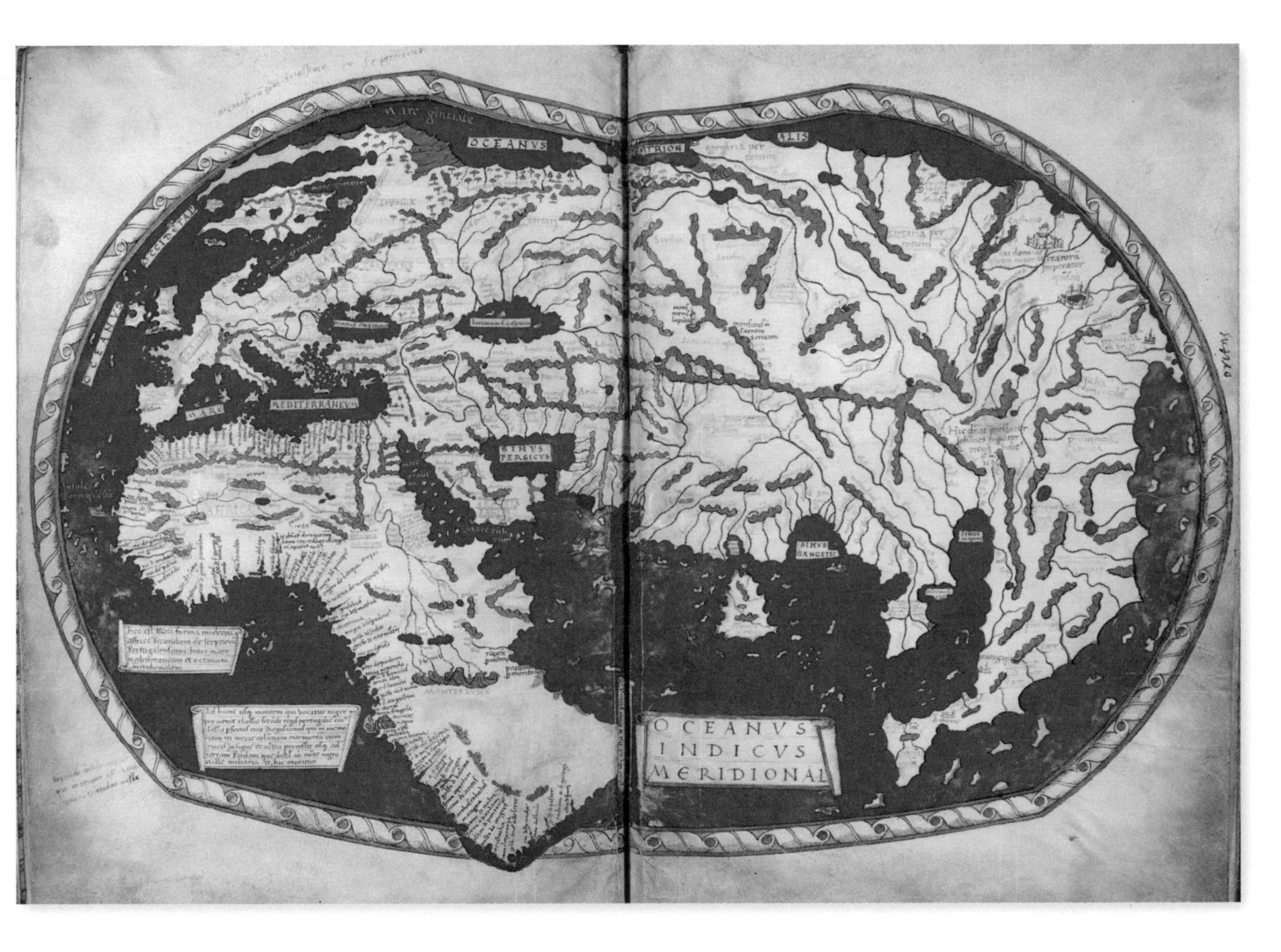

Henricus Martellus Germanus
WORLD MAP, CA. 1489

A world map contemporary with Columbus's voyage, created by German cartographer Heinrich Hammer, which may have been taken from a map by Columbus's brother Bartolomeo.

structure for the centre of the universe required there to be a sphere of earth, surrounded by a sphere of water, surrounded by a sphere of air, surrounded by a sphere of fire – each having less of a tendency to be at the centre of everything than the sphere it contained.

If the universe was perfectly centred, the Earth would have no land at all above the surface of the water, which would have been inconvenient to say the least. It was therefore accepted that the sphere of earth was off-centre, making it possible for a chunk of earth to stand out above the waters, forming the land. If that were the case, though, apart from small local cracks such as the English Channel, there had to be a single contiguous land mass. As it became clearer that the New World was widely separated from Europe, Aristotle's model seemed increasingly unlikely, laying the foundations for easier acceptance of one of the most famous Renaissance titles, *De Revolutionibus Orbium Coelestium* by Copernicus (see page 87).

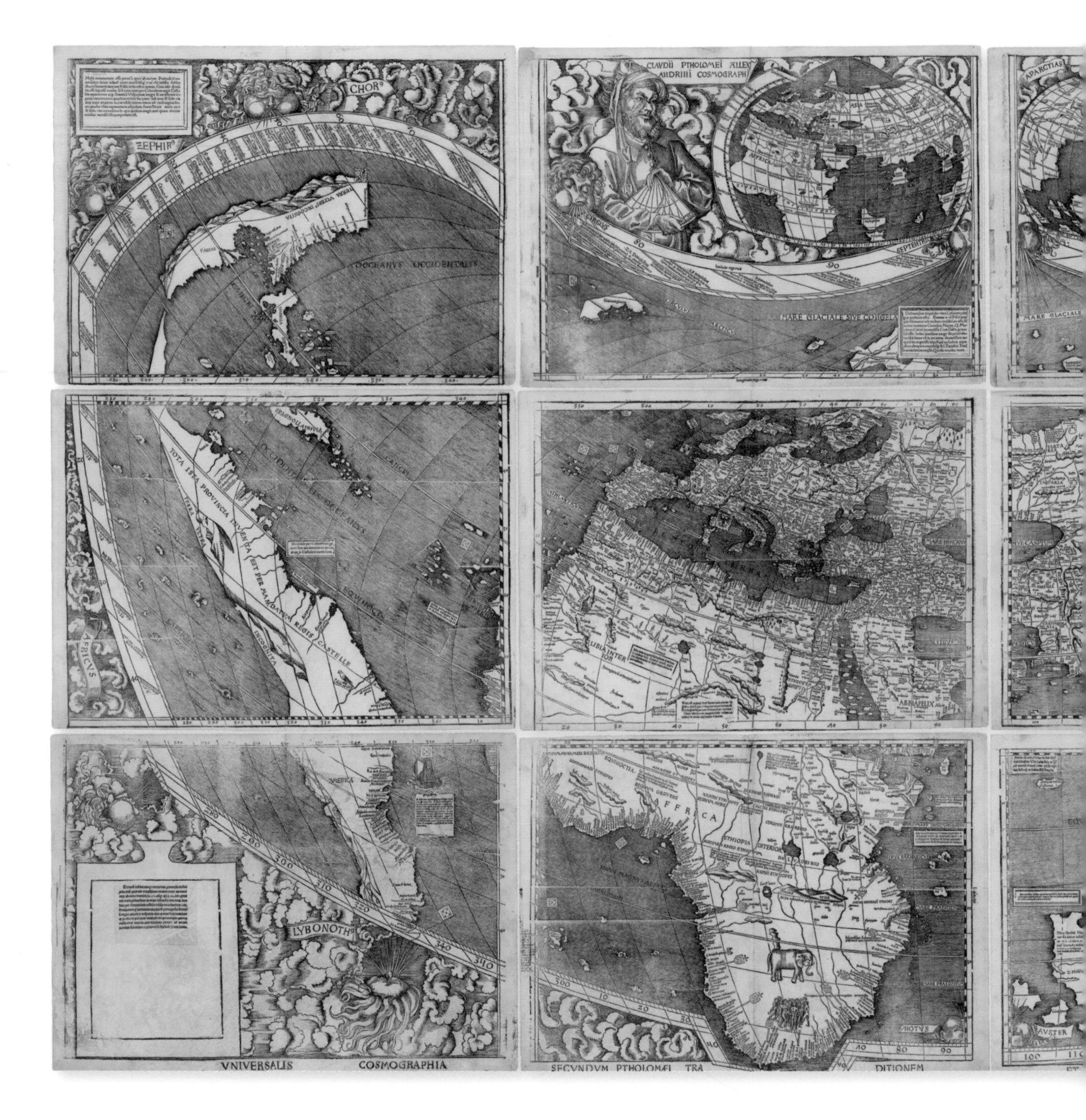

CHOR
ZEPHIR
OCEANVS OCCIDENTALIS
CLAVDII PTHOLOMEI ALEX
ANDRINI COSMOGRAPHI
SEPTENTRIO
MARE GLACIALE SIVE CONGELA
APARCTIAS
MARE GLACIALE
TOTA ISTA PROVINCIA INVENTA EST PER MANDATVM REGIS CASTELLE
TERRA VLTRA INCOGNITA
AFRICVS
LIBIA INTERIOR
AFFRICA
ETHIOPIA INTERIOR
LYBONOTH
NOTVS
AVSTER
VNIVERSALIS COSMOGRAPHIA
SECVNDVM PTHOLOMAEI TRA DITIONEM

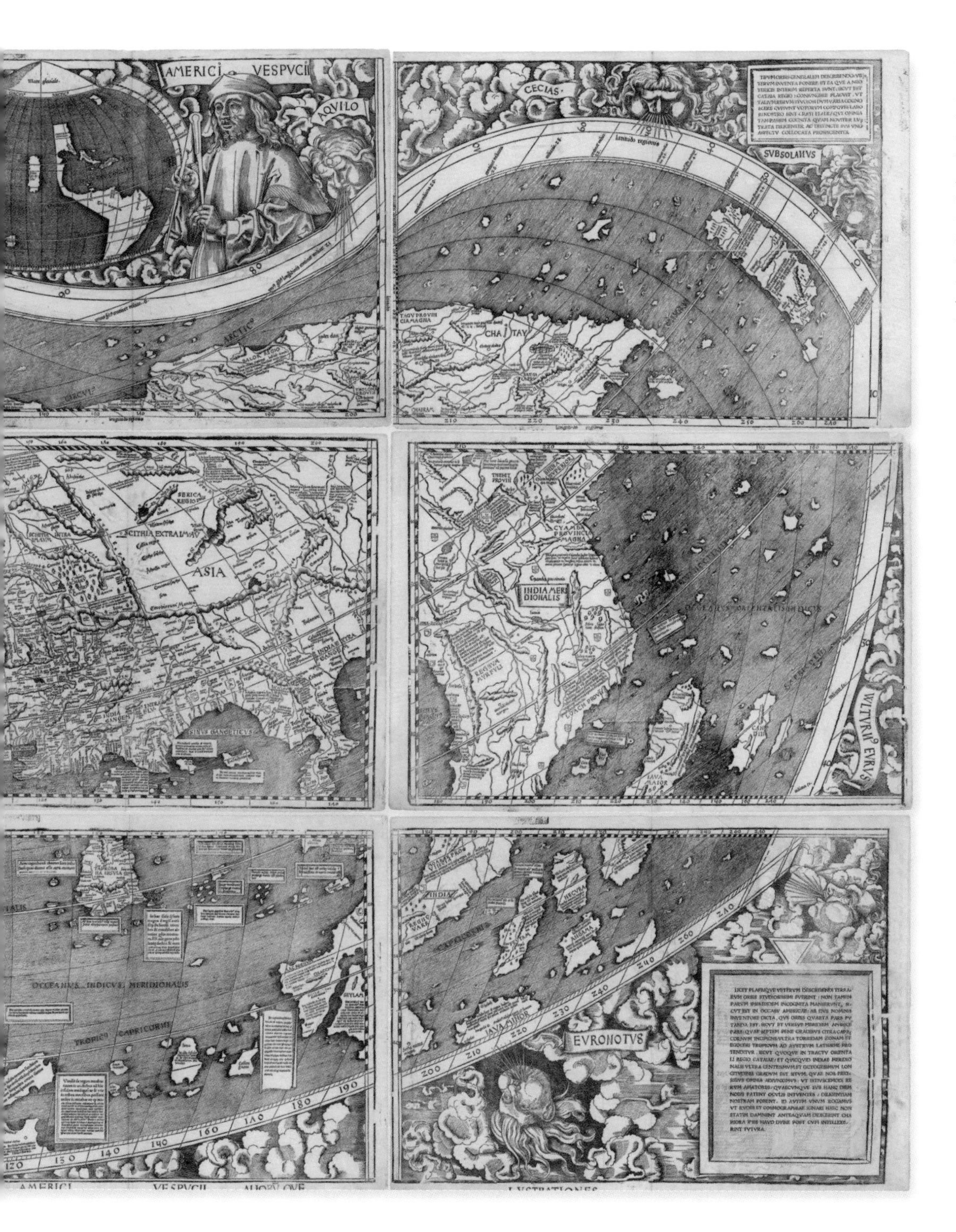

Martin Waldseemüller
UNIVERSALIS COSMOGRAPHIA,
VOSGEAN GYMNASIUM,
CA. 1507

A wall map by the German cartographer, the first to use the name America – its full title was *Universalis cosmographia secundum Ptholomaei traditionem et Americi Vespucii aliorumque lustrationes.*

Master of invention

First, though, we need to take a step back to a collection of old-style handwritten books that introduced no new scientific facts, but that are iconic in the history of science and technology. These are the notebooks of Leonardo da Vinci. His work marks perfectly the transitional period between the old and the new – the notebooks were produced after moveable type was invented, but were handwritten and never intended for publication – yet Leonardo's work was amongst the first where the visual nature of some of the content makes it as attractive to the non-technical reader as is it is to the engineer. This is doubly ironic as Leonardo appeared to go out of his way to make his texts inaccessible, often working in mirror writing and making notes that were intended for his eyes only.

In some ways, Leonardo's approach is reminiscent of the view put across by Roger Bacon, the thirteenth-century English friar. As Bacon stated, 'The cause of the obscurity in the writings of all wise men has been that the crowd derides and neglects the secrets of wisdom and knows nothing of the use of these exceedingly important matters. And, if by chance, any magnificent truth falls to its notice, it seizes upon it and abuses it to the manifold disadvantage of persons and of the community.' This is quoted from a short work with a long title, thought to be written about 1250: *De Mirabile Potestate Artis et Naturae* (in full, its title translates as 'Letter Concerning the Marvellous Power of Art and of Nature and Concerning the Nullity of Magic'). The main content of the letter is a critique of fraudsters who pretend to do magic, contrasting this with the wonders of nature and science that Bacon mostly claimed to have observed. However, he also described a series of mechanisms for ensuring that knowledge is kept to the elite few and not made available to the masses.

Some such writing, Bacon said, is 'hidden under characters and symbols, others in enigmatical and figurative expressions'. So, he was suggesting, it was possible to use ciphers, or such elliptical language that only those in the know could follow what was being said (a familiar approach to Bacon from the parables in the Bible). He then, as illustration, went on to describe the manufacture of gunpowder several times, using a range of wording that can be hard to follow and finished with an encrypted phrase that has never been satisfactorily decoded. We can see in Leonardo's notebooks that same urge to keep the secrets of his discoveries and inventions from ordinary people. And what a remarkable series of books this produced.

Born in the town of Vinci, near Florence in Italy, in 1452, Leonardo started out in 1476 as an assistant to the artist Andrea del Verroccio, and would go on to have the definitive Renaissance career as an artist, inventor and engineer. His notebooks were voluminous. Through to his death in 1519, he wrote the equivalent of around 20 books. When he was not sketching humans, animals, plants and geological features, he was devising technology. Leonardo first came to this through producing mechanisms for the stage, which at the time featured increasingly complex mechanisms to put on spectacles where actors floated through the air and new vistas opened up before the enthralled audience.

Leonardo's notebooks range from detailed guidance on painting techniques (his pupil Francesco Melzi extracted material from the notebooks to produce a *Treatise on Painting*)

Leonardo da Vinci
SELF-PORTRAIT, CA. 1512

This red chalk picture is widely (but not universally) accepted to be a self-portrait by Leonardo, aged around 60.

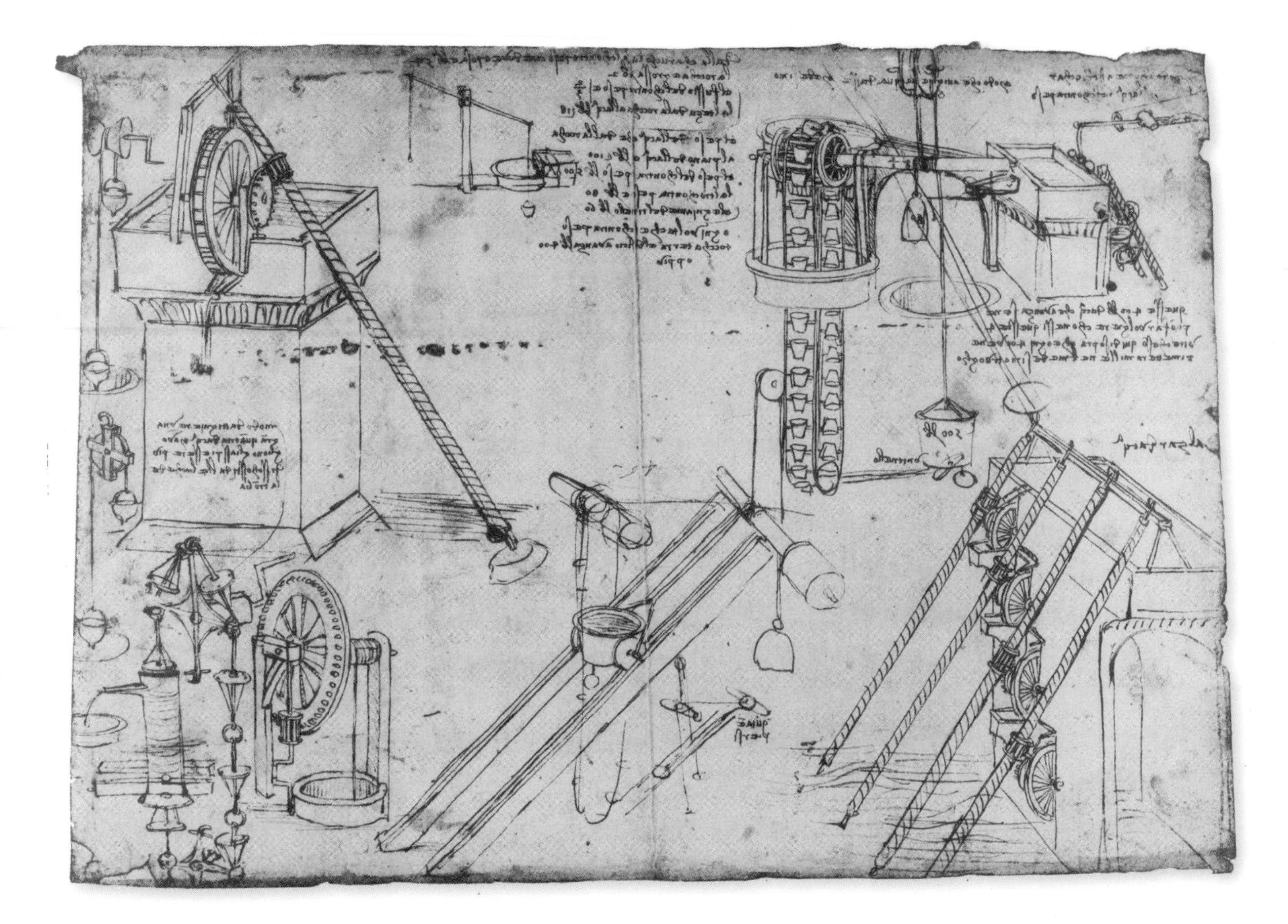

Leonardo da Vinci
CODEX ATLANTICUS,
1478–1519

A spread from one of Leonardo's notebooks, showing hydraulic machines with wheels and gears that exploit water energy.

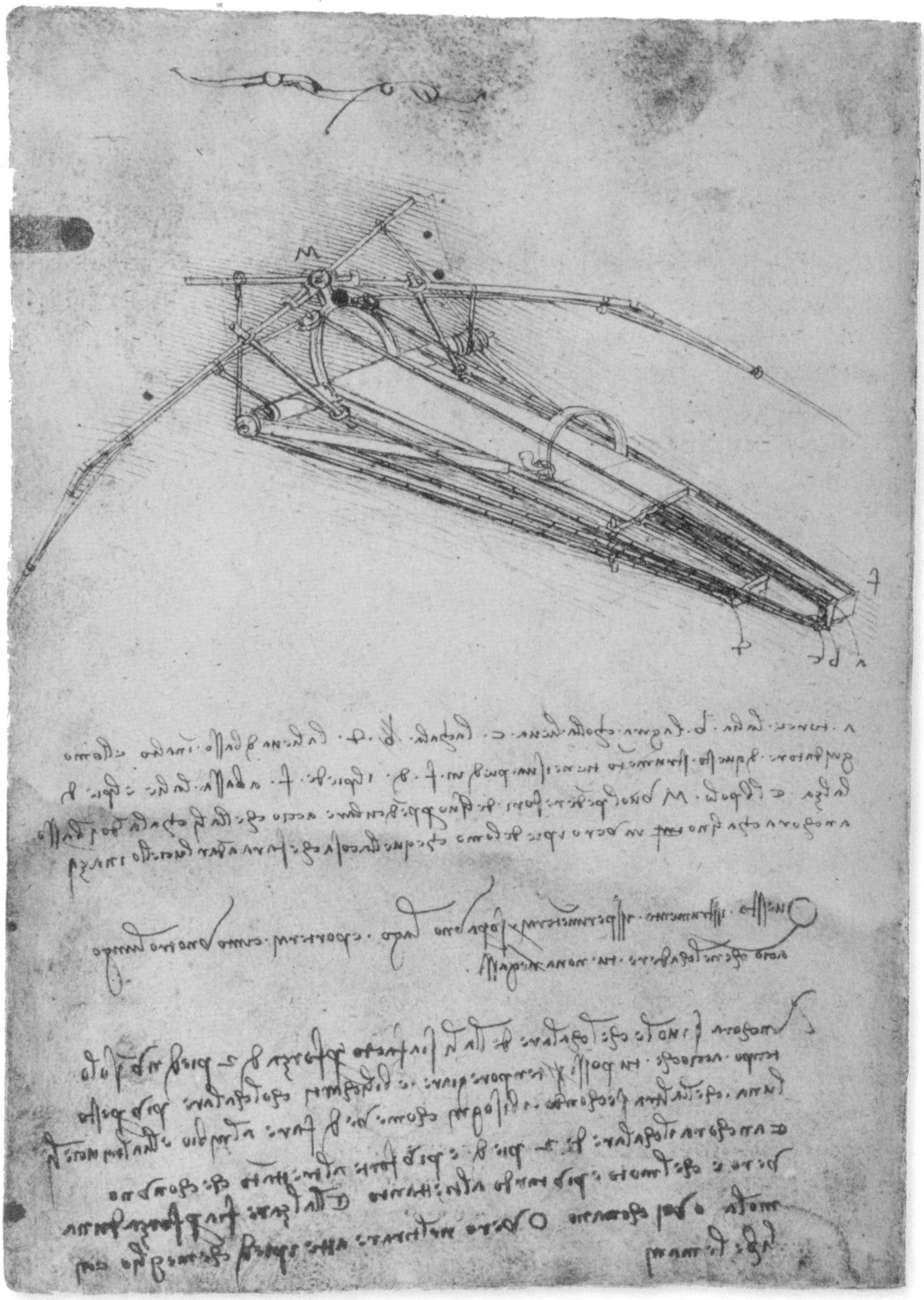

Leonardo da Vinci
CODEX ATLANTICUS,
1478–1519

A page showing 'the machine for flying', which looks very similar to a modern-day hang glider.

Leonardo da Vinci
CÓDICE MADRID I, 1493

Leonardo's sketches of chains, links and counterweights.

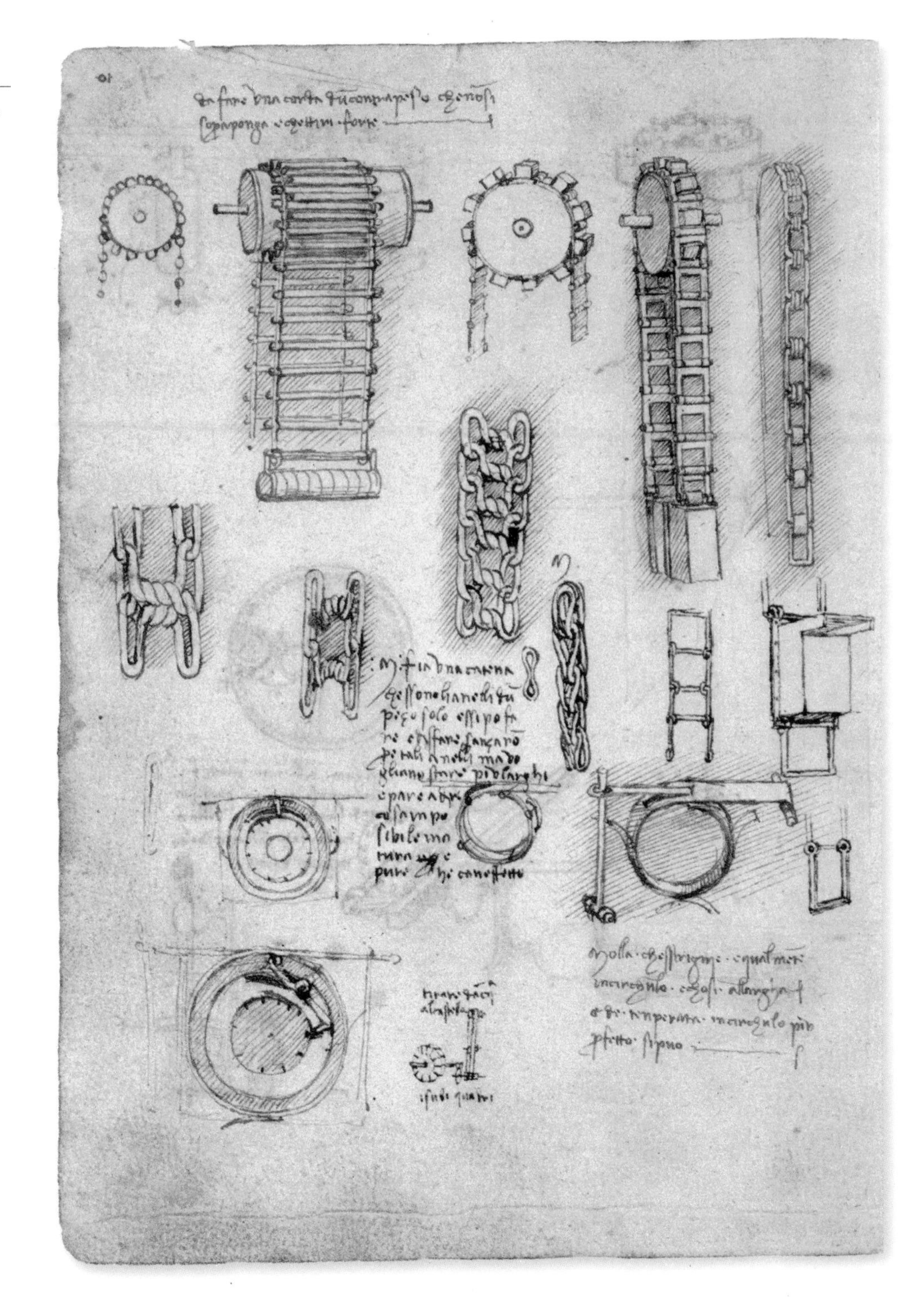

Leonardo da Vinci
CODEX ARUNDEL, 1508

Leonardo's study of breathing apparatus for a diver.

to exploded diagrams for his mechanical devices. These included automata, such as a mechanical knight and a 'self-driving' cart with a mechanism to gradually increase its steer to the right during its movement. Biologically, he drew from dissections, took a particularly modern-feeling mechanical approach to his understanding of the workings of the body and dissected eyes to understand the optics of vision before pouring all of this into his notebooks. He even included many examples from Euclid's *Elements*, illustrated in his own style.

Some of the most intricate and delightful illustrations in the notebooks show the workings of gears. These were in their relative infancy at the time, but Leonardo made use of everything from simple wheels with pegs around to sophisticated worm gears. The inventions he showed in his notebooks range from a diving suit to a form of tank, and his civil engineering included everything from canals to the design of sophisticated bridges. While Leonardo's physics may have been little different from that of Aristotle, the presentation of it in his work transformed the subject.

A revolution in the heavens

The European introduction of moveable-type printing presses in the 1440s would see a series of transforming titles that might not yet have thrown off Bacon's view that science should be kept from the masses, but that would spread far and wide among the educated, carrying the scientific message. Of these, the first to have a dramatic (if distinctly slow)

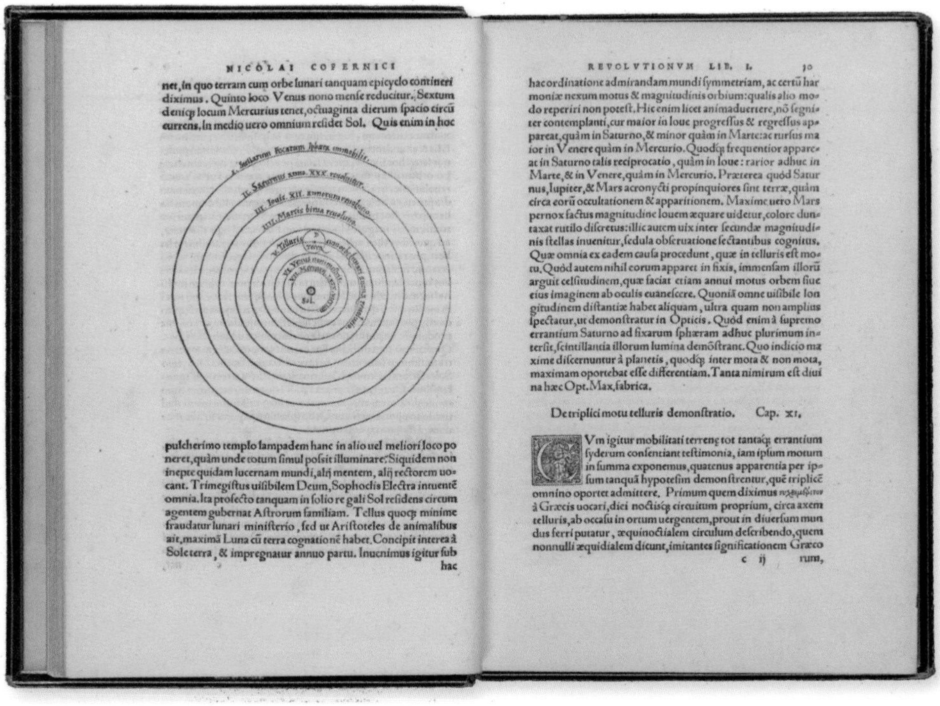

Nicolaus Copernicus
DE REVOLUTIONIBUS ORBIUM COELESTIUM,
HENRICUS PETRUS, 1566

Early edition of Copernicus's *De Revolutionibus* from 23 years after its first publication in 1543, showing the orbits of the planets around the sun.

impact was *De Revolutionibus Orbium Coelestium* (On the Revolutions of Heavenly Spheres) by Nicolaus Copernicus.

Copernicus, more accurately Mikolaj Kopernik, was born in Thorn (now Toruń) in Poland in 1473. Nominally a canon in the church, Copernicus never became an active cleric, but he did make use of part of his lengthy university education (he studied until he was 30), acting as physician to his uncle. Over the years, he developed a significant interest in astronomy and made many observations. Astronomers were still making use of the epicycles from Ptolemy's day (see page 54), and Copernicus gradually became convinced that the only way to improve on this explanation featuring the mess of strange motions in the heavens, was to redraw the model of the universe (the solar system in modern terms) putting the Sun rather than Earth at the centre. Copernicus did this in *De Revolutionibus*.

The manuscript of this famous title was pretty well complete in the 1530s, but Copernicus hesitated to publish it, and it would only be printed shortly before his death in 1543. In the final version, the Lutherian minister Andreas Osiander, who acted as publisher, added an introductory letter which made it clear that this was a model that made calculation easier, rather than being necessarily a more accurate version of the truth than Ptolemy's model. (It's often thought that resistance to the Copernican theory came solely from the Catholic Church, which banned the title from 1616 to 1835, but early Protestant leader Martin Luther was also scathing about the book, and called Copernicus a fool. In both cases, the problem was that the Bible mentions the Sun moving in the sky, rather than the Earth rotating.)

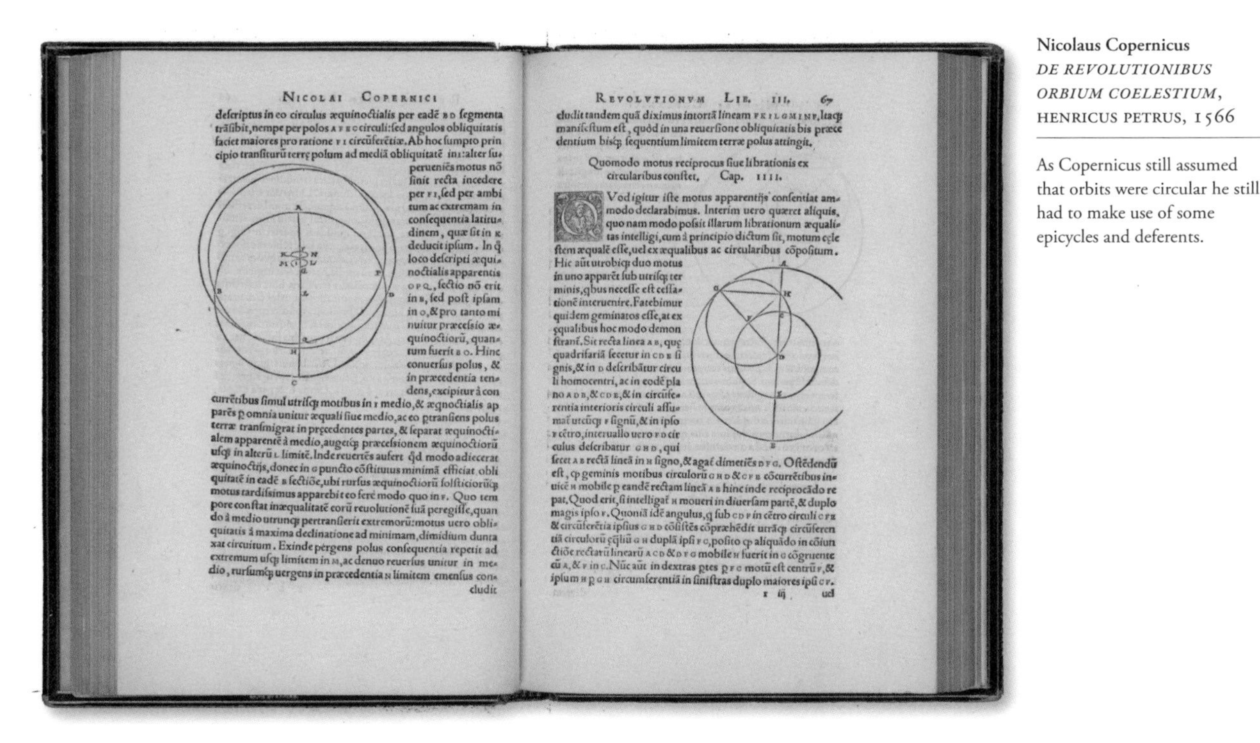

NICOLAI COPERNICI

descriptus in eo circulus æquinoctialis per eadẽ B D segmenta trãsibit, nempe per polos A F E G circuli: sed angulos obliquitatis faciet maiores pro ratione F I circũferẽtiæ. Ab hoc sumpto principio transiturũ terrę polum ad mediã obliquitatẽ in I: alter superueniẽs motus nõ sinit recta incedere per F I, sed per ambitum ac extremam in consequentia latitudinem, quæ sit in K deducit ipsum. In q̃ loco descripti æquinoctialis apparentis O P Q, sectio nõ erit in B, sed post ipsam in O, & pro tanto minuitur præcessio æquinoctiorũ, quantum fuerit B O. Hinc conuersus polus, & in præcedentia tendens, excipitur à concurrẽtibus simul utrisq; motibus in I medio, & æquinoctialis apparẽs p omnia unitur æquali siue medio, ac eo ptransiens polus terræ transmigrat in pręcedentes partes, & separat æquinoctialem apparentẽ à medio, augetq; præcessionem æquinoctiorũ usq; in alterũ L limitẽ. Inde reuertẽs aufert q̃d modo adiecerat æquinoctijs, donec in G puncto cõstitutus minimã efficiat obliquitatẽ in eadẽ B sectiõe, ubi rursus æquinoctiorũ solsticiorũq; motus tardissimus apparebit eo ferè modo quo in F. Quo tempore constat inæqualitatẽ eorũ reuolutionẽ suã peregisse, quando à medio utrunq; pertransierit extremorũ: motus uero obliquitatis à maxima declinatione ad minimam, dimidium duntaxat circuitum. Exinde pergens polus consequentia repetit ad extremum usq; limitem in M, ac denuo reuersus unitur in medio, rursumq; uergens in præcedentia N limitem emensus concludit

REVOLVTIONVM LIB. III. 67

duxit tandem quã diximus intortã lineam F K I L G M I N F, Itaq; manifestum est, quòd in una reuersione obliquitatis bis præcedentium bisq; sequentium limitem terræ polus attingit.

Quomodo motus reciprocus siue librationis ex circularibus constet. Cap. IIII.

QVod igitur iste motus apparentijs consentiat amodo declarabimus. Interim uero quæret aliquis, quo nam modo possit illarum librationum æqualitas intelligi, cum à principio dictum sit, motum cęlestem æqualẽ esse, uel ex æqualibus ac circularibus cõpositum. Hic aũt utrobiq; duo motus in uno apparẽt sub utrisq; terminis, q̃bus necesse est cessationẽ interuenire. Fatebimur quidem geminatos esse, at ex ęqualibus hoc modo demonstrari. Sit recta linea A B, quę quadrifariã secetur in C D E signis, & in D describãtur circuli homocentri, ac in eodẽ plano A D B, & C D E, & in circũferentia interioris circuli assumatur utcũq; F signũ, & in ipso F cẽtro, interuallo uero F D circulus describatur G H D, qui secet A B rectã lineã in H signo, & agat dimetiẽs D F G. Ostẽdendũ est, q geminis motibus circulorũ G H D & C F E cõcurrẽtibus inuicẽ H mobile p eandẽ rectam lineã A B hinc inde reciprocãdo repat. Quod erit, si intelligat H moueri in diuersam partẽ, & duplo magis ipso F. Quoniã idẽ angulus, q sub C D F in cẽtro circuli C F E & circũferẽtia ipsius G H D cõsistẽs cõpræhẽdit utrãq; circũferentiã circulorũ ęq̃liũ G H duplã ipsi F C, posito q aliquãdo in cõiunctiõe rectarũ linearũ A C D & D F G mobile H fuerit in G cõgruente cũ A, & F in C. Nũc aũt in dextras ptes p F C motũ est centrũ F, & ipsum H p G H circumferentiã in sinistras duplo maiores ipsi C F.

E iij   uel

Nicolaus Copernicus
DE REVOLUTIONIBUS ORBIUM COELESTIUM,
HENRICUS PETRUS, 1566

As Copernicus still assumed that orbits were circular he still had to make use of some epicycles and deferents.

Das erst general/inhaltend die beschreibung vn
Chorus
Zephyrus
Africus Lips
Terra Florida
Cuba
Europa
Africa
Aethiopia
America vel Brasilij insu

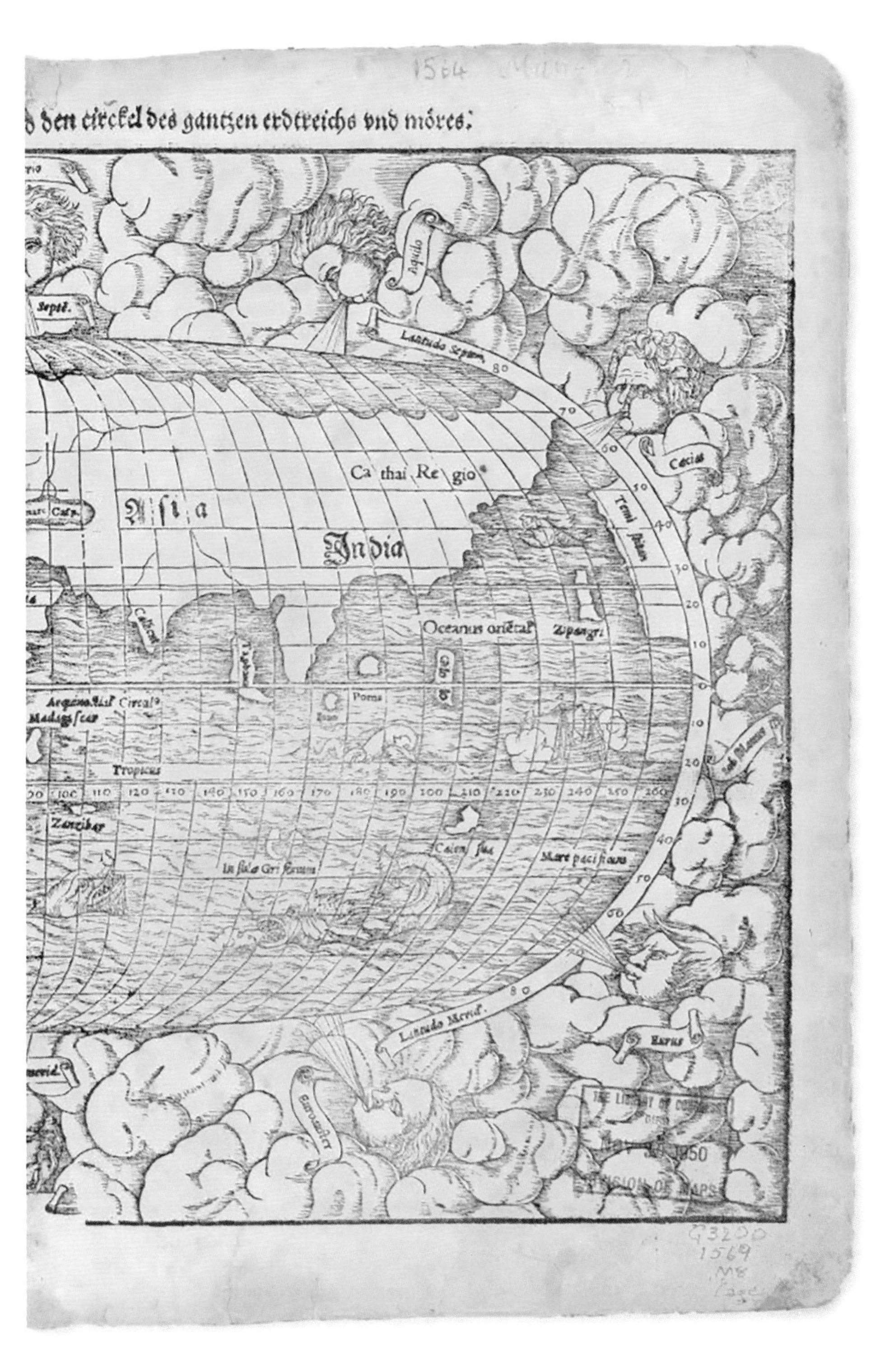

Sebastian Münster
COSMOGRAPHIA,
HENRICHUM PETRI, 1564

An early world map from Münster's far-reaching geographical, political and scientific encyclopaedia, in an edition from around 20 years after its original printing in 1544.

It's easy to picture Copernicus making the leap from Ptolemy's model to our current one in a single bound, but in reality, he still retained the idea that planets should be attached to perfect spheres, giving them circular orbits. Although his approach of putting the Sun at the centre did away with the strange retrograde motion of the outer planets, circular orbits simply didn't work properly, and Copernicus still had to make use of Ptolemy's idea of eccentrics in order to match his theory with observation.

Astronomy and cosmology also featured in another book from around the same period – a title destined to outsell not only Copernicus, but pretty well every other book published in the sixteenth century apart from the Bible. This was *Cosmographia* by Sebastian Münster. Born in Ingelheim am Rhein in Germany in 1488, Münster only had an amateur interest in natural philosophy – his day job was as professor of Hebrew – but his enthusiasm for the subject and ability to communicate it made his book a bestseller.

Perhaps one other aspect of the book hints at a reason why it sold so well, and reflects a change in science publishing that would take well over 100 years to complete. Published in 1544, Münster's book had the full title of *Cosmographia. Beschreibung aller Lender: in welcher begriffen aller Voelker, Herrschaften, Stetten, und namhafftiger Flecken, herkommen: Sitten, Gebreüch, Ordnung, Glauben, Secten und Hantierung durch die gantze Welt und fürnemlich Teütscher Nation* (roughly: 'Cosmographia. Describing all lands: covering all peoples, sovereignties, states and named locations including customs, faiths and laws throughout the world and for a whole German nation') – Münster believed in giving value for money in his titles. But note that this book was written in German. At a time when the majority of learned scientific books were published in Latin, Münster made his work more accessible to the wider public. It would still be translated into Latin (and a number of other languages), but it was written for the people.

In reality, Münster was more the contributing editor than the author of this scientific encyclopaedia, which had over 100 contributors. The focus was not primarily scientific, with many sections devoted to the geography of a wide range of countries and regions,

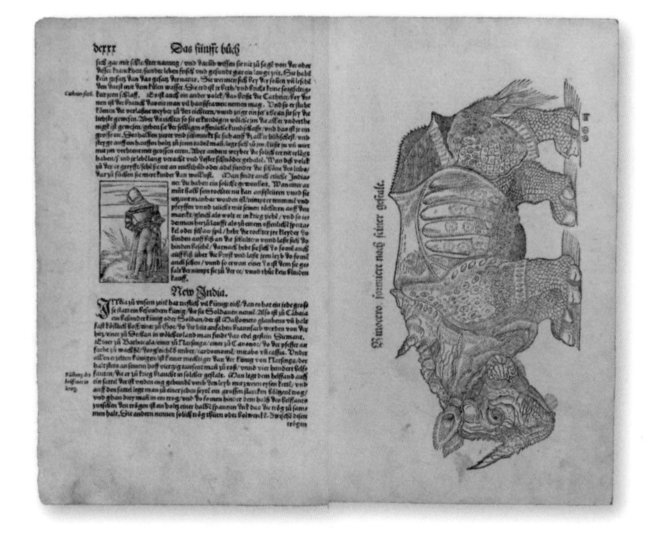

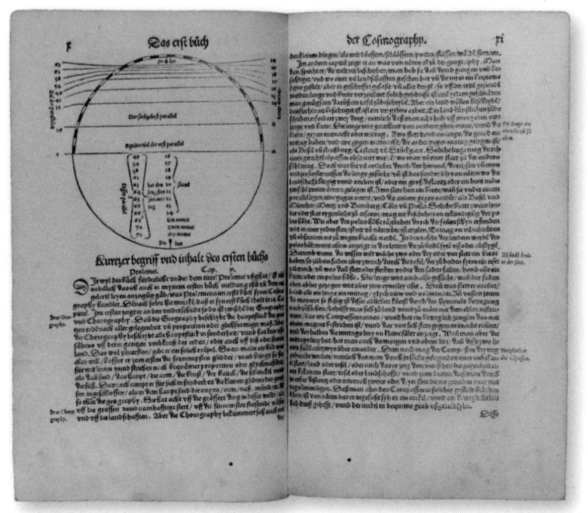

Sebastian Münster
COSMOGRAPHIA,
HENRICHUM PETRI, 1544

Interior pages of the first edition of *Cosmographia*.

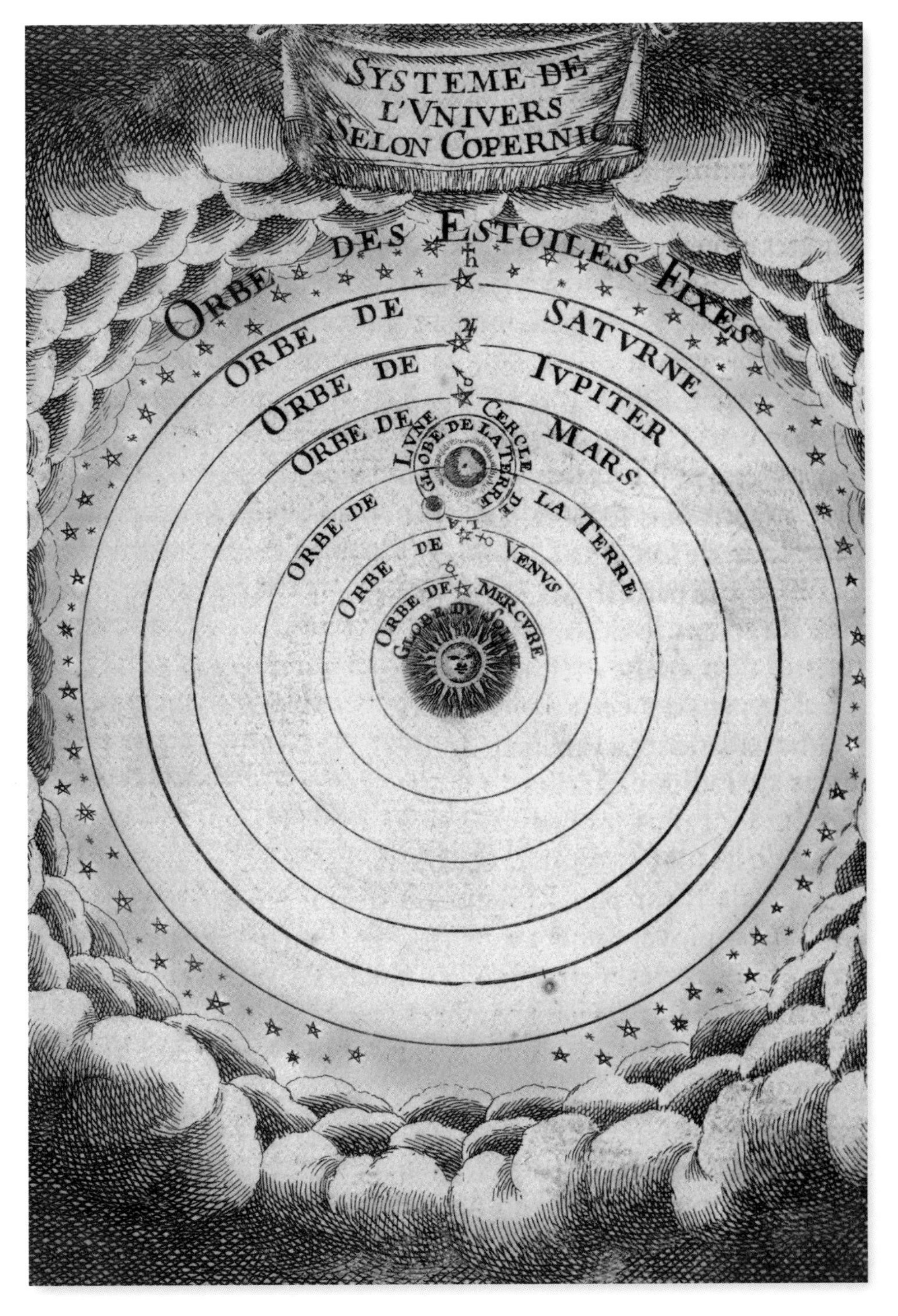

Alain Manesson Mallet
SYSTEME DE L'UNIVERS SELON COPERNICUS,
CA. 1683

A later, coloured illustration of the universe as propounded by Copernicus (see page 86), from Mallet's 1683 book *Description de l'Univers.* The book resembled Münster's *Cosmographia* both in being published in the country's language, rather than in Latin, and in covering a combination of geographical and astronomical topics.

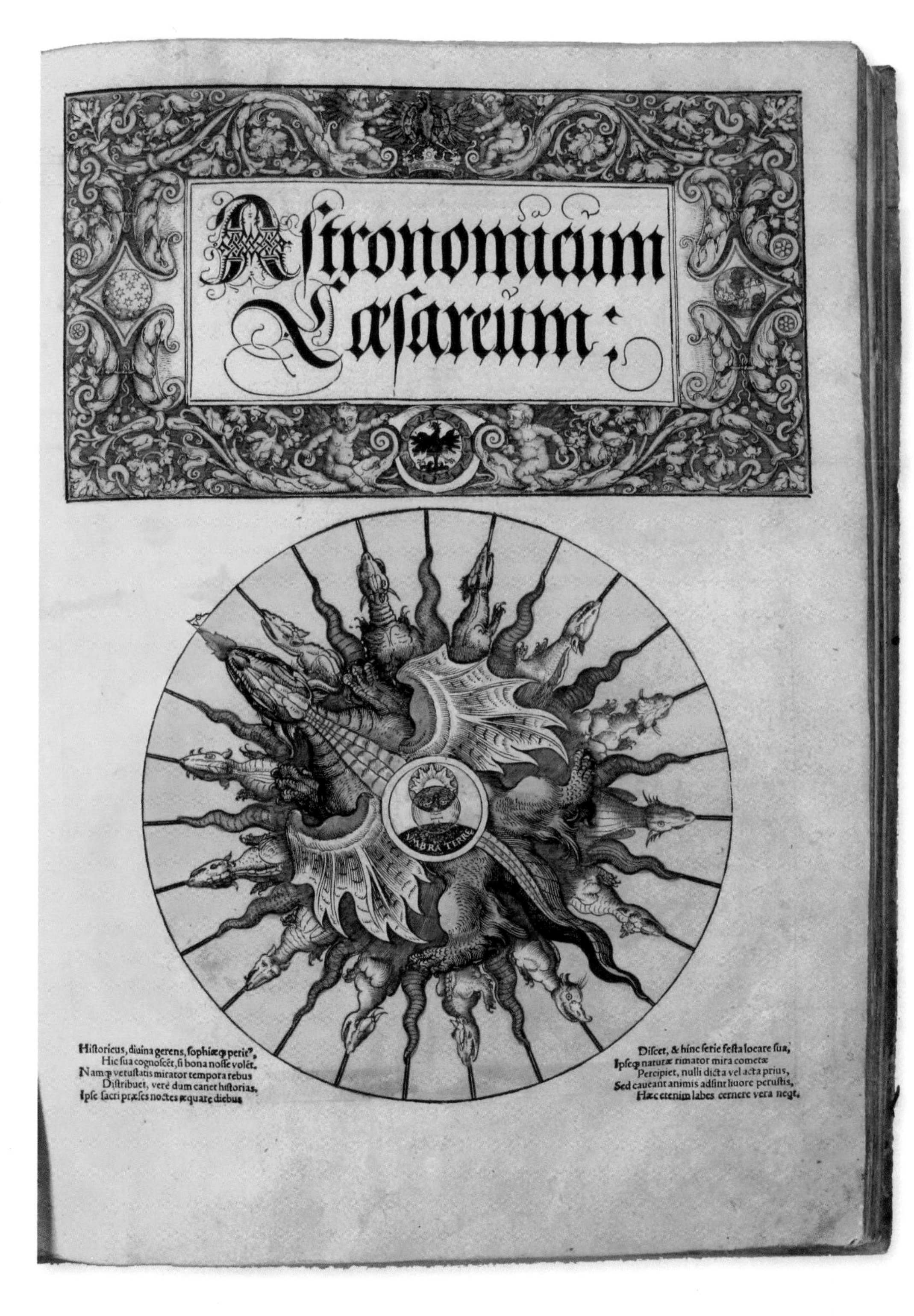

Astronomicum Cæsareum:

VMBRA TERRE

Hiſtoricus, diuina gerens, ſophiæq; perit',
Hic ſua cognoſcét, ſi bona noſſe volét.
Namq; vetuſtatis mirator tempora rebus
Diſtribuet, veré dum canet hiſtorias,
Ipſe ſacri præſes noctes æquare diebus

Diſcet, & hinc ſerie feſta locare ſua,
Ipſeq; naturæ rimator mira cometæ
Percipiet, nulli dicta vel acta prius,
Sed caueant animis adſint liuore peruſtis,
Hæc etenim labes cernere vera negt.

Peter Bienewitz
ASTRONOMICUM CAESAREUM,
PETER BIENEWITZ, 1540

The title page of this remarkable illustrated book, published by Bienewitz's own press. Though based on Ptolemy's astronomy, the book was revolutionary in its use of interactive diagrams.

but *Cosmographia* lived up to its name by starting off on astronomy and mathematics before focusing on more detailed geography and many beautiful maps.

The lasting influence of Münster's book was clear over 100 years later, when the French military engineer and mapmaker Alain Manesson Mallet wrote his five-volume *Description de L'Univers* (Description of the Universe). Like Münster, Mallet made the decision to work in his native language, though this was significantly more common by 1683 when *Description de L'Univers* (which features on this book's cover) was published. But the book was also similar in format, combining astronomical information with a large amount of geographical material, including both maps and information on customs, religions and laws, just as *Cosmographia* did.

It's interesting to contrast Münster's bestseller, which ran to at least 50,000 copies in German alone, with a contemporary volume that probably sold fewer than 200 – yet also managed to be revolutionary in a strangely literal sense. This was *Aſtronomicum Caesareum* (Caesarean Astronomy) from 1540 by the German mathematician Peter Bienewitz, who was given the Latin nickname Apianus (the Latin for 'bee', as *Biene* means bee in German). The astronomy presented in *Aſtronomicum Caesareum* was purely that of Ptolemy, untainted by new-fangled ideas, but what was remarkable about the book is that for each planet there were beautiful calculator dials with revolving moving parts, which allowed the reader to work out the planet's location at a given time, with calculators for eclipses, phases of the Moon and a universal calendar thrown in.

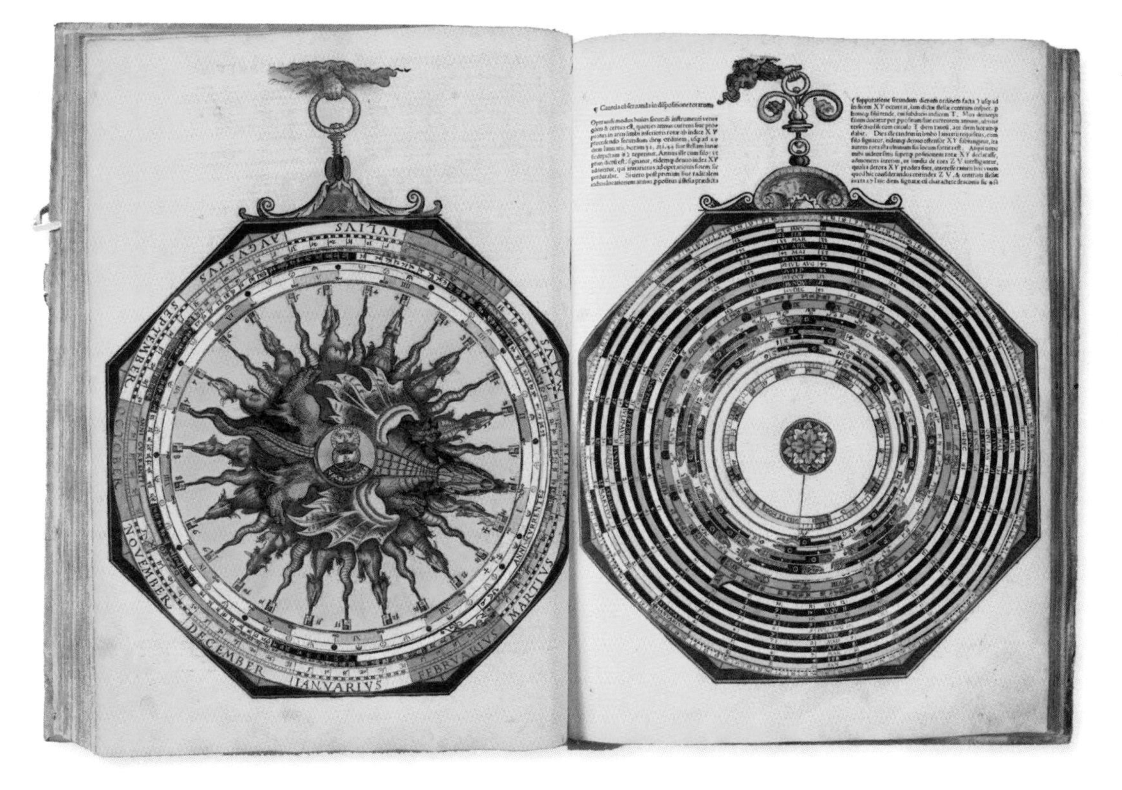

Peter Bienewitz
ASTRONOMICUM CAESAREUM,
PETER BIENEWITZ, 1540

Two of the 35 beautiful dial-like illustrations, known as volvelles, from this rare book.

ASTRONOMICVM

ENVNCTATVM DECIMVM.

Cursum Martis secundum Zodiaci longitudinem, quo cũque tempore, celeriter & citra negotium deprehendere.

ARS PLANETA TERTIVS à Saturno, calidæ siccæq; temperaturę est, ideoq; bellorum dominus putatus, quod flauæ bili non absimile quid, in corporib⁹ humanis efficiat. Hinc enim est, ꝙ cum græci ΠΥΡΟΙΔΑ id est, ignitum vocent. Huius motum qui est scitu rus, Rotas singulatim pspiciat hortor. Quas quidem non ita multum à prioribus differre videbit, nisi ꝙ in Marte venter Draconis rectè in auge deferentis habetur, semperq; aquilonẽ versus, ab ecliptica spectat. Venter aũt meridionalis ex aduerso augis situatur, seduloq; in latitudine meridionali perdurat. MARTIS cursum quæsituro, propositi temporis centenarius ꝓximè ꝓcedens minor, ex tabula accipiendus erit, & hoc, si tempus sit post Christum natum. Radices præterea medii motus & argumenti, sub titulis suis capiendi. Si vero ante Saluatoris ortum, tempus proponitur, iam centenarius ꝓpositi numeri proximus sequens seu minor, eligendus est. Habita ex tabula radice medii motus & argumẽti, eadem In Zodiaco quæratur, indice M eidem superposito. Annos deinde centenarium superantes, in limbo rotæ illius inspice, per quos filum A tensum cũ fuerit, index M rursus subordinetur. Filo A, diebus & horis electi temporis adiũcto, index denuo substituatur. Eo facto, simulac rota pro voluntate operaturi instituta est. In Zodiaco aũt medius motus Martis cernitur. Ostensorem P augi Martis (quam 4 enunctiatũ inuenire docuit) adhibens, rotam eandem rite locaueris. Ad epiciclum veniens, nota cõtactũ lineæ indicis M & circuli G H I, eundem locum in circulo K L M, secundum lineam sequutus obserua, per quem filum E ducatur, epicicli centro cum cruce ✠ augis subornato. Epiciclo sic ordinato, quæratur argumenti radix ante per tabulam accepta in limbo inferiori, cui indicem Y superioris circuli superapta. In superiore quoque circulo, annos superfluos numera, cum filoq; epicicli nota, indice Y eodem ducto. Præterea cum diebus similiter agenti, epiciclus perfectè locatus erit.

Quomodo verus Martis locus, argumentum & centrum inueniantur.

¶ Verum locum Martis in Zodiaco, filo A per stellam epicicli ductovidebis, Idem filum per centrum epicicli tensum, in G H I circulo centrum verum dat. A sectione exterioris epicicli vsq; ad indicem Y superioris, numerans. Argumentum verum habebit. Centro & Argumento custoditis, infra ad latitudinem opus esse scito.

Hactenus dicta, triplici cõpendio libet repetere, quorũ primum Imperatoris CAROLI natiuitas, illustrabit.

Exemplũ CAROLI Imperatoris.

¶ Natiuitatis Imperatoriæ tempus æquatum, ut sæpe auditum est, 1500 annos, 23 dies, 16 hor. 20 mi, continet. Hoc tempus, quia post Christum est, proximè antecedens minor centenari⁹, qui est 1400 accipiendus erit. Illius radix medii motus, ♈ 6 ♉ 3 ♊ 23, Argumẽtum verò ♈ 3 ♉ 15 ♊ 3, sub titulis accõmodis habentur. Signa ergo, gradus & minuta medii motus in Zodiaco inuestiganda, inuestigataq; cum indice M notanda. Quandoquidem vero à 1400 vsq; ad 1500 natiuitatis tempus, adhuc 99 anni supersunt, iccirco iidem in limbo eiusdem quærendi sunt. Quæsitis filum A addendum, indice sub idẽ reducto. Eodem modo cum diebus & horis, rationabiliter distinctis, agendum. Ibidem igitur filo & indice firmato, medius motus Martis in Zodiaco ad horam ꝓpositam, ♈ 2 ♉ 23 ♊ 56 cernetur. Aux postea per 4 enunctiatum habita cum est, quæ ♉ 14 ♊ 57 ♌ habet, eadem in Zodiaco requiratur, & cũ P monstratore firmetur. Locus insuper circuli G H I, ab indice M tactus, obseruetur. Transeũtiq; in K L M circulum, punctus ibidem inuentus notetur, per quem filo æquantis educto, centrum epicicli cum cruce ✠ augis supponatur. A cruce ✠ post, radix argumenti ex tabula sumpta, vbi numerata fuerit, cum indice Y signetur. Filum epicicli rursus per 99 annos (& hoc in limbo superioris epicicli) ostensore Y admoto extendatur. Vltimo filum per dies 23 hor 16 mi 20 Februarii ducatur. Cui filo, si denuo indicem substituis, in limbo inferioris epicicli, medium argmẽtum Martis Signorũ 8 graduum 20 minutorum 31 patet. Iam filum A repete, quod per stellam

TABVLA MEDII MOTVS MARTIS — TABVLA MEDII ARGVMENTI MART

Centenarii annorum ante & post Christi aduentum. Annus	Radices post Christum. S.	G.	M.	Radices ante Christum. S.	G.	M.	Centenarii annorum ante & post Christi aduentum. Ann⁹	Radices post Christum S	G	M	Radices ante Christum S	G	M
Christi	1	11	24	1	11	24	Christi	7	26	54	7	26	54
100	3	12	58	11	9	49	100	5	26	3	9	27	44
200	5	14	33	9	8	14	200	3	25	12	11	28	[illegible]
300	7	16	8	7	6	39	300	1	24	21	1	29	[illegible]
400	9	17	43	5	5	4	400	11	23	31	4	0	[illegible]
500	11	19	18	3	3	29	500	9	22	40	6	1	[illegible]
600	1	20	53	1	1	55	600	7	21	49	8	1	[illegible]
700	3	22	27	11	0	20	700	5	20	59	10	2	[illegible]
800	5	24	2	8	28	45	800	3	20	8	0	3	[illegible]
900	7	25	37	6	27	10	900	1	19	17	2	4	[illegible]
1000	9	27	12	4	25	35	1000	11	18	26	4	5	[illegible]
1100	11	28	47	2	24	1	1100	9	17	36	6	6	[illegible]
1200	2	0	21	0	22	26	1200	7	16	45	8	7	[illegible]
1300	4	1	56	10	20	51	1300	5	15	54	10	7	[illegible]
1400	6	3	31	8	19	16	1400	3	15	3	0	8	44
1500	8	5	6	6	17	41	1500	1	14	13	2	9	[illegible]
1600	10	6	41	4	16	6	1600	11	13	22	4	10	[illegible]
1700	0	8	16	2	14	32	1700	9	12	31	6	11	[illegible]
1800	2	9	50	0	12	57	1800	7	11	41	8	12	[illegible]
1900	4	11	25	10	11	22	1900	5	10	50	10	12	[illegible]
2000	6	13	0	8	9	47	2000	3	9	59	0	13	48
2100	8	14	35	6	8	12	2100	1	9	8	2	14	[illegible]
2200	10	16	10	4	6	38	2200	11	8	18	4	15	[illegible]
2300	0	17	44	2	5	3	2300	9	7	27	6	16	[illegible]
2400	2	19	19	0	3	28	2400	7	6	36	8	17	[illegible]
2500	4	20	54	10	1	53	2500	5	5	45	10	18	[illegible]
2600	6	22	29	8	8	18	2600	3	4	55	0	18	[illegible]
2700	8	24	3	5	28	43	2700	1	4	4	2	19	44
2800	10	25	38	3	27	9	2800	11	3	13	4	20	[illegible]
2900	0	27	13	1	25	34	2900	9	2	23	6	21	[illegible]
3000	1	28	48	11	23	59	3000	7	1	31	8	22	[illegible]
3100	5	0	23	9	22	24	3100	5	0	41	10	23	[illegible]
3200	7	1	58	7	20	49	3200	2	29	50	0	23	[illegible]
3300	9	3	33	5	19	15	3300	0	28	59	2	24	[illegible]
3400	11	5	7	3	17	40	3400	10	28	9	4	25	[illegible]
3500	1	6	42	1	16	5	3500	8	27	18	6	26	[illegible]
3600	3	8	17	11	14	30	3600	6	26	27	8	27	[illegible]
3700	5	9	52	9	12	55	3700	4	25	36	10	28	[illegible]
3800	7	11	27	7	11	20	3800	2	24	46	0	29	[illegible]
3900	9	13	2	5	9	46	3900	0	23	55	2	29	[illegible]
4000	11	14	36	3	8	11	4000	10	23	4	5	0	[illegible]
4100	1	16	11	1	6	36	4100	8	22	14	7	1	[illegible]
4200	3	17	46	11	5	1	4200	6	21	23	9	2	[illegible]
4300	5	19	21	0	3	25	4300	4	20	32	11	3	[illegible]
4400	7	20	56	7	1	52	4400	2	19	41	1	4	[illegible]
4500	9	22	30	5	0	17	4500	0	18	50	3	4	[illegible]
4600	11	14	5	2	28	42	4600	10	18	0	5	5	[illegible]
4700	1	25	40	0	27	7	4700	8	17	9	7	6	[illegible]
4800	3	27	15	10	25	32	4800	6	16	18	9	7	[illegible]
4900	5	28	50	8	23	57	4900	4	15	28	11	8	[illegible]
5000	8	0	25	6	22	23	5000	2	14	37	1	9	[illegible]
5100	10	1	59	4	20	48	5100	0	13	46	3	10	[illegible]
5200	0	3	34	2	19	13	5200	10	12	55	5	10	[illegible]
5300	2	5	9	0	17	38	5300	8	12	5	7	11	[illegible]
5400	4	6	44	10	16	3	5400	6	11	14	9	12	[illegible]
5500	6	8	19	8	14	29	5500	4	10	23	11	13	[illegible]
5600	8	9	53	6	12	54	5600	2	9	32	1	14	[illegible]
5700	10	11	28	4	11	19	5700	0	8	42	3	15	[illegible]
5800	0	13	3	2	9	44	5800	10	7	51	5	15	[illegible]
5900	2	14	38	0	8	9	5900	8	7	0	7	16	[illegible]
6000	4	16	13	10	6	34	6000	6	6	9	9	17	[illegible]
6100	6	17	48	8	5	0	6100	4	5	19	11	18	[illegible]
6200	8	19	22	6	3	25	6200	2	4	28	1	19	[illegible]
6300	10	20	57	4	1	50	6300	0	3	37	3	20	[illegible]
6400	0	22	32	2	0	15	6400	10	2	47	5	21	[illegible]
6500	2	24	7	11	28	40	6500	8	1	56	7	21	[illegible]
6600	4	25	42	9	27	5	6600	6	1	5	9	22	[illegible]
6700	6	27	16	7	25	31	6700	4	0	14	11	23	[illegible]
6800	8	28	51	5	23	56	6800	1	29	24	1	24	[illegible]
6900	11	0	26	3	22	21	6900	11	28	33	3	25	[illegible]
7000	1	2	1	1	20	46	7000	9	27	42	5	26	[illegible]

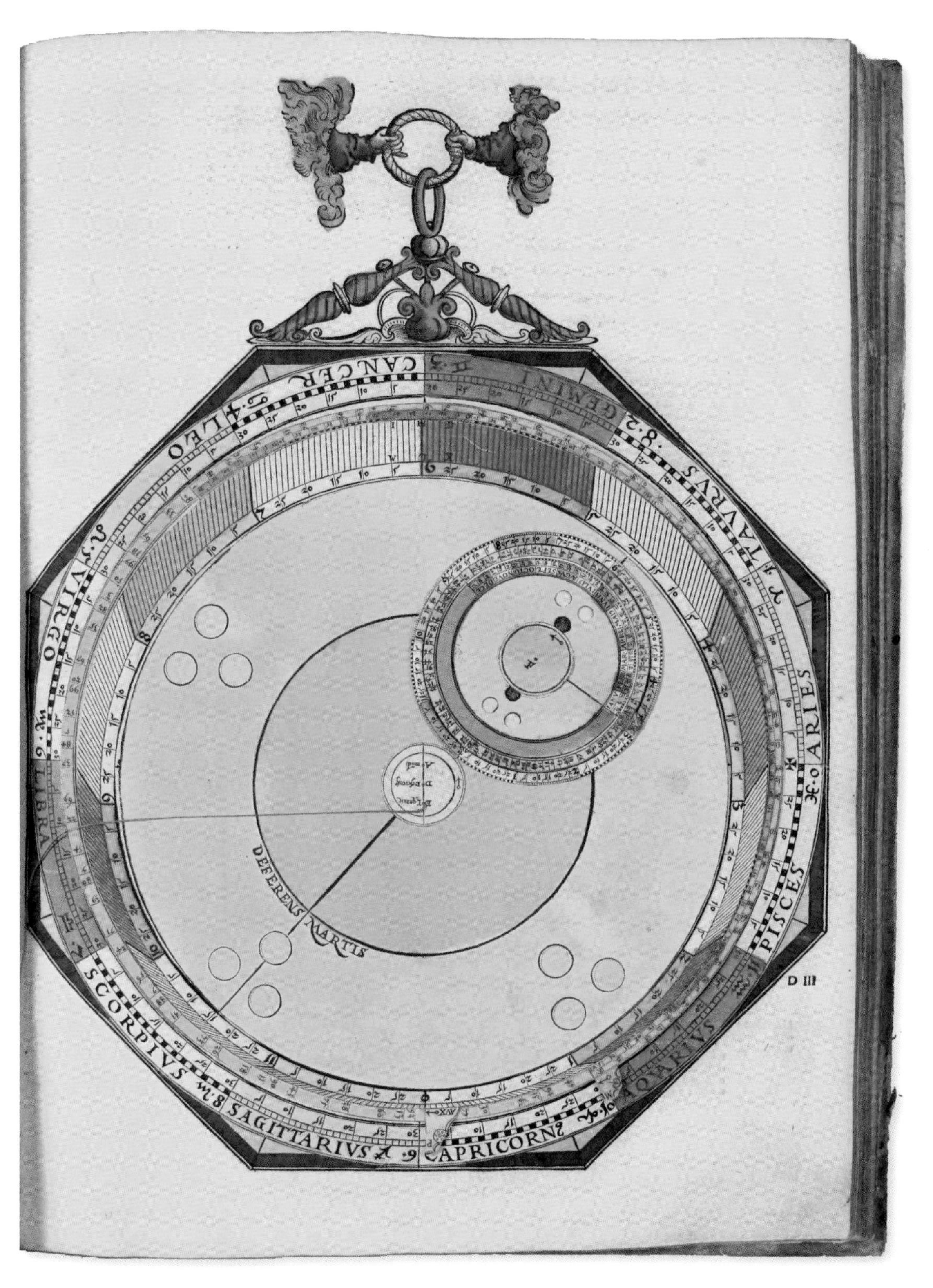

Peter Bienewitz
ASTRONOMICUM CAESAREUM,
PETER BIENEWITZ, 1540

Another illustrated spread from this rare book, showing a calculator for the position of Mars against the zodiac.

From mines to maths to minds

If Münster took his readers on a journey from the heavens to terrestrial geography, it was down to another German writer to carry on down into the depths of the Earth. Generally recognised as the first significant mineralogist, Georgius Agricola, born in the German town of Glauchau in 1494, would not live to see a finished copy of his book *De Re Metallica* (On the Nature of Metals). (It might seem a little odd that a mineralogist should be called 'Farmer George' in Latin – but Agricola was born Georg Pawer, and Pawer (now Bauer) is simply the German for 'farmer'.)

De Re Metallica is more an engineering book that one of pure science, in that it is a practical guide for mining engineers, describing how to find appropriate minerals and giving guidance on digging up, crushing and smelting the ore and providing methods for separating out mixed metals. It would remain an important guide right into the eighteenth century. Agricola had seen a number of earlier works published, notably *De Ortu et Causis Subterraneorum*, which was more scientific as a geology text, but these had far less lasting impact. However, after completing *De Re Metallica* in around 1550 there was a considerable delay in getting the many illustrations carved into wooden blocks ready for printing. It seems likely it was the production of these woodcuts that delayed the publication until 1556, the year after Agricola's death.

The sixteenth century was a period of a new awakening of mathematical ideas. These would not fully come to fruition until the following century, when the focus would be in Germany and England, but in the sixteenth century, Italy was the home of two of the mathematical greats. The first, Gerolamo Cardano, was born in Pavia in 1501. Cardano qualified as a physician but was not able to practise medicine as his illegitimate birth and lack of social graces turned those who could have awarded certification against him. Medicine's loss was mathematics' gain. Cardano was a prolific writer on maths and science, publishing over 200 works.

Georgius Agricola
DE RE METALLICA,
HIERONYMUM FROBENIUM ET NICOLAUM EPISCOPIUM, 1556

Illustrations of a machine for drawing water (left) mining techniques (middle) and a smelting furnace (right) from the woodcuts that delayed publication of the title.

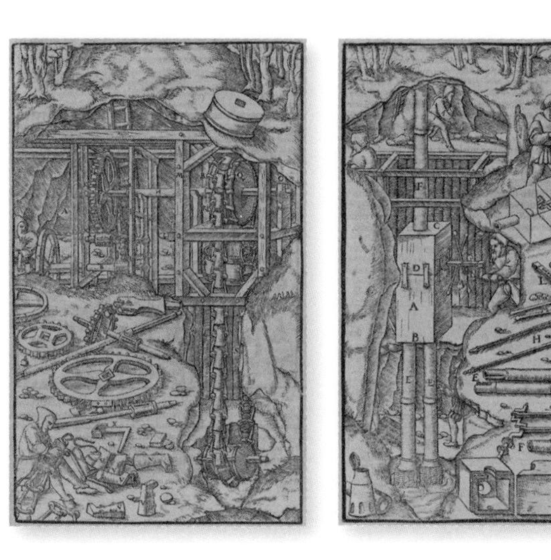

Georgius Agricola
DE RE METALLICA,
HIERONYMUM FROBENIUM
ET NICOLAUM EPISCOPIUM,
1556

From mining to smelting, the whole process laid out from one of the wooden blocks that took so much time to produce.

Out of Cardano's prolific output, two books stand out. *Ars Magna* (The Great Art), published in 1545, was a masterwork on algebra, pulling together solutions to equations that had never been seen before (including, controversially, solutions to cubic equations discovered by another mathematician, Niccolò Tartaglia, who had told Cardano in confidence and asked him not to publish them). The book also made wide use of negative numbers – rarely seen in mathematics up to this point – and started to address the idea of imaginary numbers based on the square root of negative numbers. However, it is Cardano's other masterpiece that seems far more significant now. This was *Liber de Ludo Aleae* (Book on Games of Chance). Although it was written in the 1560s, it would not be published until 1663, long after Cardano's death. This late work was important because it was the first to take on probability in a systematic way, opening up a whole new field of mathematics.

Although a lesser figure in some ways, the other Italian great of the period was Rafael Bombelli, born in Bologna around 1525. Bombelli covered the same area as Cardano's *Ars Magna* with his simply titled *Algebra*. This was most notable for giving what amounts to the full, modern understanding of imaginary numbers, including providing us with the modern symbol for the square root of -1, *i*.

The sixteenth century would prove to be an important time for writing on the philosophy of science – analysing the scientific method itself. As Roger Bacon had

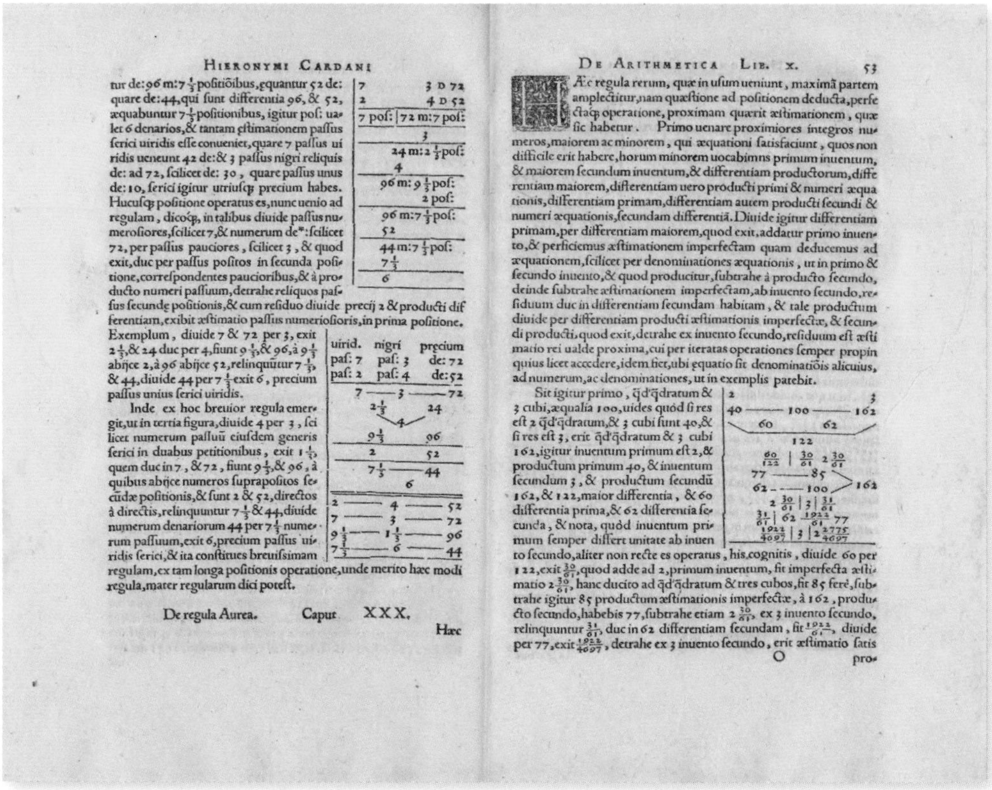

HIERONYMI CARDANI

tur de:96 m:7 ⅓ poſitiōibus, ęquantur 52 de: quare de:44, qui ſunt differentia 96, & 52, æquabuntur 7 ⅓ poſitionibus, igitur poſ: ualet 6 denarios, & tantam ęſtimationem paſſus ſerici uiridis eſſe conueniet, quare 7 paſſus uiridis ueneunt 42 de: & 3 paſſus nigri reliquis de: ad 72, ſcilicet de: 30, quare paſſus unus de: 10, ſerici igitur utriuſq; precium habes. Hucuſq; poſitione operatus es, nunc uenio ad regulam, dicoq;, in talibus diuide paſſus numeroſiores, ſcilicet 7, & numerum de^rum: ſcilicet 72, per paſſus pauciores, ſcilicet 3, & quod exit, duc per paſſus poſitos in ſecunda poſitione, correſpondentes paucioribus, & à producto numeri paſſuum, detrahe reliquos paſſus ſecundę poſitionis, & cum reſiduo diuide precij 2 & producti differentiam, exibit æſtimatio paſſus numerioſioris, in prima poſitione.

7 3 D 72
2 4 D 52
7 poſ: | 72 m:7 poſ:
3
24 m:2 ⅓ poſ:
4
96 m: 9 ⅓ poſ:
2 poſ:
96 m:7 ⅓ poſ:
52
44 m:7 ⅓ poſ:
7 ⅓
6

Exemplum, diuide 7 & 72 per 3, exit 2 ⅓, & 24 duc per 4, fiunt 9 ⅓, & 96, à 9 ⅓ abijce 2, à 96 abijce 52, relinquũtur 7 ⅓, & 44, diuide 44 per 7 ⅓ exit 6, precium paſſus unius ſerici uiridis.

uirid.	nigri	pręcium
paſ: 7	paſ: 3	de: 72
paſ: 2	paſ: 4	de: 52

Inde ex hoc breuior regula emergit, ut in tertia figura, diuide 4 per 3, ſcilicet numerum paſſuũ eiuſdem generis ſerici in duabus petitionibus, exit 1 ⅓, quem duc in 7, & 72, fiunt 9 ⅓, & 96, à quibus abijce numeros ſupra poſitos ſecũdæ poſitionis, & ſunt 2 & 52, directos à directis, relinquuntur 7 ⅓ & 44, diuide numerum denariorum 44 per 7 ⅓ numerum paſſuum, exit 6, precium paſſus uiridis ſerici, & ita conſtitues breuiſsimam regulam, ex tam longa poſitionis operatione, unde merito hæc modi regula, mater regularum dici poteſt.

7 — 3 — 72
2 ⅓ 24
4
9 ⅓ 96
2 52
7 ⅓ — 44
6

2 — 4 — 52
7 — 3 — 72
9 ⅓ — 1 ⅓ — 96
7 ⅓ — 6 — 44

De regula Aurea. Caput XXX.

Hæc

DE ARITHMETICA LIB. X. 53

HÆc regula rerum, quæ in uſum ueniunt, maximã partem amplectitur, nam quæſtione ad poſitionem deducta, perfectaq; operatione, proximam quærit æſtimationem, quæ ſic habetur. Primo uenare proximiores integros numeros, maiorem ac minorem, qui æquationi ſatisfaciunt, quos non difficile erit habere, horum minorem uocabimus primum inuentum, & maiorem ſecundum inuentum, & differentiam productorum, differentiam maiorem, differentiam uero producti primi & numeri æquationis, differentiam primam, differentiam autem producti ſecundi & numeri æquationis, ſecundam differentiã. Diuide igitur differentiam primam, per differentiam maiorem, quod exit, addatur primo inuento, & perficiemus æſtimationem imperfectam quam deducemus ad æquationem, ſcilicet per denominationes æquationis, ut in primo & ſecundo inuento, & quod producitur, ſubtrahe à producto ſecundo, deinde ſubtrahe æſtimationem imperfectam, ab inuento ſecundo, reſiduum duc in differentiam ſecundam habitam, & tale productum diuide per differentiam producti æſtimationis imperfectæ, & ſecundi producti, quod exit, detrahe ex inuento ſecundo, reſiduum eſt æſtimatio rei ualde proxima, cui per iteratas operationes ſemper propinquius licet accedere, idem fiet, ubi ęquatio ſit denominatiõis alicuius, ad numerum, ac denominationes, ut in exemplis patebit.

Sit igitur primo, q̄d'q̄dratum & 3 cubi, æqualia 100, uides quód ſi res eſt 2 q̄d'q̄dratum, & 3 cubi ſunt 40, & ſi res eſt 3, erit q̄d'q̄dratum & 3 cubi 162, igitur inuentum primum eſt 2, & productum primum 40, & inuentum ſecundum 3, & productum ſecundũ 162, & 122, maior differentia, & 60 differentia prima, & 62 differentia ſecunda, & nota, quód inuentum primum ſemper differt unitate ab inuento ſecundo, aliter non recte es operatus, his cognitis, diuide 60 per 122, exit 30/61, quod adde ad 2, primum inuentum, fit imperfecta æſtimatio 2 30/61, hanc ducito ad q̄d'q̄dratum & tres cubos, fit 85 ferè, ſubtrahe igitur 85 productum æſtimationis imperfectæ, à 162, producto ſecundo, habebis 77, ſubtrahe etiam 2 30/61, ex 3 inuento ſecundo, relinquuntur 31/61, duc in 62 differentiam ſecundam, fit 1922/61, diuide per 77, exit 1922/4697, detrahe ex 3 inuento ſecundo, erit æſtimatio ſatis

2 3
40 — 100 — 162
60 62
122
60/122 | 30/61 2 30/61
77 — 85
62 — 100 162
2 30/61 | 3 | 31/61
31/61 | 62 1922/61 77
1922/4697 | 3 | 2 2775/4697

O pro-

Gerolamo Cardano
ARS MAGNA,
JOHANN PETREIUS, 1545

The Italian mathematician's masterpiece on algebra, first published in 1545, included a number of previously unseen solutions.

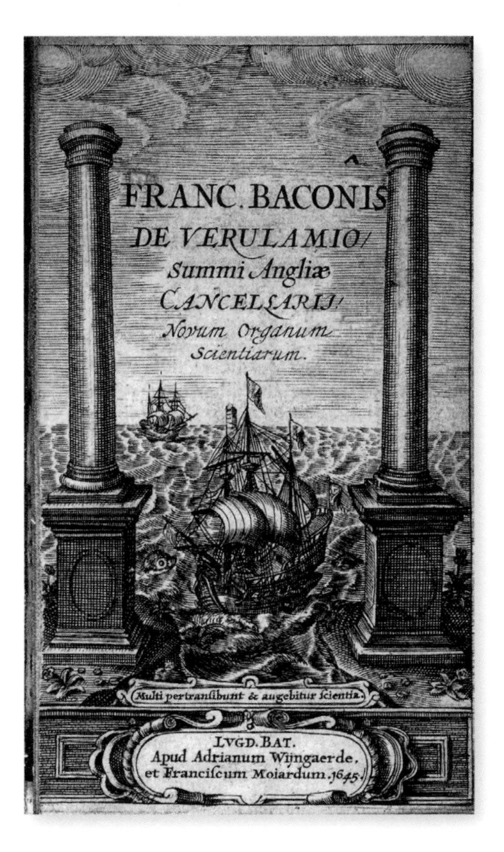

Francis Bacon
NOVUM ORGANUM SCIENTIARUM,
FRANCISCUM MOIARDUM ET ADRIANUM WIJNGAERDE,
1645

The frontispiece illustration from Bacon's best-known title in an edition from 25 years after its original publication in 1620.

suggested should be the case in the thirteenth century, proto-scientists such as Copernicus had tried to take an approach to natural philosophy that was driven more by observation than based on philosophical theorising alone. This approach would be clarified and described effectively, if sometimes obscurely, by the English politician Francis Bacon (no relation to Roger, as far as we know), born in London in 1561. Bacon wrote a number of books, most notably *Novum Organum Scientiarum* (The New Instrument of Science) from 1620. His writing style was odd and full of hyperbole, and it was more his approach than the detail of his books that made a difference. He argued that natural philosophers should be sceptical and should build knowledge by combining observation with the logical process of induction.

Bacon is often described as the father of the scientific method, though modern historians of science tend to play down his significance. His work was certainly a notable influence on the founders of England's Royal Society, which was established in 1660, the oldest surviving, continuously existing, scientific institution in the world. One of Bacon's most important observations was to do away with the traditional distinction between artificial and natural. Natural philosophers had held it impossible to learn from, say, an artificially produced rainbow about a natural rainbow – but Bacon argued there was no difference 'in form or essence' between nature and artifice.

Hans von Aachen
JOHANNES KEPLER,
OIL ON CANVAS, CA. 1612

Portrait of the German mathematician.

The magnetic universe

Bacon's methodology was slow to fully take root: if we look at the work of the man who finally fixed the problem of epicycles, the German mathematician Johannes Kepler, we see a gradual mental shift in his work from an ancient Greek-style philosophical approach to a more Baconian, data-driven method – from an inward looking approach to more outward looking science. Kepler wrote three significant books on astronomy and cosmology. The earliest, *Myſterium Cosmographicum* (Cosmographic Mystery) from 1596, supported Copernicus, but in doing so came up with a visually attractive but flawed philosophical argument for the structure of the solar system.

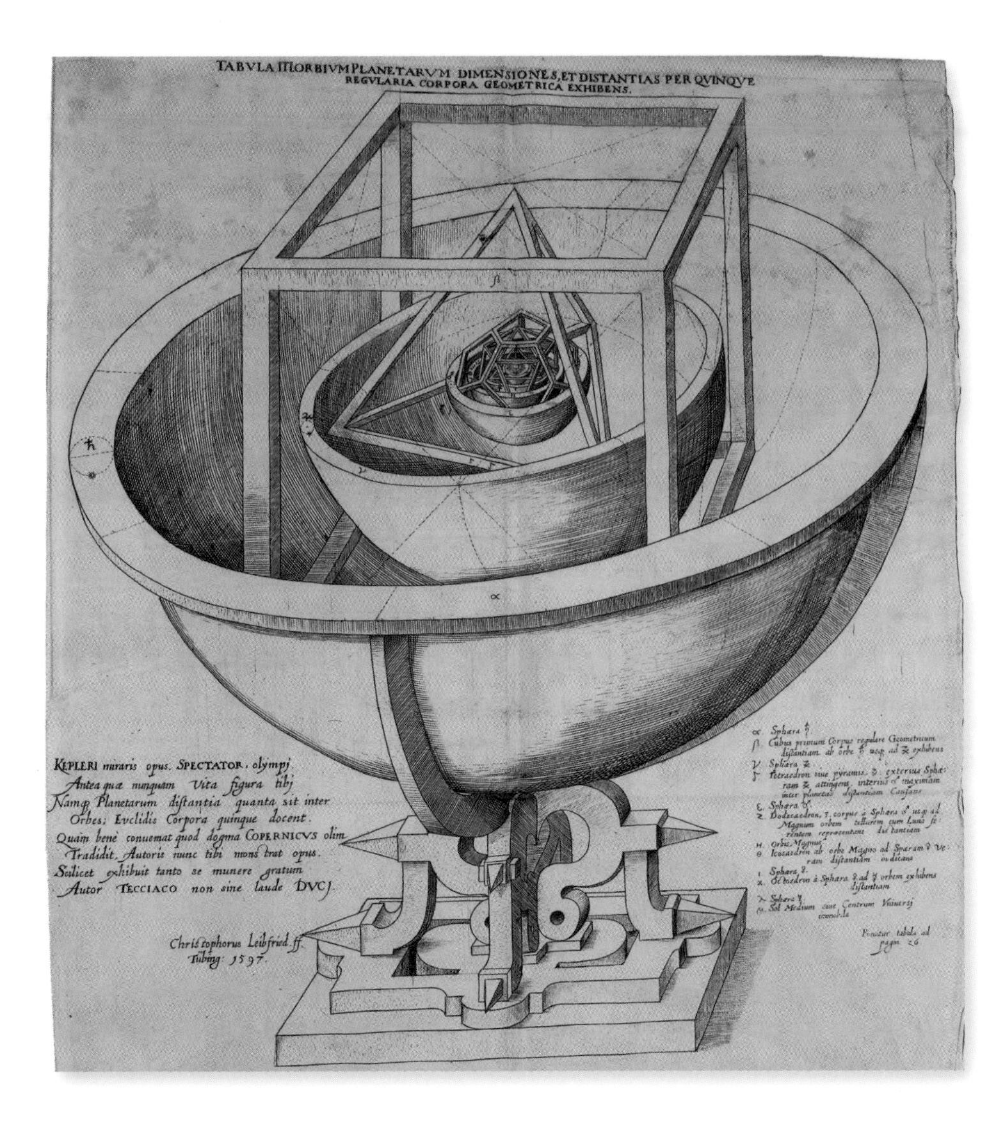

Johannes Kepler
MYSTERIUM COSMOGRAPHICUM,
ERASMI KEMPFERI/
GODEFRIDI TAMPACHII,
1621

The Platonic solids that Kepler envisaged structuring the solar system in this edition of his 1596 title.

Kepler argued that each of the six planets known at the time, from Mercury to Saturn, could be considered to exist on spheres, separated by the five Platonic solids – the solids the ancient Greeks had identified that could be constructed with sides of the same regular shape, such as triangles or squares. His basis for this model was more theological and philosophical than based on any scientific reasoning. However, it did enable him to start the move away from the epicycles that had continued to plague the model that Copernicus used (see page 87).

Kepler's second astronomical title was *De Stella Nova* (On a New Star). This was a description of the supernova of 1604, which appeared as a new bright star that gradually faded away. Kepler was able to argue that, because the new star had no parallax

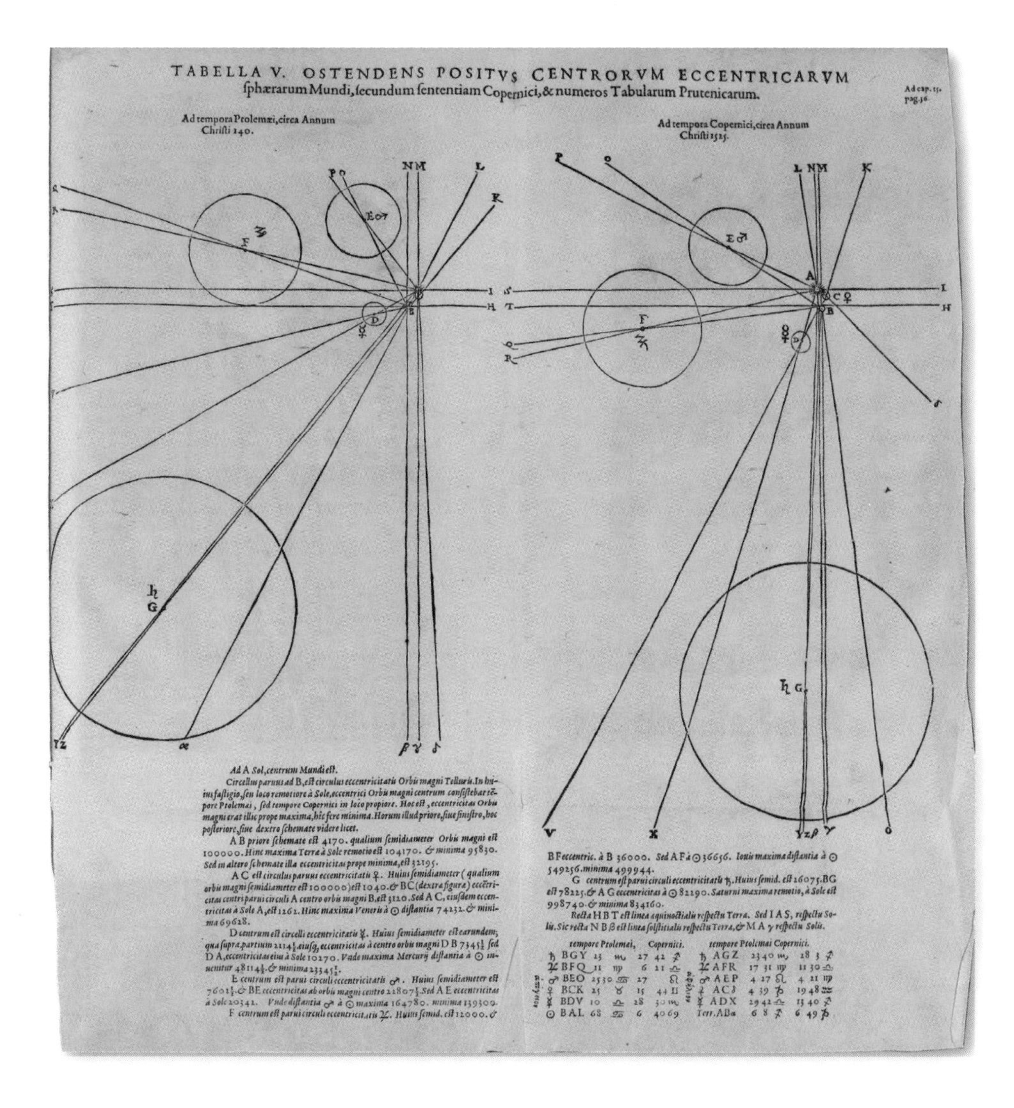

Johannes Kepler
MYSTERIUM COSMOGRAPHICUM,
ERASMI KEMPFERI/GODEFRIDI TAMPACHII, 1621

Table and diagrams showing the 'centres of eccentrics' in this title predating Kepler's doing away with the need for these structures by using elliptical orbits.

Johannes Kepler
DE STELLA NOVA,
PAUL SESSIUS, 1606

An illustration showing the location of the 'new star' in the constellation of Serpens – the serpent.

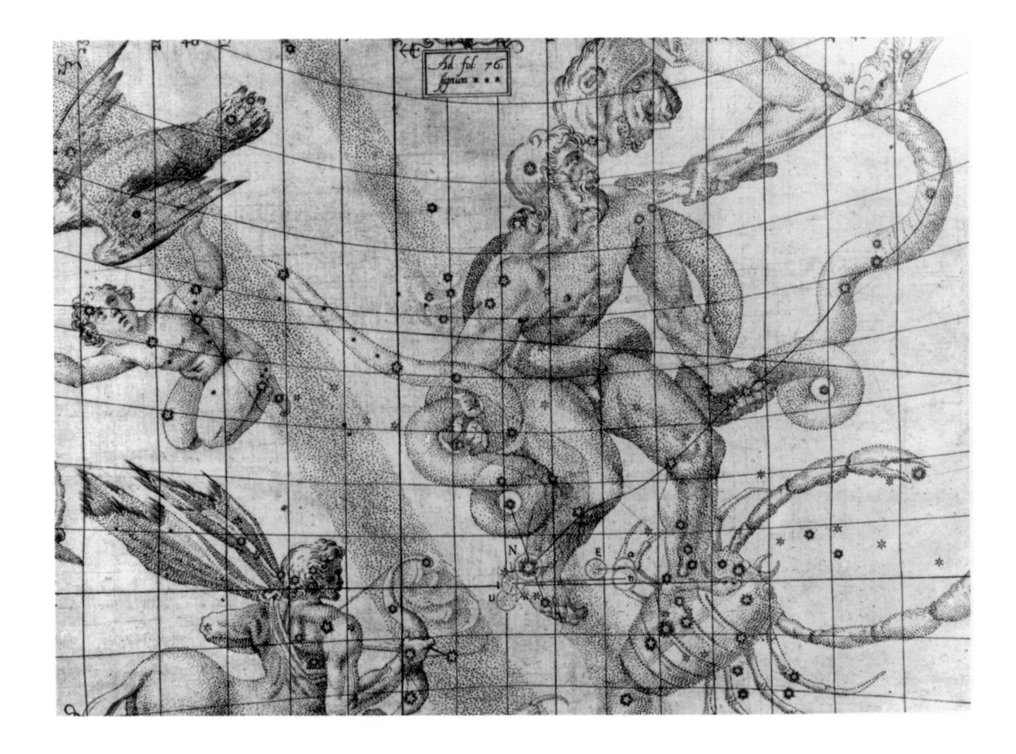

Johannes Kepler
ASTRONOMIA NOVA,
GOTTHARD VÖGELIN, 1609

The book in which Kepler made the move away from circular orbits and hence was able to simplify the structure of the solar system.

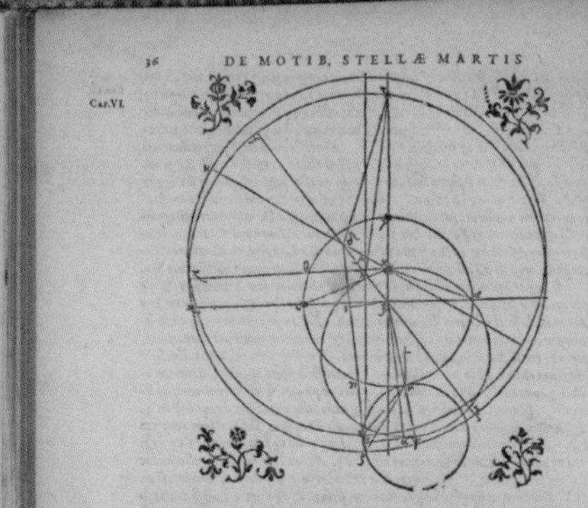

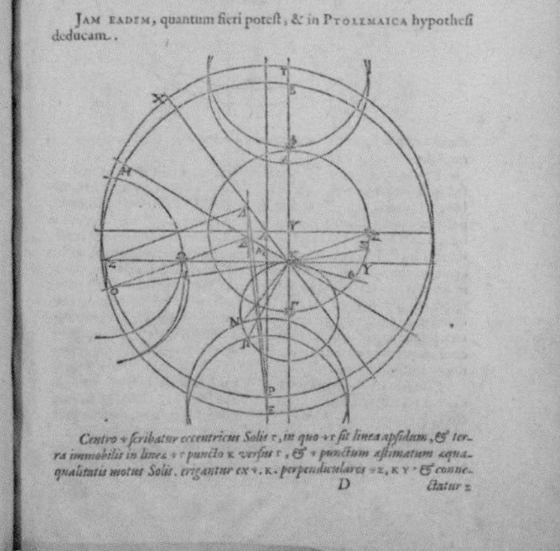

movement (the movement we see when, for example, we look at an object with first one eye and then the other), it had to be far enough away to be well outside the orbit of the Moon. As Aristotle's model required everything from the Moon's orbit outwards to be unchanging, this was another piece of evidence weakening the Aristotelian view.

However, Kepler's astronomical masterwork was realising that he could fully do away with epicycles and accurately model what was observed by removing the heavenly spheres entirely and instead having the planets move on paths that he called 'orbits'. These orbits were not circular, but followed the egg-like shape of an ellipse (though some, such as that of the Earth, were close to being circular). This model and Kepler's first two laws of planetary motion – that planets travelled in an ellipse with the Sun at one focus, and that a line from a planet to the Sun would sweep out equal areas in equal amounts of time – were published in his book *Astronomia Nova* (The New Astronomy) in 1609 and were fundamental to the acceptance of the Copernican model.

Ironically, Kepler based a lot of his argument on the high-quality observational work of Danish astronomer Tycho Brahe. Yet the two had not agreed on the structure of the heavens. Brahe realised the benefits of having the Moon rotate around the Earth, and the planets around the Sun. But he made a special case for the Earth, having the Sun (and its linked planets) rotate around the Earth, thereby managing to keep in line with much of Aristotle's physics (and biblical commentary), even though he was happy to dispute the idea of an unchanging space above the Moon's orbit.

It is worth mentioning two further books of Kepler's. First was *Harmonices Mundi* (Harmony of the World), published in 1619. Like his *Mysterium Cosmographicum*, this also made use of what we would now see as philosophical rather than scientific arguments in suggesting that the spacing of the planets obeyed the same sort of relationship as the musical notes that form harmonies (producing the fanciful concept of the 'music of the spheres', though Kepler did not believe that the planets generated actual musical notes). However, the final section of the book also brought in his third law of planetary motion, relating the size of the orbit to the time taken to complete it. The second book, the *Tabulae Rudolphinae* (Rudolphine Tables), published in 1627 and named after the Emperor Rudolf II, was a star catalogue, based largely on the data of Brahe, which was not only scientifically state of the art, but also included an impressive map of the world.

Many philosophers wondered exactly how the planets were kept in place if there were no crystal spheres to hold them there. One popular possible answer was magnetism – the natural force known to have the ability to influence solid objects at a distance. Kepler thought this was the case, basing his belief on a book by the English natural philosopher William Gilbert, *De Magnete, Magneticisque Corporibus, et de Magno Magnete Tellure* (On the Magnet and Magnetic Bodies, and on That Great Magnet the Earth), published in 1600.

Although Gilbert was wrong about gravity, his book was still an important text as it was the first detailed scientific study of magnets. It described a range of experiments he carried out, including producing spherical magnets called 'terrellas', which he used to show how the effect of the Earth's magnetism would vary depending on your position on Earth. This was a clear example of Bacon's approach of learning about the natural from the artificial, an essential for the experimental method to be able to fully take hold. In his approach, Gilbert produced the first true science book in the modern sense.

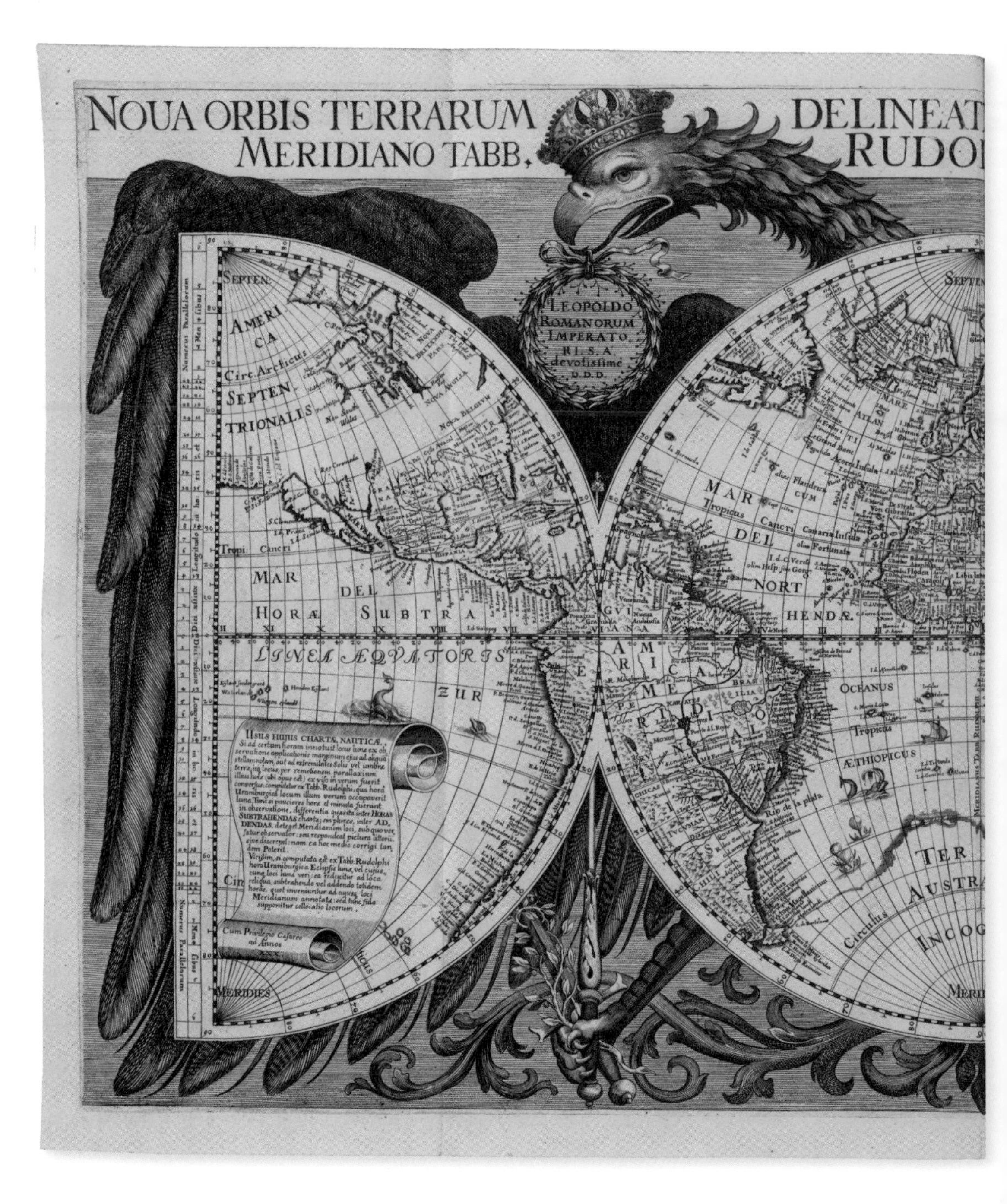
NOUA ORBIS TERRARUM DELINEAT
MERIDIANO TABB. RUDO
LEOPOLDO
ROMANORUM
IMPERATO
RI S. A.
devotissimé
D.D.D.
AMERICA
SEPTENTRIONALIS
MAR DEL NORT
MAR DEL ZUR
LINEA ÆQVATORIS
OCEANUS ÆTHIOPICUS

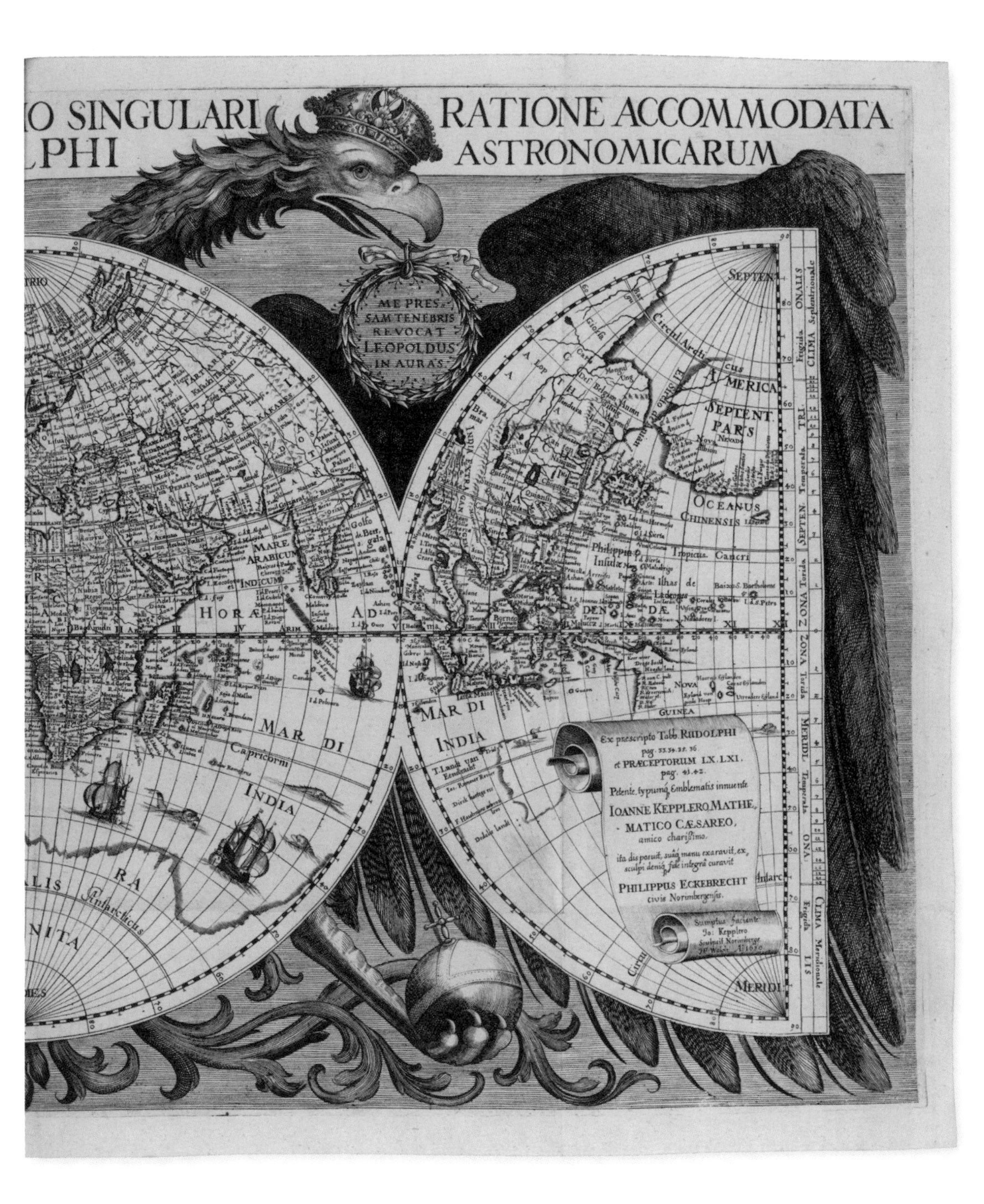

Johannes Kepler
TABULAE RUDOLPHINAE,
J. SAUR, 1627/1658

The impressive world map commissioned by Kepler, drawn by Philip Eckebrecht and engraved by J.P. Walsh. It was a later addition to Kepler's book, which was first published in 1627; a proof exists from 1630, but it's likely that the map was first published in 1658.

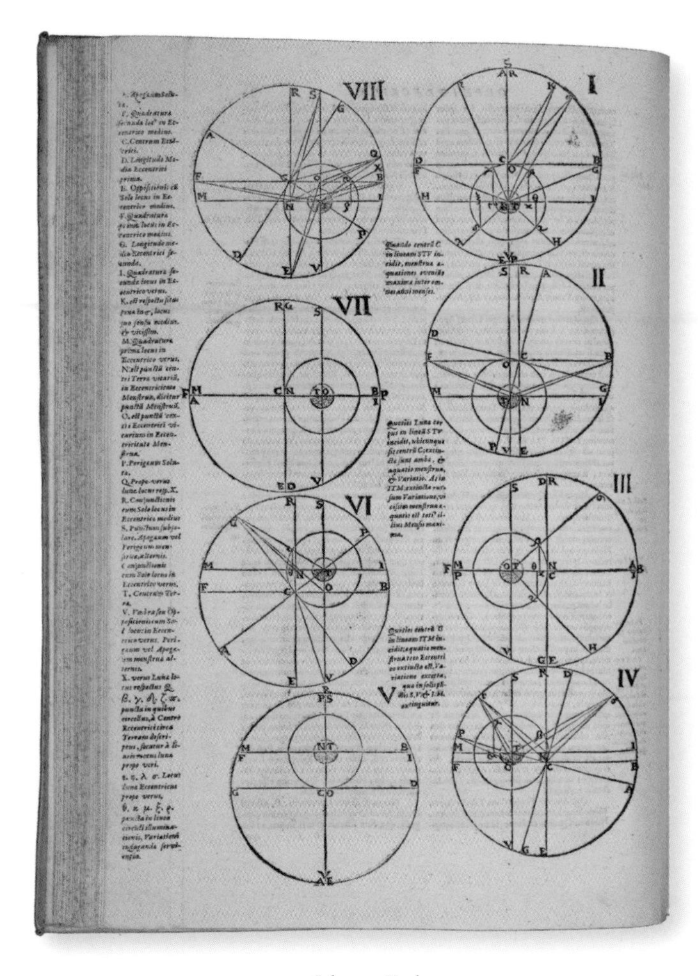

TABULÆ

RUDOLPHINÆ,

QUIBUS ASTRONOMICÆ SCIENTIÆ, TEMPOrum longinquitate collapſa RESTAURATIO *continetur;*

A Phœnice illo Aſtronomorum

TYCHONE

Ex Illuſtri & Generoſa BRAHEORUM *in Regno Daniæ familiâ oriundo Equite,*

PRIMUM ANIMO CONCEPTA ET DESTINATA ANNO CHRISTI MDLXIV: EXINDE OBSERVATIONIBUS SIDERUM ACCURATISSIMIS, POST ANNUM PRÆCIPUE MDLXXII, QUO SIDUS IN CASSIOPEJÆ CONSTELLATIONE NOVUM EFFULSIT, SERIÒ AFFECTATA; VARIISQUE OPERIBUS, cùm mechanicis, tùm librariis, impenſo patrimonio ampliſſimo, accedentibus etiam ſubſidiis FRIDERICI II. DANIÆ REGIS, regali magnificentia dignis, tracta per annos XXV, potiſſimùm in Inſula freti SUNDICI HUENNA, & arce URANIBURGO, in hos uſus à fundamentis extructâ:

TANDEM TRADUCTA IN GERMANIAM, INQUE AULAM ET Nomen RUDOLPHI IMP. *anno* MDIIC.

TABULAS IPSAS, JAM ET NUNCUPATAS, ET AFFECTAS, SED MORTE AUTHORIS SUI ANNO MDCI DESERTAS,

JUSSU ET STIPENDIIS FRETUS TRIUM IMPPP.

RUDOLPHI, MATTHIÆ, FERDINANDI,

ANNITENTIBUS HÆREDIBUS BRAHEANIS; EX FUNDAMENTIS obſervationum relictarum; ad exemplum ferè partium jam exſtructarum: continuis multorum annorum ſpeculationibus, & computationibus, primùm PRAGÆ *Bohemorum continuavit; deinde* LINCII, *ſuperioris Auſtriæ Metropoli, ſubſidiis etiam Ill. Provincialium adjutus, emendavit, perfecit, abſolvit; adq; cauſarum & calculi perennis formulam traduxit*

IOANNES KEPLERUS,

TYCHONI *primùm à* RUDOLPHO II. *Imp. adjunctus calculi miniſter; indéq; trium ordine Imppp. Mathematicus:*

Qui idem de ſpeciali mandato FERDINANDI II. IMP. petentibus inſtantibúsq; Hæredibus,

Opus hoc ad uſus præſentium & poſteritatis, typis, numericis propriis, cæteris & prælo JONÆ SAURII, *Reip. Ulmanæ Typographi, in publicum extulit, & Typographicis operis* ULMÆ *curator affuit.*

Cum Privilegiis, IMP. & Regum Rerúmq; publ. vivo TYCHONI ejúsq; Hæredibus, & ſpeciali Imperatorio, ipſi KEPLERO conceſſo, ad annos XXX.

ANNO M. DC. XXVII.

Johannes Kepler
TABULAE RUDOLPHINAE,
J. SAUR, 1627

Internal page and title page from Kepler's star catalogue.

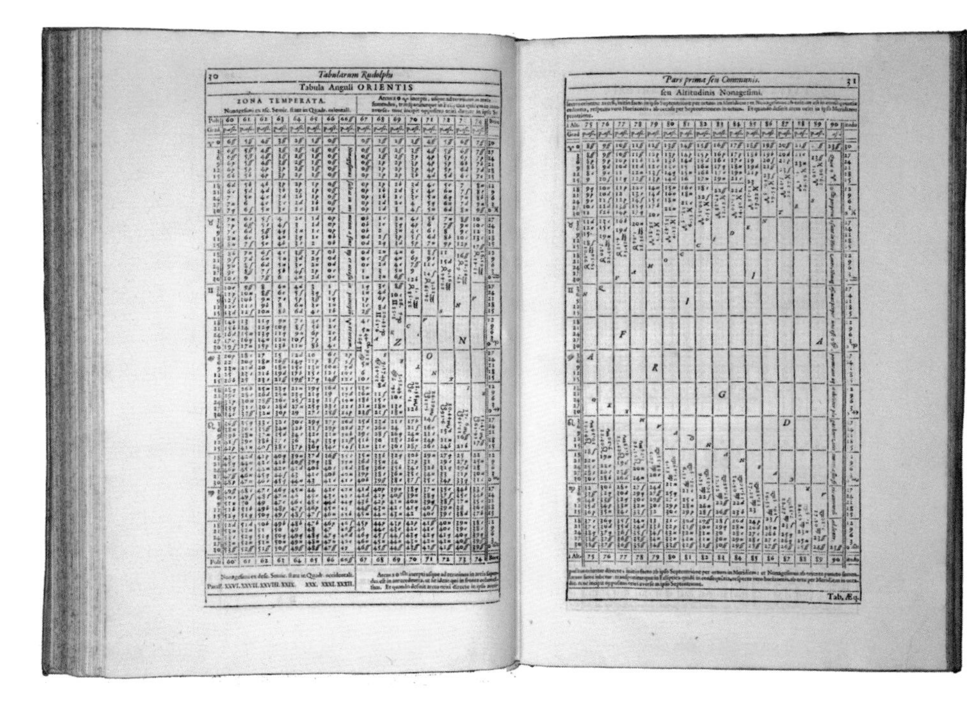

30 Tabularum Rudolphi
Tabula Anguli ORIENTIS
ZONA TEMPERATA.

Pars prima seu Communis. 31
seu Altitudinis Nonagesimi.

Johannes Kepler
TABULAE RUDOLPHINAE,
J. SAUR, 1627

One of the large number of Kepler's astronomical tables based on Tycho Brahe's data.

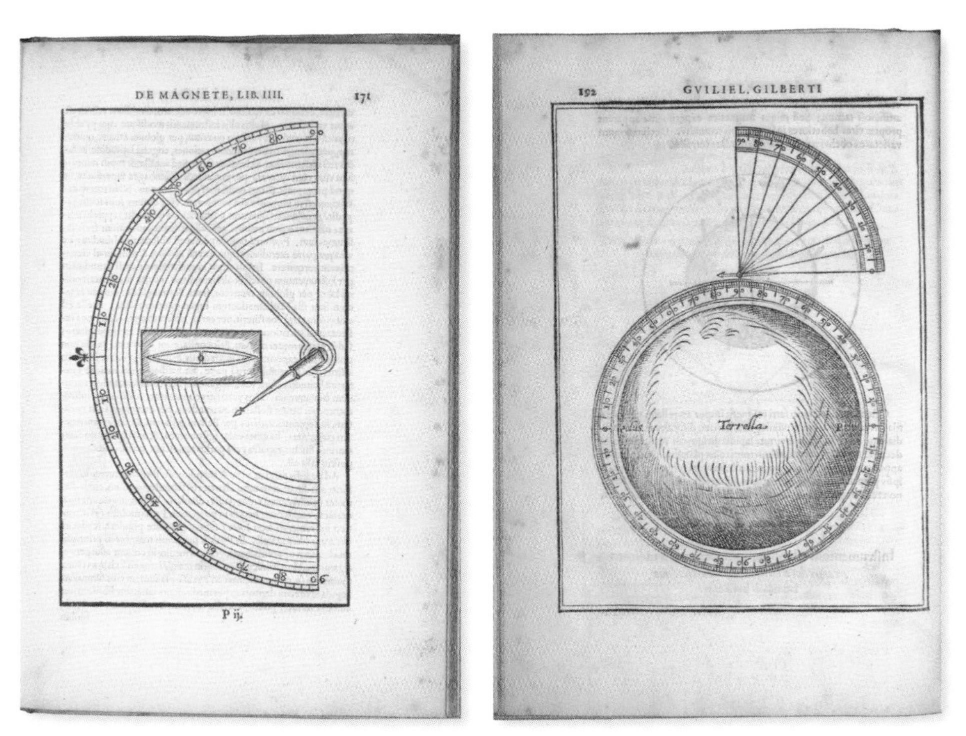

DE MAGNETE, LIB. IIII. 171

P ij.

192 GVILIEL. GILBERTI

William Gilbert
DE MAGNETE,
PETER SHORT, 1600

Pages from Gilbert's book on magnets, with the illustration on the right showing one of his spherical magnetic 'terrellas'.

Jan Collaert I
THE INVENTION OF BOOK PRINTING, CA. 1600

An early illustration of men working at a book printer's from a collection of prints entitled 'New Inventions of Modern Times'.

By the time Gilbert was writing his book, the printing press was beginning to have a noticeable influence on science and the distribution of information. In fact, Gilbert was the first to complain of a problem that has troubled scientists and readers in general ever since: information overload. He wrote that the intellectual now faced 'so vast an Ocean of Books by which the minds of studious men are troubled and fatigued'.

However, the impact of the printing press on science communication was, without doubt, more good than bad. It meant that new ideas could be shared between natural philosophers, enabling them to build on each other's inspiration. Where once the words of Aristotle were considered sacrosanct, now it was possible to challenge ideas in print. The book was not only a medium for the distribution of scientific information, but it also fostered debate in a way that isolated individuals in far-flung universities never could.

Inside the human body

Although astronomy (and to a degree magnetism) was a major factor in the new science, it was not the only area that was seeing rapid development. In 1543, the Flemish physician Andreas Vesalius had written *De Humani Corporis Fabrica Libri Septem* (On the Human Body in Seven Volumes), a book that bettered even Leonardo da Vinci in its illustrations of human anatomy. Vesalius broke away from the domination of the work

Andreas Vesalius
DE HUMANI CORPORIS,
JOHANNES OPORINUS, 1543

The colourful woodcut frontispiece illustration from book seven of the first edition of Vesalius's work on the human body, showing Vesalius teaching at a medical school. This is the only completely coloured copy known.

Andreas Vesalius
DE HUMANI CORPORIS,
JOHANN OPORINUS, 1555

Interior illustrations of the human body in a Swiss printed edition from 12 years after the original publication in 1543.

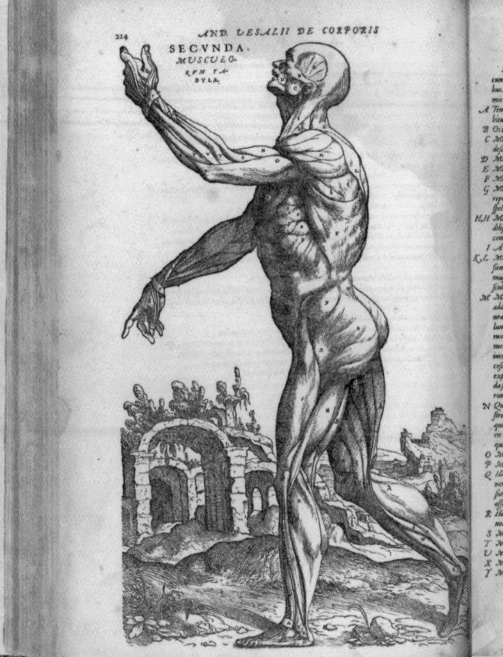

214 AND. VESALII DE CORPORIS

SECUNDA MUSCULORUM TABULA.

HUMANI FABRICA LIBER II. 215

SECUNDAE MUSCULORUM TABULAE
characterum Index.

SECUNDA tabula primae, quod ad sectionem spectat, omnino correspondet, & in latus versa, eosdem cum illa commonstrat musculos, una cum ossium sedibus, in superficie prominentibus, quodammodoque excarnibus. Facies itaque propemodum universa, & caput quoque, quum paulo minus toto sit osseum, literis nisi paucissimis in hunc modum insignitur.

A Temporalis musculus. Quicquid autem supra semicirculum hunc, quo temporalis musculi exortus circumscribitur, consistit, prorsus excarne est.

B Os iugale. Foramen illud ad ossis huius terminum sese offerens, auditorius est meatus.

C Musculus, à mandendo mansorius praecipuè nuncupatus, & inter maxillam inferiorem mouentes secundus describendus.

D Musculus, buccarum motibus subseruiens, qui ex una maxilla in alteram pronascitur.

E Musculus similiter, ab inferiori maxilla in os υ referens productus.

F Musculus, à pectoris osse in os υ referens insertus.

G Musculus similiter, à superiori pectoris ossis sede principium ducens, & in humiliorem cartilaginis scuti imaginem repraesentantis regionem insertus. Is enim, quum sub musculo F insignito lateat, non prorsus in latere hic conspectum fugit.

H, H [illegible]

[illegible]

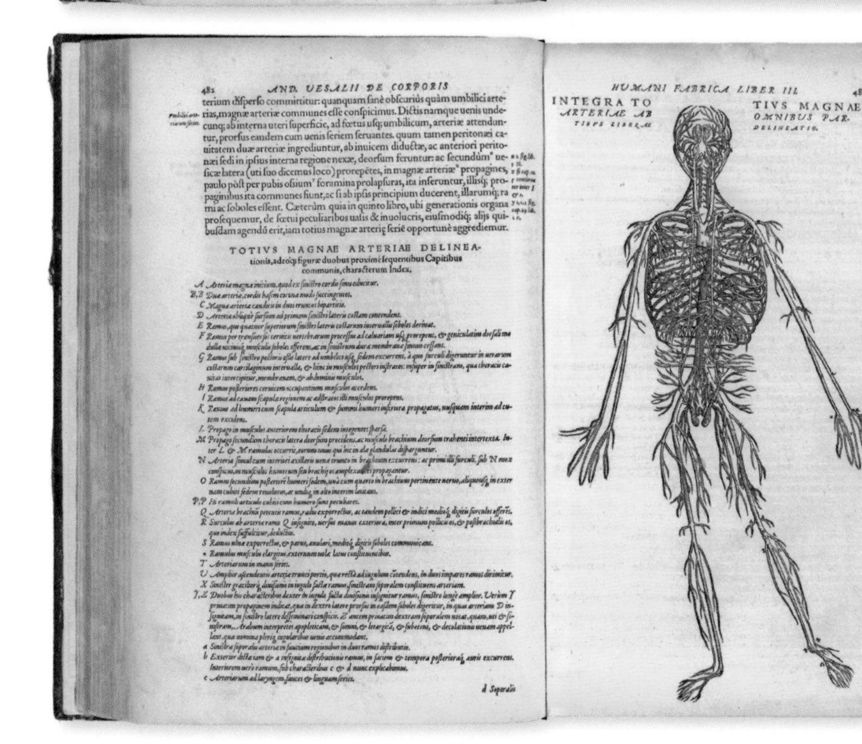

482 AND. VESALII DE CORPORIS

terium disperso committitur: quanquam sanè obscuriùs quàm umbilici arterias, magnae arteriae communes esse conspicimus. Dictis namque uenis undecunque ab interna uteri superficie, ad foetus usque umbilicum, arteriae attenduntur, prorsus eandem cum uenis seriem seruantes. quum tamen peritonaei cauitatem duae arteriae ingrediuntur, ab inuicem diductae, ac anteriori peritonaei sedi in ipsius interna regione nexae, deorsum feruntur: ac secundùm uesicae latera (uti suo dicemus loco) prorepētes, in magnae arteriae propagines, paulo pòst per pubis ossium foramina prolapsuras, ita inseruntur, illisque propaginibus ita communes fiunt, ac si ab ipsis principium ducerent, illarumque rami ac soboles essent. Caeterùm quia in quinto libro, ubi generationis organa prosequemur, de foetui peculiaribus uasis & inuolucris, eiusmodique alijs quibusdam agendū erit, iam totius magnae arteriae seriē opportunè aggrediemur.

TOTIUS MAGNAE ARTERIAE DELINEAtionis, adeoque figurae duobus proximè sequentibus Capitibus communis, characterum Index.

A Arteriae magnae initium, quod ex sinistro cordis sinu educitur.

B, B Duae arteriae, cordis basim coronae modo succingentes.

C Magnae arteriae caudicis in duos truncos bipartitio.

D Arteria obliquè sursum ad primam sinistri lateris costam contendens.

E Ramus, qui quatuor superiorum sinistri lateris costarum interuallis soboles deriuat.

F Ramus per transuersos ceruicis uertebrarum processus ad caluariam usque prorepens, & geniculatim dorsali medullae uicinisque musculis soboles offerens, ac in sinistrum durae membranae sinum cessans.

G Ramus sub sinistro pectoris ossis latere ad umbilici usque sedem excurrens, à quo surculi digeruntur in uerarum [illegible]

[illegible]

HUMANI FABRICA LIBER III. 483

INTEGRA TOTIUS MAGNAE ARTERIAE AB OMNIBUS PARTIBUS LIBERAE DELINEATIO.

T 2

of Galen and Hippocrates, whose theories were as limiting as Aristotle's were in physics and cosmology, pointing out many anatomical errors and giving a new understanding of anatomy based on far better observation of dissections.

Less than a hundred years later, the English physician William Harvey, born in Folkestone in 1578, published a landmark medical book in 1628. Just 72 pages long, *Exercitatio Anatomica de Motu Cordis et Sanguinis in Animalibus* (Anatomical Exercise of the Motion of the Heart and Blood in Organisms) gives the first detailed analysis of the way that blood is circulated around the body. Harvey's work was based on careful observation and experiments on animals, and by using ligatures in humans to temporarily restrict the blood flow. Until this point the role of the heart was largely seen as more spiritual than physical, but Harvey clearly identified it as a pump that produced the blood flow. He identified one-way valves and showed that circulation was necessary to account for the sheer volume of blood that the heart pumped. Because Harvey's model went against medical ideas dating back to Galen, it took several decades before this new hypothesis was widely accepted.

It is worth briefly jumping further forward a little in time to contrast Harvey's work with another famous medical book of the period, Nicolas Culpeper's *Complete Herbal* (originally published as *The English Physitian*). Although Harvey's work demonstrated an improving knowledge of human anatomy, medicine was still mostly in the dark ages, tied to unscientific and often harmful ancient Greek ideas. What Culpeper, born in London in 1616, would do is focus on the one aspect of the medicine of the time that had some potential for actually improving the patient's health – the pharmacopeia.

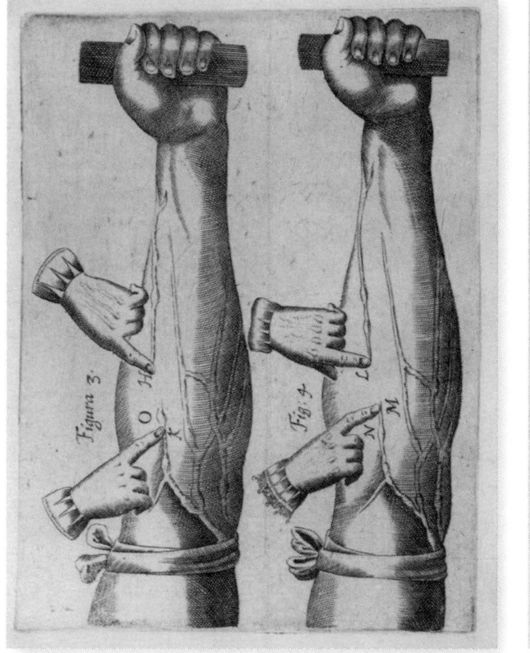

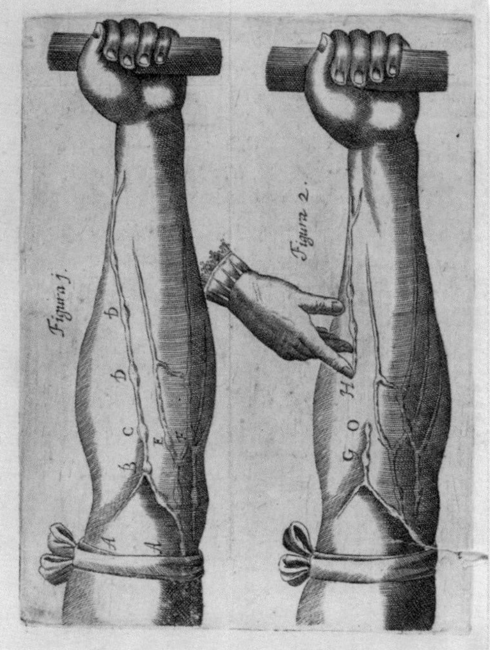

William Harvey
DE MOTU CORDIS,
DOMINICI RICCIARDI, 1643

Illustrations on the use of ligatures to restrict blood flow from a 1643 edition, 15 years after its original publication in 1628.

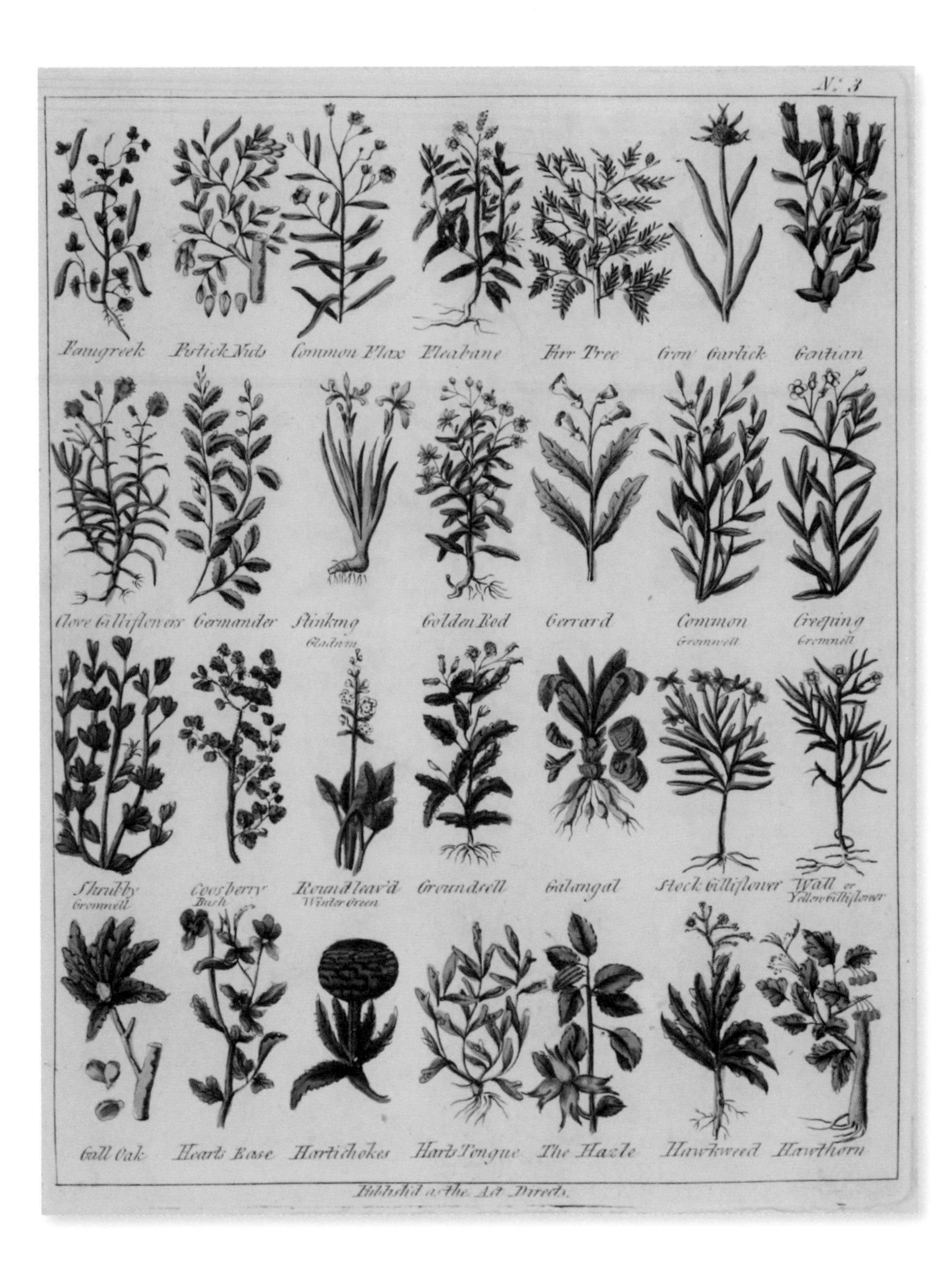
N.° 3
Fenugreek
Pistick Nuts
Common Flax
Fleabane
Firr Tree
Gron Garlick
Gentian
Clove Gilliflowers
Germander
Stinking
Gladwin
Golden Rod
Gerrard
Common
Gromwell
Creeping
Gromwell
Shrubby
Gromwell
Goosberry
Bush
Round leav'd
Winter Green
Groundsell
Galangal
Stock Gilliflower
Wall or
Yellow Gilliflower
Gall Oak
Hearts Ease
Hartichokes
Harts Tongue
The Hazle
Hawkweed
Hawthorn
Publishd as the Act Directs.

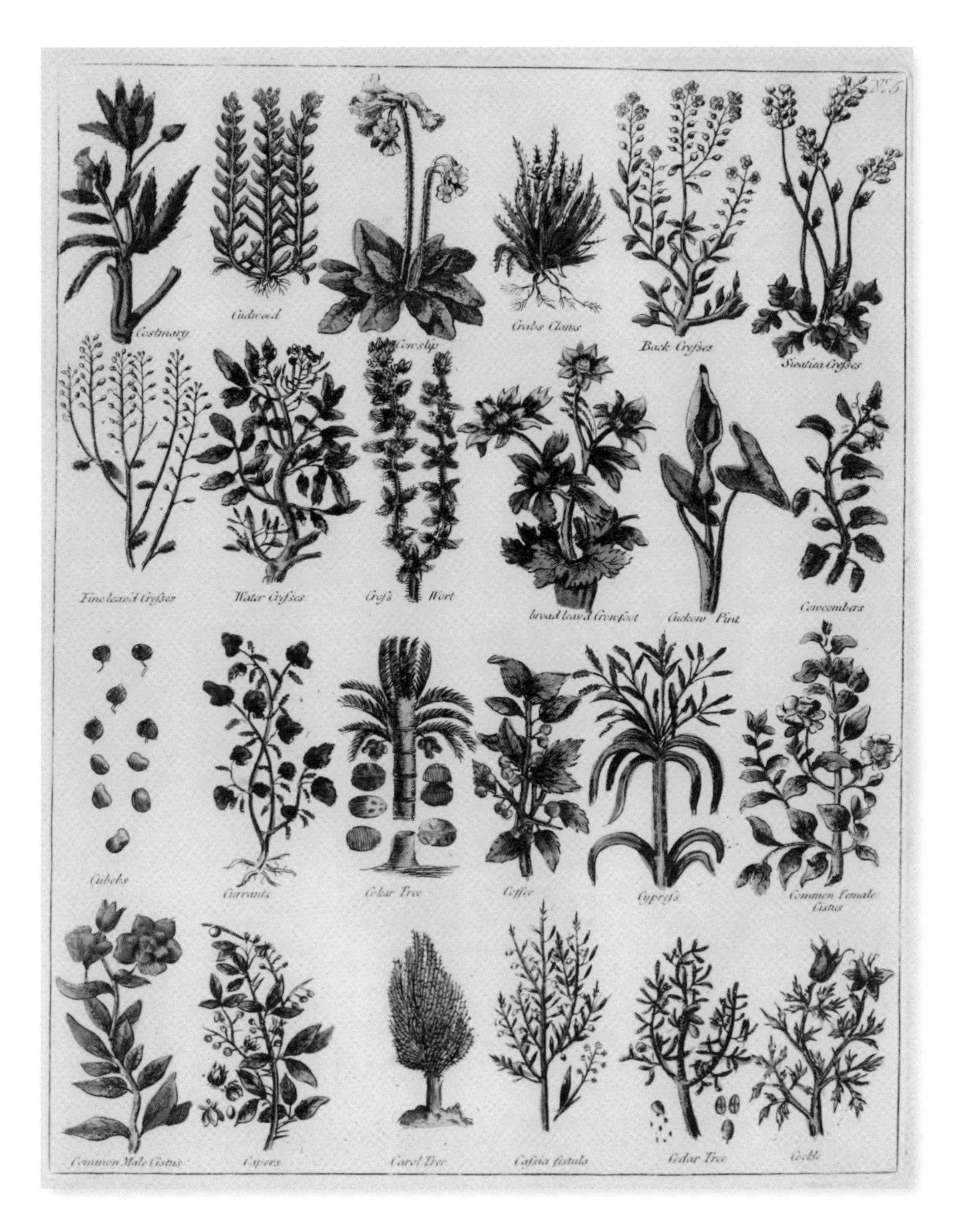

Nicholas Culpeper
COMPLETE HERBAL,
EBENEZER SIBLEY, 1789

This edition, published over 100 years after the first edition in 1652, makes use of rich illustrations of the herbs to help with identification in the field.

The English Physitian was first published in 1652, and as its better-known name from 1653 suggests, was a pharmaceutical guide concentrating primarily on herbal medicines. One reason the book is fascinating is that despite Culpeper's enthusiasm for the benefits of herbal treatments, some of which genuinely would have a positive medical effect, he was from an age that had yet to throw off the non-scientific influence of astrology, and therefore combined folk knowledge of what actually worked with fictional reasoning that paired plants with the supposed influence of the planets.

So good they named him twice

Medicine, however, was making only slow steps forward, and the most frequent breakthroughs in this period were in physics and cosmology, which we return to for one of the most famous names from this era, Galileo Galilei. Galileo was born in Pisa in 1564, the son of a professional musician who was fascinated by the science behind musical instruments and who influenced the young Galileo's analytical view of the world.

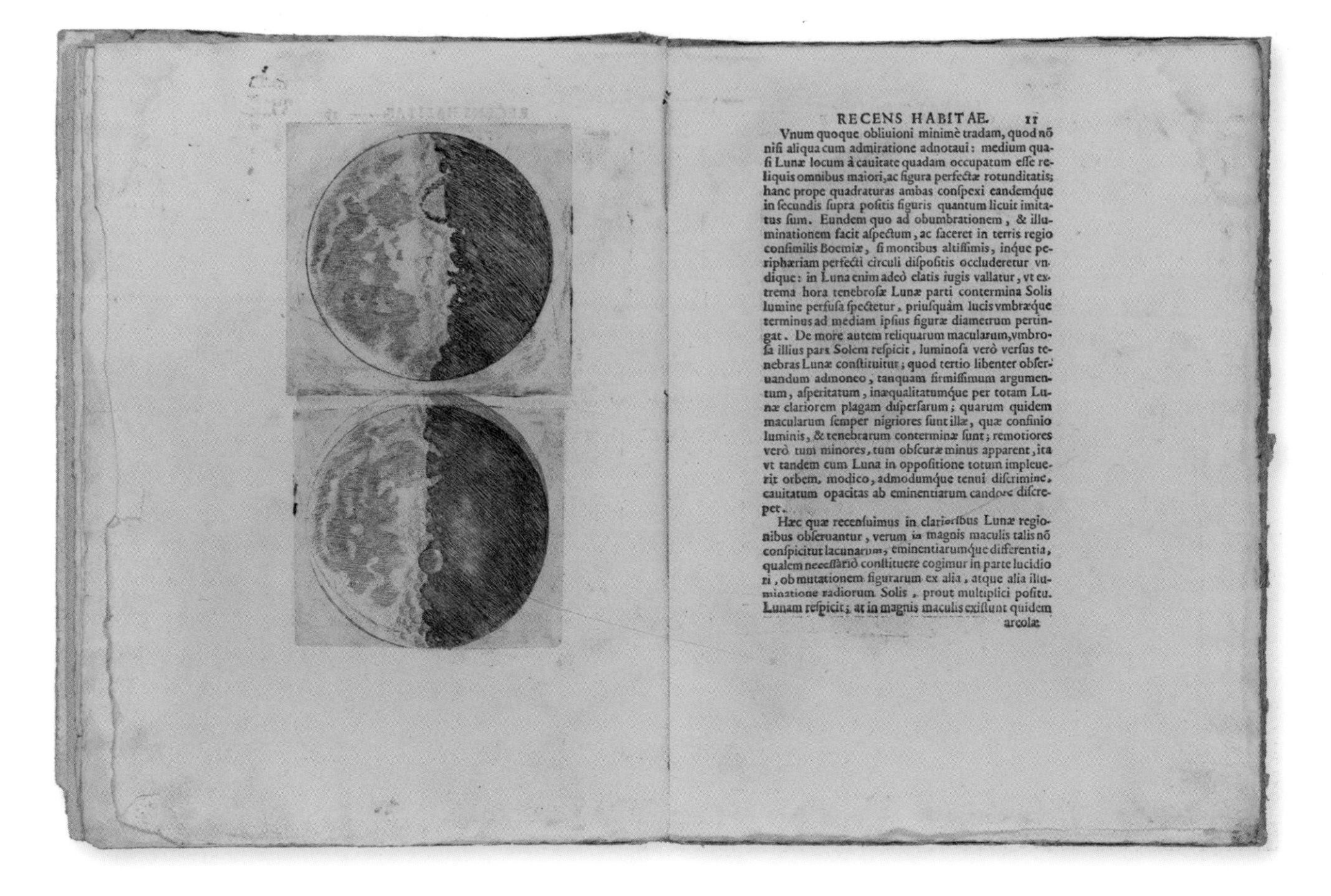

RECENS HABITAE. 11

Vnum quoque obliuioni minimè tradam, quod nõ nisi aliqua cum admiratione adnotaui: medium quasi Lunæ locum à cauitate quadam occupatum esse reliquis omnibus maiori, ac figura perfectæ rotunditatis; hanc prope quadraturas ambas conspexi eandemque in secundis supra positis figuris quantum licuit imitatus sum. Eundem quo ad obumbrationem, & illuminationem facit aspectum, ac faceret in terris regio consimilis Boemiæ, si montibus altissimis, inque peripheriam perfecti circuli dispositis occluderetur vndique: in Luna enim adeò elatis iugis vallatur, vt extrema hora tenebrosæ Lunæ parti contermina Solis lumine perfusa spectetur, priusquàm lucis vmbræque terminus ad mediam ipsius figuræ diametrum pertingat. De more autem reliquarum macularum, vmbrosa illius pars Solem respicit, luminosa verò versus tenebras Lunæ constituitur; quod tertio libenter obseruandum admoneo, tanquam firmissimum argumentum, asperitatum, inæqualitatumque per totam Lunæ clariorem plagam dispersarum; quarum quidem macularum semper nigriores sunt illæ, quæ confinio luminis, & tenebrarum conterminæ sunt; remotiores verò tum minores, tum obscuræ minus apparent, ita vt tandem cum Luna in oppositione totum impleuerit orbem, modico, admodumque tenui discrimine, cauitatum opacitas ab eminentiarum candore discrepet.

Hæc quæ recensuimus in clarioribus Lunæ regionibus obseruantur, verum in magnis maculis talis nõ conspicitur lacunarum, eminentiarumque differentia, qualem necessariò constituere cogimur in parte lucidiori, ob mutationem figurarum ex alia, atque alia illuminatione radiorum Solis, prout multiplici positu Lunam respicit; at in magnis maculis existunt quidem

areolæ

Galileo Galilei
SIDEREUS NUNCIUS,
TOMMASO BAGLIONI, 1610

Two of Galileo's sketches of the Moon, showing the terminator separating the light and dark sides.

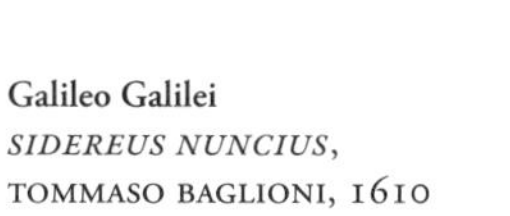

Galileo Galilei
SIDEREUS NUNCIUS, TOMMASO BAGLIONI, 1610

Galileo's drawing of the Pleiades constellation brings in stars that had not previously been seen, thanks to his observation through his newly constructed telescope.

Stefano della Bella,
FRONTISPIECE TO *OPERE DI GALILEO GALILEI*, ETCHING, 1656

The frontispiece from a collection of Galileo's works, showing Galileo with a telescope.

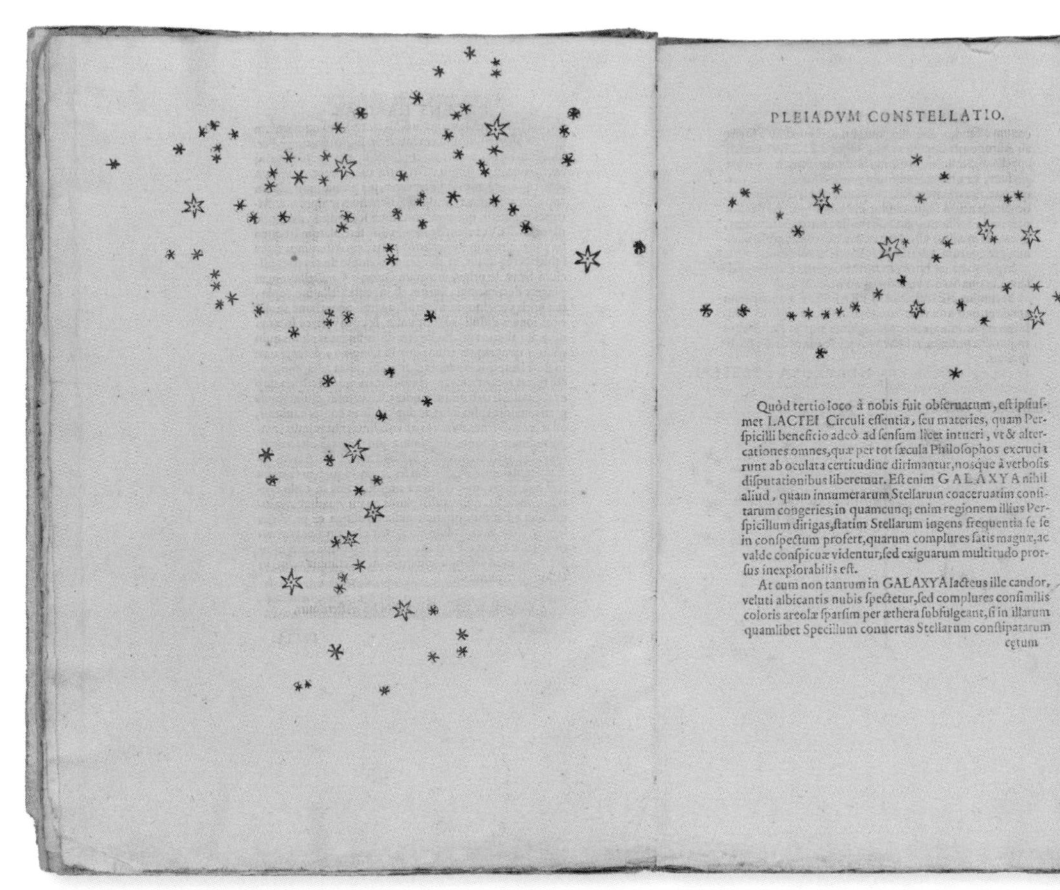

PLEIADVM CONSTELLATIO.

Quòd tertio loco à nobis fuit obſeruatum, eſt ipſiuſmet LACTEI Circuli eſſentia, ſeu materies, quam Perſpicilli beneficio adeò ad ſenſum licet intueri, vt & altercationes omnes, quæ per tot ſæcula Philoſophos excruciarunt ab oculata certitudine dirimantur, noſque à verboſis diſputationibus liberemur. Eſt enim GALAXYA nihil aliud, quam innumerarum Stellarum coaceruatim conſitarum congeries; in quamcunq; enim regionem illius Perſpicillum dirigas, ſtatim Stellarum ingens frequentia ſe ſe in conſpectum profert, quarum complures ſatis magnæ, ac valde conſpicuæ videntur; ſed exiguarum multitudo prorſus inexplorabilis eſt.

At cum non tantum in GALAXYA lacteus ille candor, veluti albicantis nubis ſpectetur, ſed complures conſimilis coloris areolæ ſparſim per æthera ſubfulgeant, ſi in illarum quamlibet Specillum conuertas Stellarum conſtipatarum cętum

Galileo worked widely in mathematics, physics and astronomy. As far as personality goes, he was something of an opportunist. When, for example, he heard that a Dutch inventor had arrived in Italy to demonstrate his telescope in Venice, Galileo arranged for a friend to hold the inventor up, to give Galileo time to assemble his own telescope and get to Venice first with it. However, Galileo was also without doubt a genius, one of the first natural philosophers who helped move the scientific viewpoint away from the strictures of ancient Greece.

Galileo wrote several notable books, though the first to reach a wide non-technical audience – *Sidereus Nuncius* from 1610 (strictly more a pamphlet than a book) – is by no means his most important. Apparently translating as 'Sidereal Messenger', *Sidereus Nuncius* is more often called the 'Starry Messenger'. In fact, 'starry' isn't a bad translation, as 'sidereal' was originally used to mean 'starlike', but over time has gained the more technical meaning of a period of time measured by the passage of the stars (i.e. the rotation of the Earth). *Sidereus Nuncius* was a summary of Galileo's early astronomical observations through his telescopes. It's often said he was the first to use a telescope for astronomy; he wasn't, but there is no doubt that his observations were significant. They notably included the discovery of the four brightest moons of Jupiter (which Galileo, hoping to win favour with the Tuscan court, named the Medicean Stars) and his detailed study of the Moon.

Galileo's description of the Moon was accompanied by beautiful engraved images from his sketches, showing that the lunar surface, which Aristotle's worldview required to be perfect and flat, was, in fact, rugged and varied. He concluded that the lunar landscape included high mountains, which Galileo could see was the case because the view through his telescope showed that the terminator – the line between the dark side and the light side of the partly lit lunar face – was not a straight line, but crinkly, as higher parts of the surface cast shadows into the sunlit parts.

Galileo Galilei
SYSTEMI DEL MONDO,
I. A. HUGUETAN, 1641

The frontispiece from a later edition of Galileo's book on models of the universe, first published in 1632, showing Aristotle, Ptolemy and Copernicus.

One technique that Galileo used in his illustrations would be frowned upon by modern science communicators. In some of his sketches of the Moon there is a large crater on the terminator, making our natural satellite look a little like the Death Star from the *Star Wars* movies. This huge crater does not exist – it's thought that Galileo was zooming in on a smaller crater to make it clear what he was seeing – but without labelling it as such, it was technically misleading. The pamphlet was by no means universally accepted. Many, who had less effective telescopes, thought the moons of Jupiter, for example, were nothing more than flaws in Galileo's lenses.

There were elements in *Sidereus Nuncius* that ran counter to the Aristotelian view of the universe, but Galileo's big spanner in the works, and his best-known work with the general public, was *Dialogo Sopra i due Massimi Sistemi del Mondo* (Dialogue Concerning the Two Chief World Systems) from 1632. This was the book that led to Galileo's trial for heresy and subsequent house arrest for life. Galileo is often portrayed as a martyr for publishing this in the face of opposition from the Catholic Church, though some of the fault was indubitably his own. In comparing the Aristotelian and Copernican systems, Galileo makes the supporter of Aristotle a character named Simplicio, whose name was suspiciously similar to the Italian word for 'simple-minded'. And, to make matters worse, he included a section which put the Pope's message – that the Copernican model was just a mathematical nicety, useful for calculations but not reflecting reality – in Simplicio's voice. It was hardly politically sensitive to imply that the Pope was a simpleton.

We need to remember that the main thrust of the book was not Galileo's original idea; rather, he was presenting evidence to support it. Sadly, the only original part of the book – a section on the tides – is not just wrong, but obviously wrong, as it put forward a theory that would only produce one tide a day. Although this book is Galileo's best known, what tends to be missed is that its most important aspect was arguably that – like Münster's *Cosmographia* (see page 90) – it was written in the language of the people (in Galileo's case, Italian), rather than in Latin. Galileo was one of the earliest significant writers of science books to realise the importance of getting the message out to a wider public, and he wrote his two most important books in the 'vulgar' tongue of his country.

Far eclipsing *Dialogo* for originality, though, was Galileo's most significant book, written in 1638 after his house arrest. This was *Discorsi e Dimostrazioni Matematiche Intorno a Due Nuove Scienze* (Discourses and Mathematical Demonstrations Relating to Two New Sciences). In this book, Galileo published his main ideas on physics. He described the motion of pendulums, the way bodies accelerated under the force of gravity, their rate being uninfluenced by their mass, the trajectories of projectiles and more. Along the way, to future mathematicians' delight, he was also one of the very first to consider the mathematical implications of infinity.

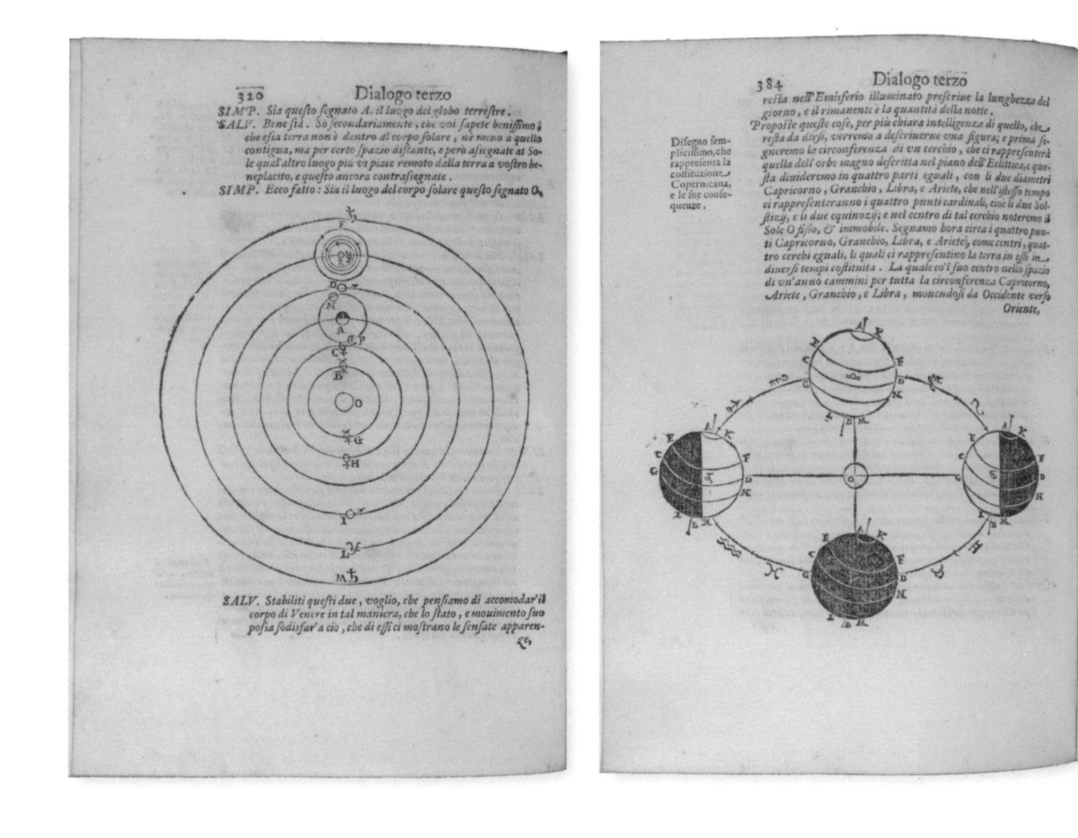

310 Dialogo terzo

SIMP. Sia questo segnato A. il luogo del globo terrestre.

SALV. Bene stà. So secondariamente, che voi sapete benissimo, che essa terra non è dentro al corpo solare, nè meno a quello contigua, ma per certo spazio distante, e però assegnate al Sole qual'altro luogo più vi piace remoto dalla terra a vostro beneplacito, e questo ancora contrassegnate.

SIMP. Ecco fatto: Sia il luogo del corpo solare questo segnato O.

SALV. Stabiliti questi due, voglio, che pensiamo di accomodar'il corpo di Venere in tal maniera, che lo stato, e movimento suo possa sodisfar'a ciò, che di essi ci mostrano le sensate apparenze,

384 Dialogo terzo

resta nell'Emisferio illuminato prescrive la lunghezza del giorno, e il rimanente è la quantità della notte.

Disegno semplicissimo, che rappresenta la costituzione Copernicana, e le sue consequenze.

Proposte queste cose, per più chiara intelligenza di quello, che resta da dirsi, verremo a descriverne una figura; e prima segneremo la circonferenza di un cerchio, che ci rappresenterà quella dell'orbe magno descritta nel piano dell'Eclittica, e questa divideremo in quattro parti eguali, con li due diametri Capricorno, Granchio, Libra, e Ariete, che nell'istesso tempo ci rappresenteranno i quattro punti cardinali, cioè li due Solstizij, e li due equinozij; e nel centro di tal cerchio noteremo il Sole O fisso, & immobile. Segnamo hora circa i quattro punti Capricorno, Granchio, Libra, e Ariete, come centri, quattro cerchi eguali, li quali ci rappresentino la terra in essi in diversi tempi costituita. La quale co'l suo centro nello spazio di un'anno cammini per tutta la circonferenza Capricorno, Ariete, Granchio, e Libra, movendosi da Occidente verso Oriente,

Galileo Galilei
SYSTEMI DEL MONDO,
BATISTA LANDINI, 1632

Diagrams of the Copernican solar system and phases of a body as it rotates around an illuminating source.

What Galileo almost certainly didn't do in researching this book was to drop balls of different weights off the Leaning Tower of Pisa, as legend has it. It would have been very difficult to accurately time the falls, and there is no evidence that the experiment ever took place. Instead, he presented an ingenious thought experiment involving tying two falling bodies together, which disproved Aristotle's view that heavier bodies naturally fell faster, and he supported his arguments on the effect of gravity with the results of experiments where he rolled balls of different mass down inclined planes, measuring their acceleration under far more controlled conditions than would have been possible by dropping balls off a tower.

Once again, his book was written in Italian and, unlike many science books from the period, it is still remarkably readable today. The book is clearly intended for a wide audience, though Galileo archly remarks in the introduction that he never intended it to be published (a service that was performed by the publisher Elsevier – who still exist today – in Protestant Holland). According to Galileo, he only intended to circulate the book to a few friends and the whole thing came as a big surprise to him. If this was intended to fool the Inquisition, he didn't think much of their abilities – but there were no repercussions this time.

Galileo Galilei
DUE NUOVE SCIENZE,
ELSEVIER, 1638

Galileo's masterpiece on physics, printed by Dutch publisher Elsevier, after Galileo's house arrest made publication in Italy impossible.

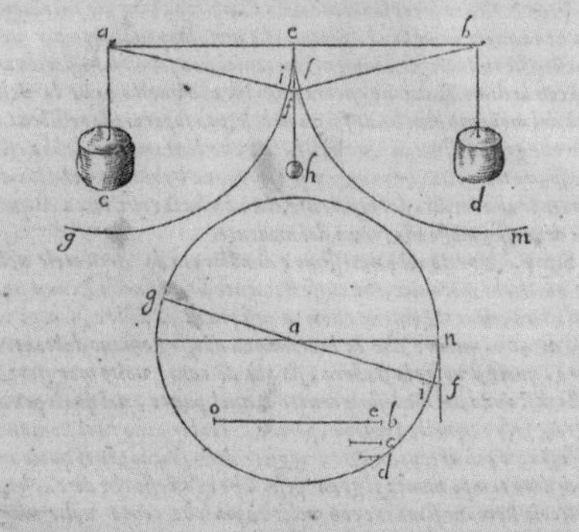

Geometry to chemistry

Someone who bridges the gap between Galileo and the other towering name of the period, Isaac Newton, is the French philosopher René Descartes. Though often best remembered for his statement 'I think, therefore I am', Descartes put forward a range of scientific theories and made a major breakthrough in mathematical technique. His most significant book was *Discours de la Méthode Pour bien conduire sa Raison, et chercher la Vérité dans les Sciences* (Discourse on the Method of Rightly Conducting One's Reason and of Seeking Truth in the Sciences) from 1637, which contains his famous 'I think' quote. The *Discours* was Descartes's attempt to give a philosophical basis for a scientific method, but its greatest significance is one of the appendices – which are effectively books in their own right – that Descartes used to illustrate the outcome of using his method.

The appendix was *La Géométrie* (The Geometry). It may sound from the title that it was little more than a re-working of Euclid, but it was far more significant: the work contained in *La Géométrie* is the reason we refer to the 'x y coordinates' of mathematical charts as 'Cartesian coordinates' (even though, perversely, Descartes didn't use x and y;

René Descartes
DISCOURS DE LA MÉTHODE,
JOANNES MAIRE, 1637

The geometry appendix of Descarte's *Discours,* which linked algebra and geometry using Cartesian coordinates.

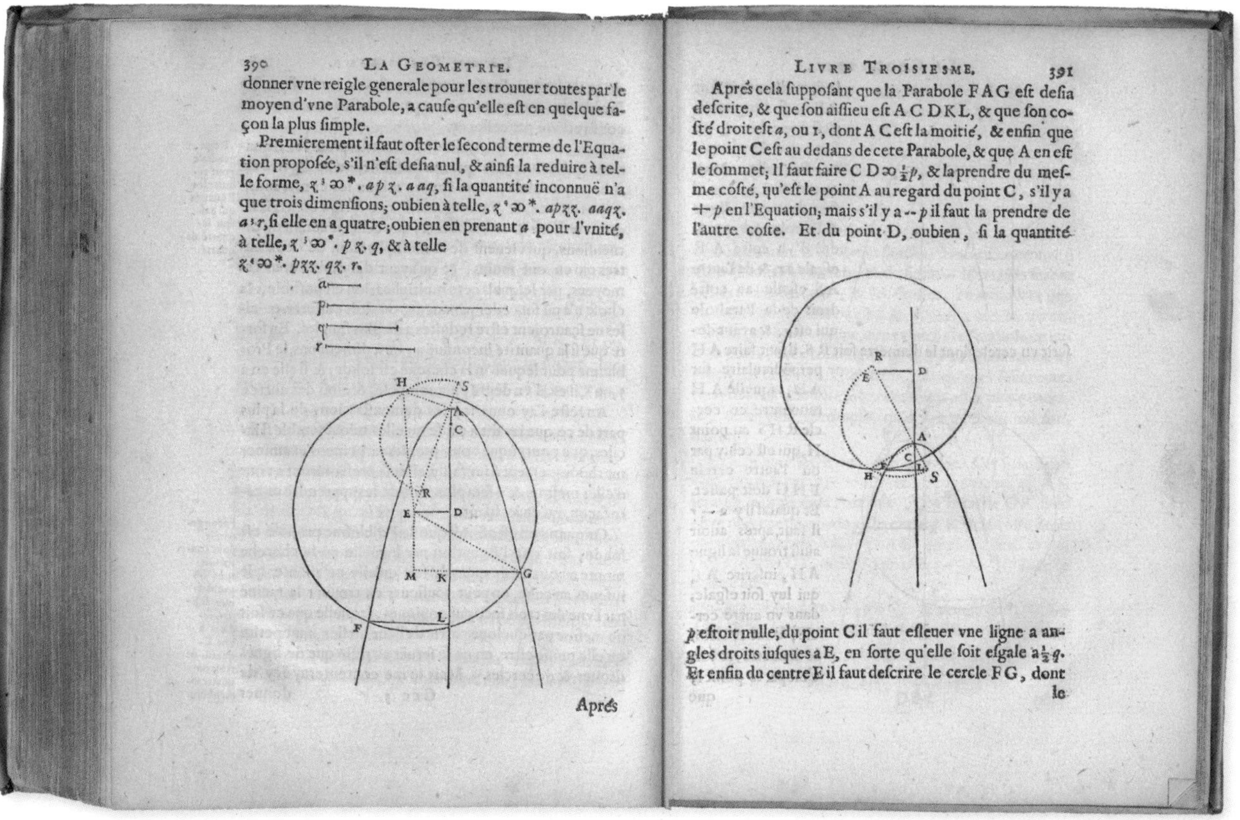

390 LA GEOMETRIE.

donner vne reigle generale pour les trouuer toutes par le moyen d'vne Parabole, a cauſe qu'elle eſt en quelque façon la plus ſimple.

Premierement il faut oſter le ſecond terme de l'Equation propoſée, s'il n'eſt deſia nul, & ainſi la reduire à telle forme, $z^3 \infty * . apz . aaq$, ſi la quantité inconnuë n'a que trois dimenſions; oubien à telle, $z^4 \infty * . apzz . aaqz . a^3r$, ſi elle en a quatre; oubien en prenant a pour l'vnité, à telle, $z^3 \infty * . pz . q$, & à telle
$z^4 \infty * . pzz . qz . r.$

Aprés

LIVRE TROISIESME. 391

Aprés cela ſuppoſant que la Parabole F A G eſt deſia deſcrite, & que ſon aiſſieu eſt A C D K L, & que ſon coſté droit eſt a, ou 1, dont A C eſt la moitié, & enfin que le point C eſt au dedans de cete Parabole, & que A en eſt le ſommet; Il faut faire C D $\infty \frac{1}{2}p$, & la prendre du meſme coſté, qu'eſt le point A au regard du point C, s'il y a $+p$ en l'Equation; mais s'il y a $-p$ il faut la prendre de l'autre coſte. Et du point D, oubien, ſi la quantité

p eſtoit nulle, du point C il faut eſleuer vne ligne a angles droits iuſques a E, en ſorte qu'elle ſoit eſgale a $\frac{1}{2}q$. Et enfin du centre E il faut deſcrire le cercle F G, dont le

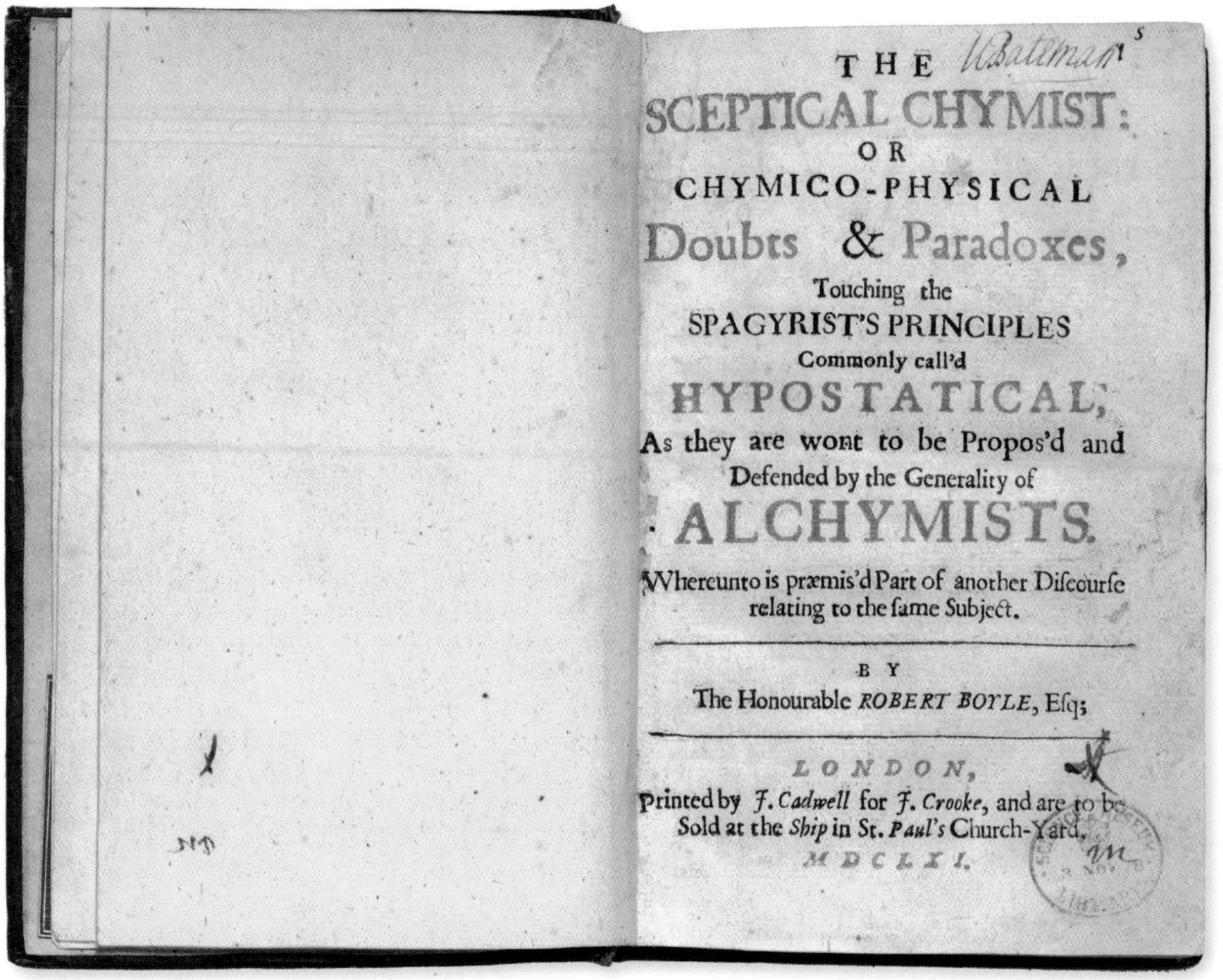

THE
SCEPTICAL CHYMIST:
OR
CHYMICO-PHYSICAL
Doubts & Paradoxes,
Touching the
SPAGYRIST'S PRINCIPLES
Commonly call'd
HYPOSTATICAL;
As they are wont to be Propos'd and
Defended by the Generality of
ALCHYMISTS.
Whereunto is præmis'd Part of another Discourse
relating to the same Subject.

BY
The Honourable *ROBERT BOYLE*, Esq;

LONDON,
Printed by *J. Cadwell* for *J. Crooke*, and are to be
Sold at the *Ship* in St. *Paul's* Church-Yard.
MDCLXI.

Robert Boyle
THE SCEPTICAL CHYMIST,
J. CADWELL, 1661

The title page from the first edition of Boyle's book, which began the separation of chemistry from alchemy as a discipline in its own right.

they were added by later mathematicians to clarify his work.) Up until Descartes, geometry and algebra were seen as totally independent disciplines. In his book, Descartes showed that geometric curves and shapes could be rendered as algebraic equations, making the manipulation of geometry something that could be handled in terms of the often-simpler algebra.

Descartes's ideas would have a huge influence on someone who could have learned a lesson or two in writing comprehensibly from Galileo. This was the great mathematician and physicist born less than a year after Galileo's death, Isaac Newton. But before we get to his book *Principia*, a landmark in the history of science books, we need to take a look at a title that effectively marks the birth of a new science: Robert Boyle's *The Sceptical Chymist*.

Boyle was born in Lismore, Ireland in 1627, the fourteenth child of the British Earl of Cork. He was technically a nobleman, but far enough down the family pecking order to require some kind of profession. Most often this would have taken the form of a career in the army or clergy, but Robert's European tour, taken when he was in his teens, seems to have sparked a fascination with science that lasted throughout his life. Like most natural philosophers of the period, his interests were wide. Boyle published a significant book in 1660, his *New Experiments Physico-Mechanical, Touching the Spring of the Air, and its Effects*, which introduced Boyle's law for gasses. Yet his work on the behaviour of gas was less original than his transformation of chemistry. Just one year later, *The Sceptical Chymist* would be published.

We now draw a very clear line between alchemy and chemistry. Alchemy involved attempts to mix and react substances with an approach that was as much spiritual as physical, hoping to achieve goals such as an elixir of life or a philosopher's stone, or the ability to transmute base metals like lead into gold. Chemistry is a scientific study of the way that elements interact. However, when Robert Boyle was active there was no such distinction.

The approach that Boyle took could not be considered 100 per cent chemistry, but he very much pushed alchemy in that direction. He was still an alchemist as well as a chemist, and certainly attempted transmutation of metals. However, his approach was far more in the spirit of modern science than traditional alchemy. In *The Sceptical Chymist* he identified chemistry as more than a tool of alchemy, as it had been seen to that date, but rather a study in its own right of the way that different substances could be combined as mixtures and compounds.

In the book, Boyle puts forward a hypothesis that seems surprisingly modern. He suggested that matter was made up of atoms, a concept that, as we have seen, had largely

School of Peter Lely
PORTRAIT OF THE HON. ROBERT BOYLE, OIL ON CANVAS, CA. 1689

A portrait of Boyle (the painter may be Johannes Kerseboom) alongside a seventeenth century painting by Hendrick Heerschop of an alchemist suffering a technical problem.

been out of favour since Aristotle (see page 38). These atoms, he believed, linked together to form compounds and were constantly in motion, with their collisions resulting in reactions. This idea was prescient, though Boyle's picture of the world was not perfect. What he wrote was more inspired guesswork than the result of detailed experimental analysis, limited as this was by the equipment and theories of the time. So, for example, he believed that all matter we experience is made of compounds, rather than ever being found as pure elements. However, his approach was far more tied to experiment than that of many of his contemporaries.

Scientific novelties

While Boyle envisaged the ultimate in small pieces of matter in the form of atoms, his contemporary, Robert Hooke, examined, and reproduced in greatly magnified drawings, the invisibly small world that was revealed by the microscope in his book *Micrographia*. In doing so, he turned genuine scientific discoveries into something close to an entertainment.

Hooke, who was born on the Isle of Wight in England in 1635, was an impressive scientist in his own right, though he tends to be remembered more for his role as curator of experiments for the Royal Society, where he traded carefully worded insults with Isaac Newton. He also undertook astronomical work, experimented in physics – producing Hooke's law, describing the elasticity of springs – and was a major player in the rebuilding of London after the great fire of 1666, acting as surveyor for the City. But as a writer, his *Micrographia* proved his most remarkable output.

With the lengthy full title of *Micrographia: or Some Physiological Descriptions of Minute Bodies Made by Magnifying Glasses. With Observations and Inquiries Thereupon*, the book, published in 1665, was the first to bring the view through the microscope to ordinary readers, in the form of what would now probably be called a coffee-table book. In what was already a large book, some of the illustrations even folded out, so that readers could be delighted and horrified by the huge rendition of his beautiful drawings of the likes of a lice, a flea and the compound eyes of a fly.

It's not that there isn't plenty of text in *Micrographia*. Hooke used the opportunity of its publication to range far and wide in his scientific interests, considering everything from the nature of light to the origins of fossils. However, it is the large copperplate engravings made from his drawings that must have enthralled the book's readers. The diarist Samuel Pepys commented that *Micrographia* was 'the most ingenious book that ever I read in my life'.

Hooke's observations also included a cross section of a piece of cork, showing how the substance, which is apparently continuous to the naked eye was made up of a collection of tiny box-like structures, which Hooke was the first to name as 'cells'. He likened the collection of cells to a honeycomb, though the word itself seems to have been taken from the name for a monk's room, which were often arranged in rows of equal-sized spaces.

Hooke was not the only one to capture the public's imagination with scientific novelties. The German scientist Otto von Guericke, born in Magdeburg in 1602, would put on a scientific demonstration that amazed the world. This would later appear in his *Experimenta Nova* (New Experiments), published in 1672, which also included a wide range of other observations on the nature of a vacuum (with work on static electricity thrown in for good measure).

We tend to take the existence of a vacuum (in space, for example) as an obvious possibility. But Aristotle had declared that a void or vacuum was abhorred by nature, and until the seventeenth century it was assumed that such a thing could simply not exist. Von Guericke carried out a vast range of experiments on vacuums with early air pumps, but the one described in *Experimenta Nova* that brought him fame was the Magdeburg hemispheres. This was a pair of copper hemispheres about 50 cm (20 inches) across. In 1654, Von Guericke arranged an elaborate demonstration, removing as much air as he could from between the pair of hemispheres. So strong was the air pressure on the outside with a vacuum inside that two teams of 15 horses, one attached to each hemisphere, could not separate them.

Both von Guericke and Hooke grasped that going beyond simple description of a theory was important if a scientific work was to attract the attention of the public.

Otto von Guericke
EXPERIMENTA NOVA,
JOHANNES JANSSONIUS,
1672

Illustration from von Guericke's book of the Magdeburg hemispheres and the teams of horses attempting to separate them after the air between was evacuated.

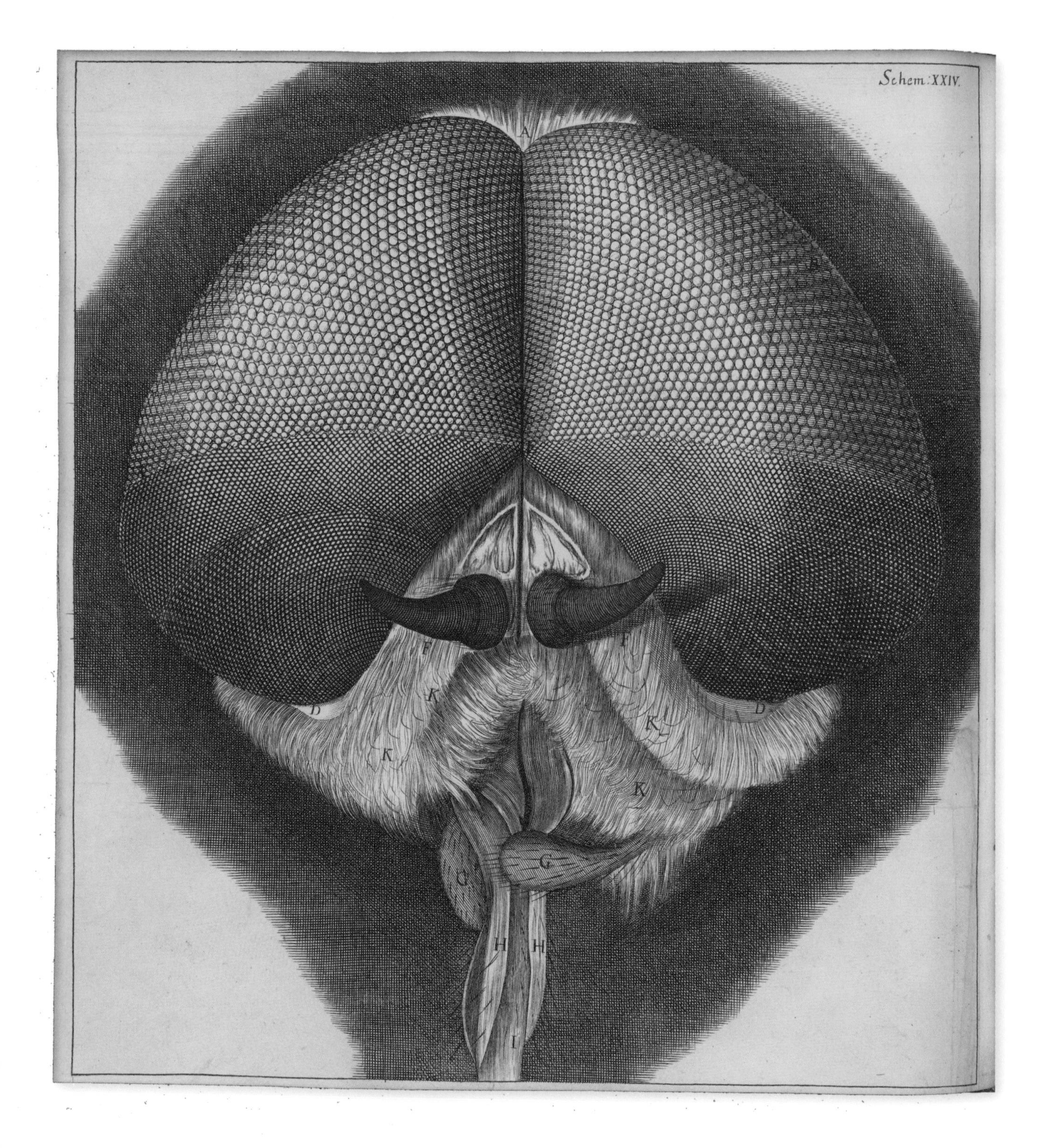
Schem: XXIV.
A
F
F
K
K
D
D
K
K
G
G
H
H
I

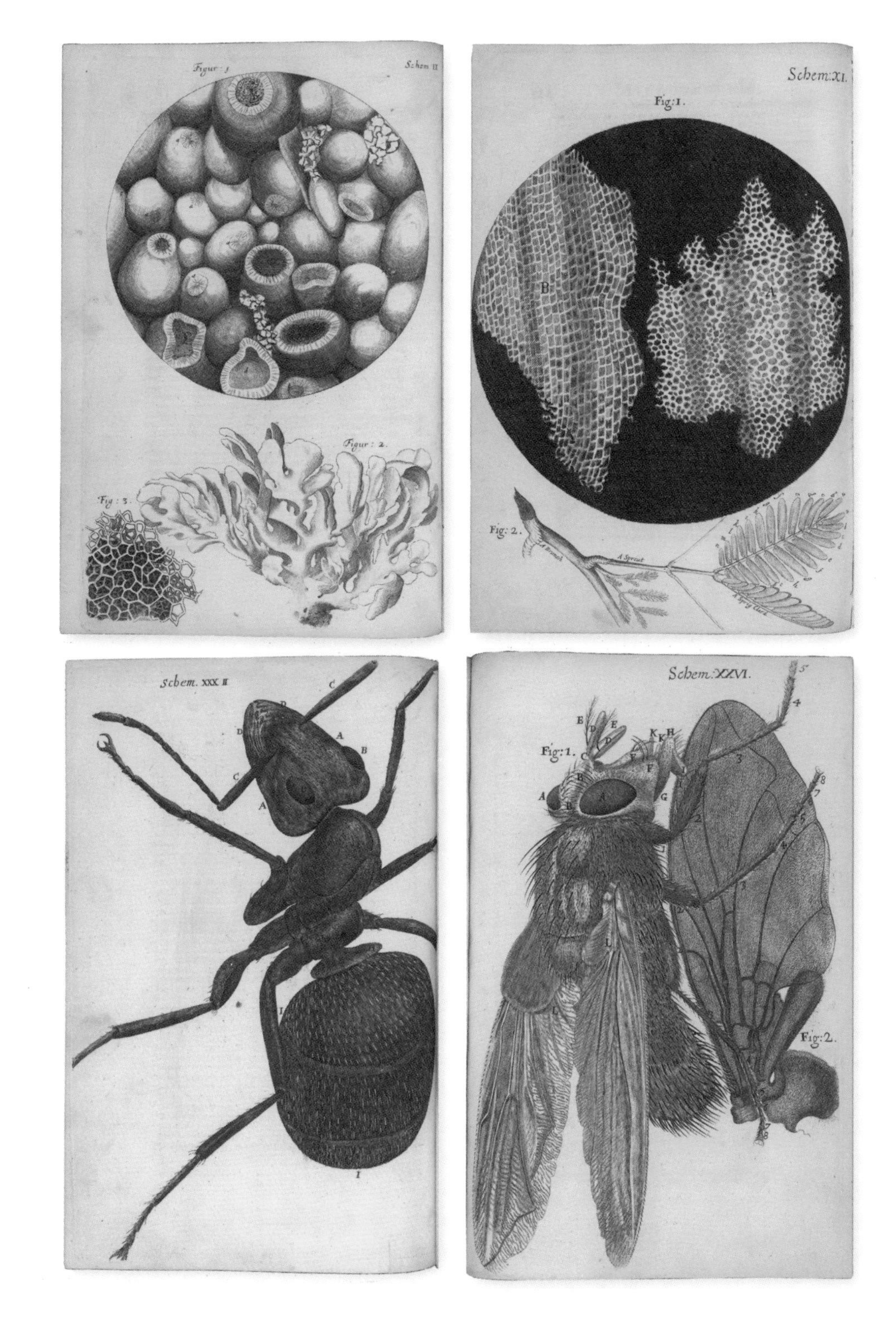

Robert Hooke
MICROGRAPHIA,
JOHN MARTYN AND
JAMES ALLESTRY, 1665

Some of the stunning illustrations from Hooke's large format title including the eyes and head of a grey drone fly (opposite), sea plants (top left), cork (top right), an ant (bottom left) and a blue fly (bottom right).

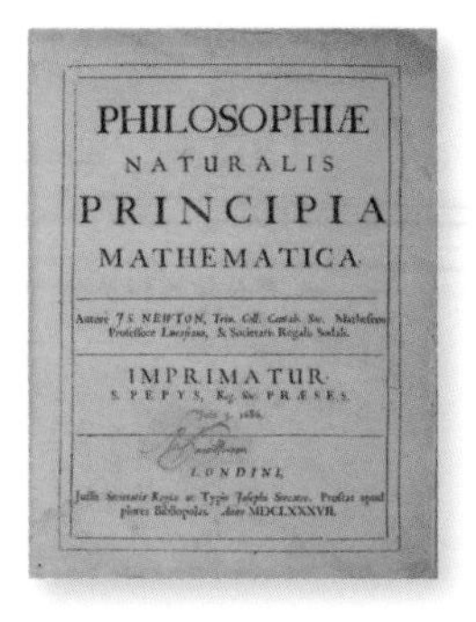
PHILOSOPHIÆ
NATURALIS
PRINCIPIA
MATHEMATICA.

IMPRIMATUR.
S. PEPYS, Reg. Soc. PRÆSES.

LONDINI,

Isaac Newton
PHILOSOPHÆ NATURALIS PRINCIPIA MATHEMATICA, JOSEPH STREATER, 1687

The title page from one of the most famous scientific works in history. This is where Newton introduced his laws of motion and his theory of universal gravitation.

The gravity of the situation

Hooke's name appeared a good few times in the original draft of one of the most influential science books of all time: Isaac Newton's *Philosophiæ Naturalis Principia Mathematica* (Mathematical Principles of Natural Philosophy), usually known as the *Principia*. Newton's poor relations with Hooke, which got worse over time, meant that before publication Newton struck out the name of his fellow scientist whenever he could.

Isaac Newton was born in Woolsthorpe in Lincolnshire in the east of England in 1643 – or 1642, depending on your preference. In the period before Britain adopted the Gregorian calendar in 1752, dates near the turn of the year can be confusing. In the Julian calendar that was used at the time, Newton was born on Christmas Day 1642, but by modern reckoning this was 4 January 1643. His date of death is even more confusing. Modern dating puts this as 31 March 1727, but in the calendar of the time it was 20 March 1726, as New Year's Day at the time was held on 25 March.

Newton's three great contributions to science were his work on light and colour (where he established, for example, that white light was composed of the colours of the rainbow mixed together), on gravitation and motion, and in developing the mathematical technique now called calculus, which he called the method of fluxions. Newton would have considered himself a mathematician rather than a natural philosopher (and he spent more time on alchemy and theology than either), but his theories on light, gravity and motion meant that the magnitude of his contributions to physics would only be rivalled by Einstein's.

Newton's work on light and colour, which included a range of experiments using both prisms and lenses and his own eyes, came first – yet he was never one to publish in a hurry and his book on light and optics, *Opticks: or, A Treatise of the Reflexions, Refractions, Inflexions and Colours of Light*, would not be published until 1704, decades after the research. This book was written in English, following the trend of using the vulgar language that Galileo had made popular. But this was not the case with his masterpiece, published in 1687. Not only was *Principia* written in Latin, it seems deliberately to have been made more obscure than it needed to be.

The book is in three volumes, the third of which Newton had intended to be a more approachable text for the general reader – but as a result of fallings out with other Fellows of the Royal Society he made it as inaccessible to the non-scientist as the rest. The *Principia* brings us everything from the concept of mass and the three laws of motion to Newton's gravitational law (though it is never shown in the modern equation form). The most inspired aspect of his work was unifying the gravitational pull we experience on the surface of the Earth with the force that keeps the Earth in orbit around the Sun and the Moon around the Earth.

To be able to achieve this remarkable feat, Newton made wide use of his new mathematical technique, the method of fluxions, though in a move that probably makes the content even harder to grasp, as much of the book as possible makes use of traditional geometric arguments. In making calculations on the effects of gravity, in the main part of the book Newton stuck to a description of what happens, rather than a hypothesis of why it happens. To emphasise this, he wrote a famous phrase in the *Principia*, 'Hypotheses

Isaac Newton
PHILOSOPHÆ NATURALIS PRINCIPIA MATHEMATICA, JOSEPH STREATER, 1687

Interior pages showing Newton's geometric approach despite his use of calculus to derive his laws. Below, proposition XXXIV looks at the hourly variation of the inclination of the lunar orbit.

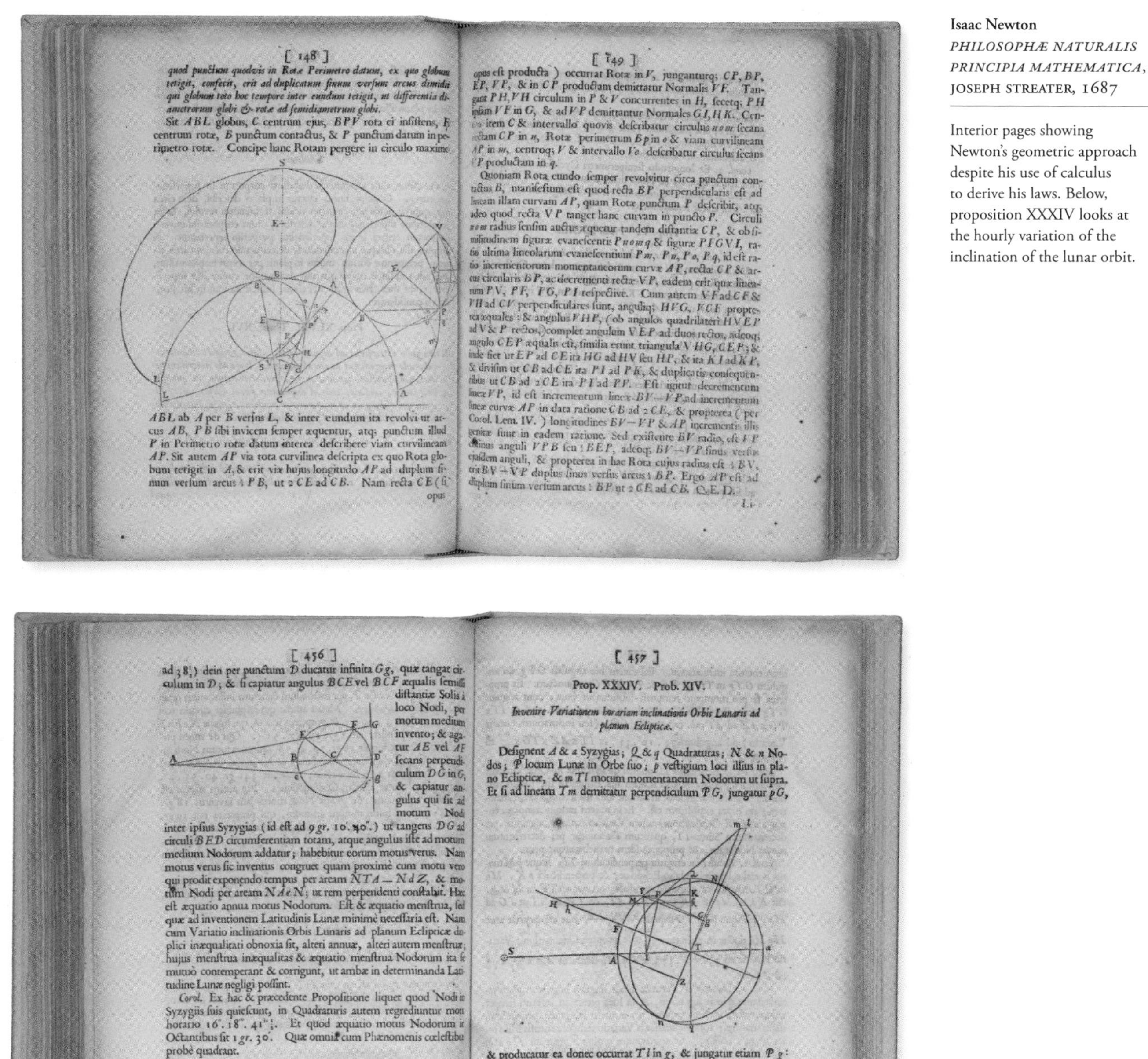

[148]

quod punctum quodvis in Rotæ Perimetro datum, ex quo globum tetigit, confecit, erit ad duplicatum sinum versum arcus dimidii qui globum toto hoc tempore inter eundum tetigit, ut differentia diametrorum globi & rotæ ad semidiametrum globi.

Sit *ABL* globus, *C* centrum ejus, *BPV* rota ei insistens, *E* centrum rotæ, *B* punctum contactus, & *P* punctum datum in perimetro rotæ. Concipe hanc Rotam pergere in circulo maximo *ABL* ab *A* per *B* versus *L*, & inter eundum ita revolvi ut arcus *AB*, *PB* sibi invicem semper æquentur, atq; punctum illud *P* in Perimetro rotæ datum interea describere viam curvilineam *AP*. Sit autem *AP* via tota curvilinea descripta ex quo Rota globum tetigit in *A*, & erit viæ hujus longitudo *AP* ad duplum sinum versum arcus ½ *PB*, ut 2 *CE* ad *CB*. Nam recta *CE* (si opus

[149]

opus est producta) occurrat Rotæ in *V*, junganturq; *CP*, *BP*, *EP*, *VP*, & in *CP* productam demittatur Normalis *VF*. Tangant *PH*, *VH* circulum in *P* & *V* concurrentes in *H*, secetq; *PH* ipsam *VF* in *G*, & ad *VP* demittantur Normales *GI*, *HK*. Centro item *C* & intervallo quovis describatur circulus *nom* secans rectam *CP* in *n*, Rotæ perimetrum *Bp* in *o* & viam curvilineam *AP* in *m*, centroq; *V* & intervallo *Vo* describatur circulus secans *VP* productam in *q*.

Quoniam Rota eundo semper revolvitur circa punctum contactus *B*, manifestum est quod recta *BP* perpendicularis est ad lineam illam curvam *AP*, quam Rotæ punctum *P* describit, atq; adeo quod recta *VP* tanget hanc curvam in puncto *P*. Circuli *nom* radius sensim auctus æquetur tandem distantiæ *CP*, & ob similitudinem figuræ evanescentis *Pnomq* & figuræ *PFGVI*, ratio ultima lineolarum evanescentium *Pm*, *Pn*, *Po*, *Pq*, id est ratio incrementorum momentaneorum curvæ *AP*, rectæ *CP* & arcus circularis *BP*, ac decrementi rectæ *VP*, eadem erit quæ linearum *PV*, *PF*, *PG*, *PI* respective. Cum autem *VF* ad *CF* & *VH* ad *CV* perpendiculares sunt, anguliq; *HVG*, *VCF* propterea æquales; & angulus *VHP*, (ob angulos quadrilateri *HVEP* ad *V* & *P* rectos,) complet angulum *VEP* ad duos rectos, adeoq; angulo *CEP* æqualis est, similia erunt triangula *VHG*, *CEP*; & inde fiet ut *EP* ad *CE* ita *HG* ad *HV* seu *HP*, & ita *KI* ad *KP*, & divisim ut *CB* ad *CE* ita *PI* ad *PK*, & duplicatis consequentibus ut *CB* ad 2 *CE* ita *PI* ad *PV*. Est igitur decrementum lineæ *VP*, id est incrementum lineæ *BV* − *VP*, ad incrementum lineæ curvæ *AP* in data ratione *CB* ad 2 *CE*, & propterea (per Corol. Lem. IV.) longitudines *BV* − *VP* & *AP* incrementis illis genitæ sunt in eadem ratione. Sed existente *BV* radio, est *VP* cosinus anguli *VPB* seu ½ *BEP*, adeoq; *BV* − *VP* sinus versus ejusdem anguli, & propterea in hac Rota cujus radius est ½ *BV*, erit *BV* − *VP* duplus sinus versus arcus ½ *BP*. Ergo *AP* est ad duplum sinum versum arcus ½ *BP* ut 2 *CE* ad *CB*. Q.E.D.

Li-

[456]

ad 38′,) dein per punctum *D* ducatur infinita *Gg*, quæ tangat circulum in *D*; & si capiatur angulus *BCE* vel *BCF* æqualis semissi distantiæ Solis à loco Nodi, per motum medium invento; & agatur *AE* vel *AF* secans perpendiculum *DG* in *G*, & capiatur angulus qui sit ad motum Nodi inter ipsius Syzygias (id est ad 9 *gr.* 10′. 40″.) ut tangens *DG* ad circuli *BED* circumferentiam totam, atque angulus iste ad motum medium Nodorum addatur; habebitur eorum motus verus. Nam motus verus sic inventus congruet quam proximè cum motu vero qui prodit exponendo tempus per aream *NTA* − *NdZ*, & motum Nodi per aream *NAeN*; ut rem perpendenti constabit. Hæc est æquatio annua motus Nodorum. Est & æquatio menstrua, sed quæ ad inventionem Latitudinis Lunæ minimè necessaria est. Nam cum Variatio inclinationis Orbis Lunaris ad planum Eclipticæ duplici inæqualitati obnoxia sit, alteri annuæ, alteri autem menstruæ; hujus menstrua inæqualitas & æquatio menstrua Nodorum ita se mutuò contemperant & corrigunt, ut ambæ in determinanda Latitudine Lunæ negligi possint.

Corol. Ex hac & præcedente Propositione liquet quod Nodi in Syzygiis suis quiescunt, in Quadraturis autem regrediuntur motu horario 16″. 18‴. 41⁗ ½. Et quod æquatio motus Nodorum in Octantibus sit 1 *gr.* 30′. Quæ omnia cum Phænomenis cœlestibus probè quadrant.

Prop

[457]

Prop. XXXIV. Prob. XIV.

Invenire Variationem horariam inclinationis Orbis Lunaris ad planum Eclipticæ.

Designent *A* & *a* Syzygias; *Q* & *q* Quadraturas; *N* & *n* Nodos; *P* locum Lunæ in Orbe suo; *p* vestigium loci illius in plano Eclipticæ, & *mTl* motum momentaneum Nodorum ut supra. Et si ad lineam *Tm* demittatur perpendiculum *PG*, jungatur *pG*, & producatur ea donec occurrat *Tl* in *g*, & jungatur etiam *Pg*: erit angulus *PGp* inclinatio orbis Lunaris ad planum Eclipticæ, ubi Luna versatur in *P*; & angulus *Pgp* inclinatio ejusdem post momentum temporis completum, adeoque angulus *GPg* Variatio

Hhh

mo-

John Quartley
A MEETING OF THE ROYAL SOCIETY IN CRANE COURT, ENGRAVING, 1878

An engraving of a Royal Society meeting in the early eighteenth century at Crane Court, London with Isaac Newton presiding: includes the mace, granted to the Society by Charles II.

non fingo' ('I frame no hypotheses'), though it has been pointed out that the word 'fingo' had negative connotations and seems to suggest that his opponents, who did try to explain how gravity worked, were just making things up.

Newton's force of gravity acted at a distance, with a force that depended only on the mass of the two bodies involved and the inverse square of the distance between them. How one thing influenced another at a distance was something he claimed not to care about – though in reality he did have a theory. This strange remote force led to some of his contemporaries, such as the Dutch natural philosopher Christiaan Huygens, being scathing about the idea, calling the action at a distance 'occult', in the sense of being hidden and inexplicable.

However, Newton's mathematics worked, and the book opened up a major aspect of physics, becoming an essential for anyone working in the field. As we have seen, for example, the eighteenth-century French mathematician and physicist Émilie du Châtelet wrote a French translation and commentary of *Principia* that was published after her death in 1759 (see page 17).

Martin Hoffman
CAROLUS LINNAEUS, OIL ON CANVAS, EIGHTEENTH CENTURY

Portrait of Carl Linnaeus in Lappish dress.

Nature, organised

The *Principia* contains many diagrams, though it is a work where the text is of primary importance. We tend to think of natural history as a more visual aspect of science than physics, but what was arguably the first truly scientific book in natural history is purely a matter of words. It is *Systema Naturae* (The System of Nature), first published in 1735, written by Swedish naturalist Carl von Linné. Appropriately known by the Latin version of his name, Linnaeus, he was born in Råshult in 1707. In a way, Linnaeus was a hangover from the past. Not only was this one of the last major science titles to be written in Latin, but the central feature of the book was the use of Latin in two-part names to identify species (such as the familiar *Homo sapiens*) – a system known as binomial nomenclature.

Syſtema Naturae ran to 12 editions (with an extra posthumous version), starting as a small collection of organisms and working up in increasing quantity and structure until, by the tenth edition, with the full title *Syſtema naturæ per regna tria naturæ, secundum classes, ordines, genera, species, cum characteribus, differentiis, synonymis, locis* (System of nature through the three kingdoms of nature, according to classes, orders, genera and species, with characters, differences, synonyms, places), Linnaeus had included the two-part (binomial) Latin structure for a whole range of animals as well as plants. The long title gives a feel for the importance of the book. As well as popularising the two-part Latin species name, which already existed but was not commonly used, it gave a whole hierarchal structure (taxonomy) to the 'kingdoms' of nature, which Linnaeus divided into animals, plants and (strangely to us now) minerals.

The basic approach taken by Linnaeus is still in use today, although his taxonomy has changed over time; we now have, for example, a higher level than 'kingdom' – the 'domain', which is split into archaea, bacteria and eukarya, with the multi-celled eukaryates then divided into kingdoms such as animals, plants and fungi. Admittedly Linnaeus's is not the only approach – his top-down structures are now rivalled by cladistics, which takes a bottom-up approach, forming groups based on common ancestors, a method that was

Carl Linnaeus
SYSTEMA NATURAE,
JOANNIS WILHELMI DE GROOT, 1735

The title page of the first edition and one of the classification diagrams from the Linnaeus book.

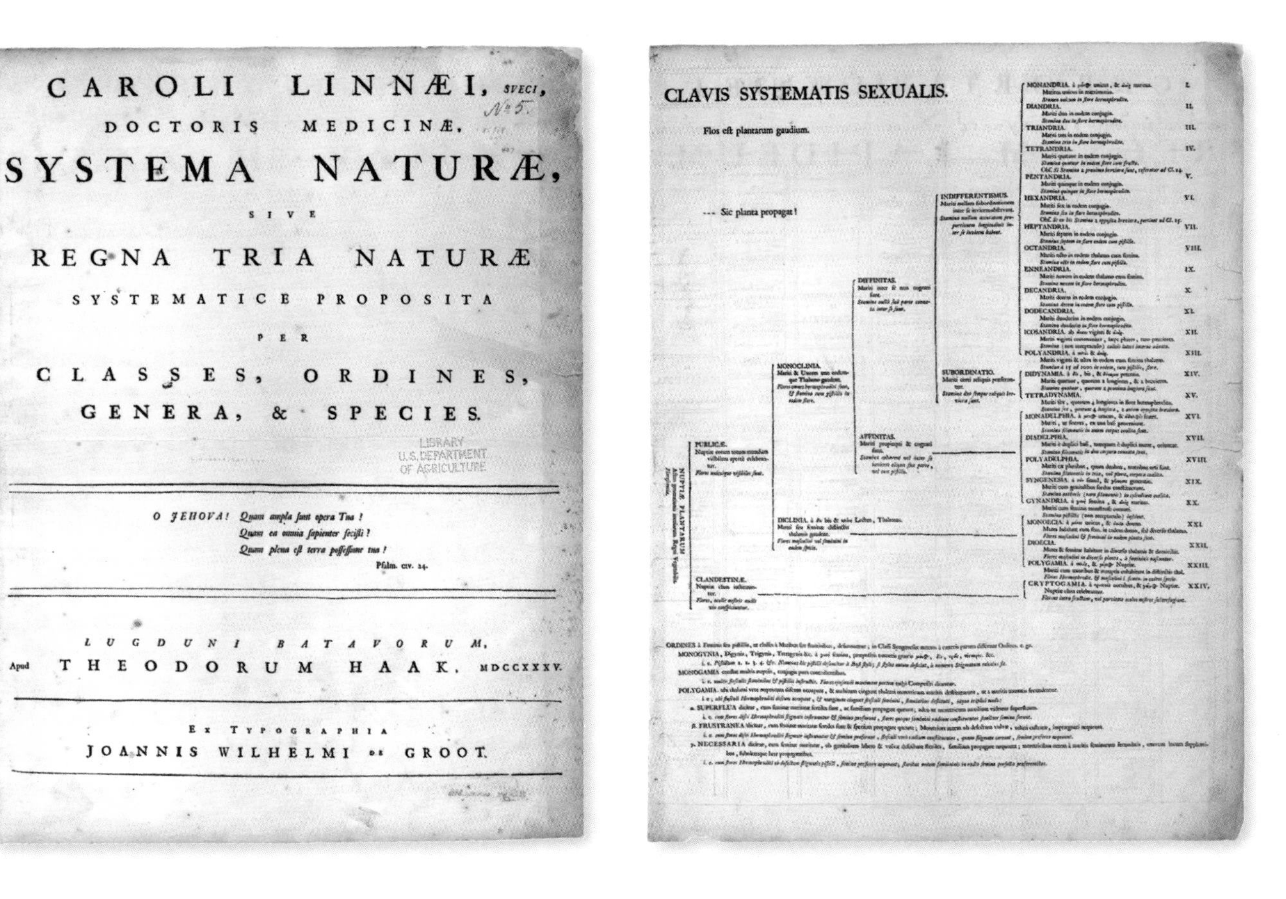
CAROLI LINNÆI, SVECI,
DOCTORIS MEDICINÆ,
SYSTEMA NATURÆ,
SIVE
REGNA TRIA NATURÆ
SYSTEMATICE PROPOSITA
PER
CLASSES, ORDINES,
GENERA, & SPECIES.

O JEHOVA! Quam ampla ſunt opera Tua! Quam ea omnia ſapienter feciſti! Quam plena eſt terra poſſeſſione tua!
Pſalm. civ. 24.

LUGDUNI BATAVORUM,
Apud THEODORUM HAAK. MDCCXXXV.

Ex Typographia
JOANNIS WILHELMI DE GROOT.

CLAVIS SYSTEMATIS SEXUALIS.

Flos eſt plantarum gaudium.

... Sic planta propagat!

Carl Linnaeus
SYSTEMA NATURAE,
GABRIEL NICOLAUS RASPE,
1773

An avian illustration from a German translation of the twelfth edition of *Systema*. This edition was the last to be produced by Linnaeus.

Tab. XXVII.
Fig. 2.
Fig. 3.
Fig. 5.

George Ehret Dionysius
METHODUS PLANTARUM SEXALIS IN SISTEMATE NATURAE DESCRIPTA, WATERCOLOUR, 1736

Original watercolour artwork for one of Linnaeus's plant catalogues, showing the sexual system of classification.

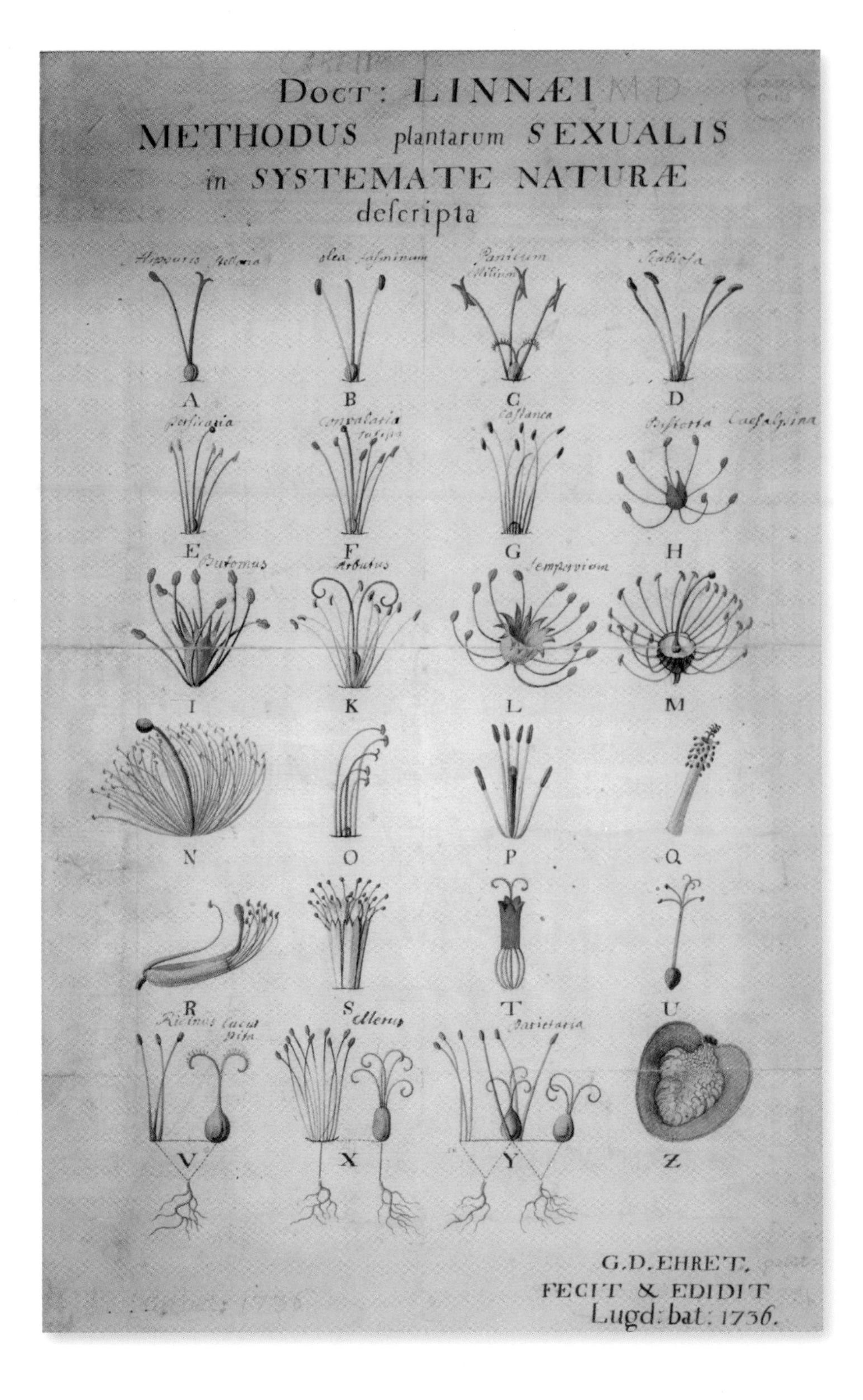

first considered around the start of the twentieth century, but has really blossomed with the ability to make DNA comparisons.

Without such technology, Linnaeus was only able to categorise animals and plants on visual similarities, which inevitably led to significant errors. This particularly occurred in plants, which he grouped together according to the number of stamens (the pollen-producing parts), even though this proved not to have any significance as a way of structuring species. Nevertheless, his tables and structural details of around 10,000 species proved the starting point of modern zoological and botanical studies.

Although the *Systema Naturae* is without doubt Linnaeus's masterpiece, he produced a number of other books, notably in botany, from his early *Flora Lapponica* (Plants of Lapland) from 1737 – written while still a student, where he first made use of his names and structures – to *Species Plantarum* (Species of Plants) published in 1753, his most comprehensive catalogue of plant species.

CAROLI LINNÆI
S:æ R:giæ M:tis Sveciæ Archiatri; Medic. & Botan. Profess. Upsal; Equitis aur. de Stella Polari; nec non Acad. Imper. Monspel. Berol. Tolos. Upsal. Stockh. Soc. & Paris. Coresp.

SPECIES PLANTARUM,

EXHIBENTES
PLANTAS RITE COGNITAS,
AD
GENERA RELATAS,
CUM
Differentiis Specificis,
Nominibus Trivialibus,
Synonymis Selectis,
Locis Natalibus,
Secundum
SYSTEMA SEXUALE
DIGESTAS.

Tomus I.

[illegible]

HOLMIÆ,
Impensis LAURENTII SALVII.
1753.

Carl Linnaeus
SPECIES PLANTARUM,
LAURENTII SALVII, 1753

The title page of the first edition of Linnaeus's plant catalogue.

Modern chemistry emerges

If Linnaeus brought order to the world of natural history, it would be the French chemist Antoine Lavoisier who performed the same function for chemistry. Where Robert Boyle's work was on the cusp between alchemy and chemistry, Lavoisier was arguably the first truly modern chemist. Born in Paris in 1743, the aristocrat would die on the guillotine during the French Revolution. His book *Traité Élémentaire de Chimie, Présenté dans un Ordre Nouveau et d'après les Decouvertes Modernes* (Elements of Chemistry in a New Systematic Order following Modern Discoveries), published in 1789, provided the foundations of chemistry as we now know it.

One essential achievement of the book was to move away from the phlogiston theory, which was a logical scientific theory that just happened to be wrong. The idea was that flammable bodies contained a substance called phlogiston, which was used up as they burned. What we now know to be oxygen (the name that Lavoisier gave to the gas) was called dephlogisticated air, meaning it was capable of absorbing phlogiston from a flammable substance, allowing it to burn.

Although the British natural philosopher Joseph Priestley had effectively noted the existence and action of oxygen, it was from this phlogiston-related viewpoint. Lavoisier turned the model on its head, regarding oxygen as an element in its own right that combined with other elements in the combustion process. *The Traité Élémentaire de Chimie* pulled together his ideas and those of a few contemporaries, giving a number of elements their familiar names, exploring a more quantitative approach to chemistry and giving the first approximation to a modern formula for a chemical reaction.

These were early days for chemistry, so there were inevitably some errors. Of the 33 elements that Lavoisier identified (out of a total of 92 found naturally on the Earth), 23 were correct – though not always carrying their now-familiar names. Nitrogen, for example, was called azote, approximately meaning 'no life' as it did not support living things. Oxygen meant 'acidifying', as it was wrongly thought to be an essential part of acids, while hydrogen, also named by Lavoisier, meant 'water making'. Among the

Antoine Lavoisier
TRAITÉ ÉLÉMENTAIRE DE CHIME, CUCHET, 1789

Illustrations of contemporary chemical laboratory equipment from Lavoisier's important chemistry text.

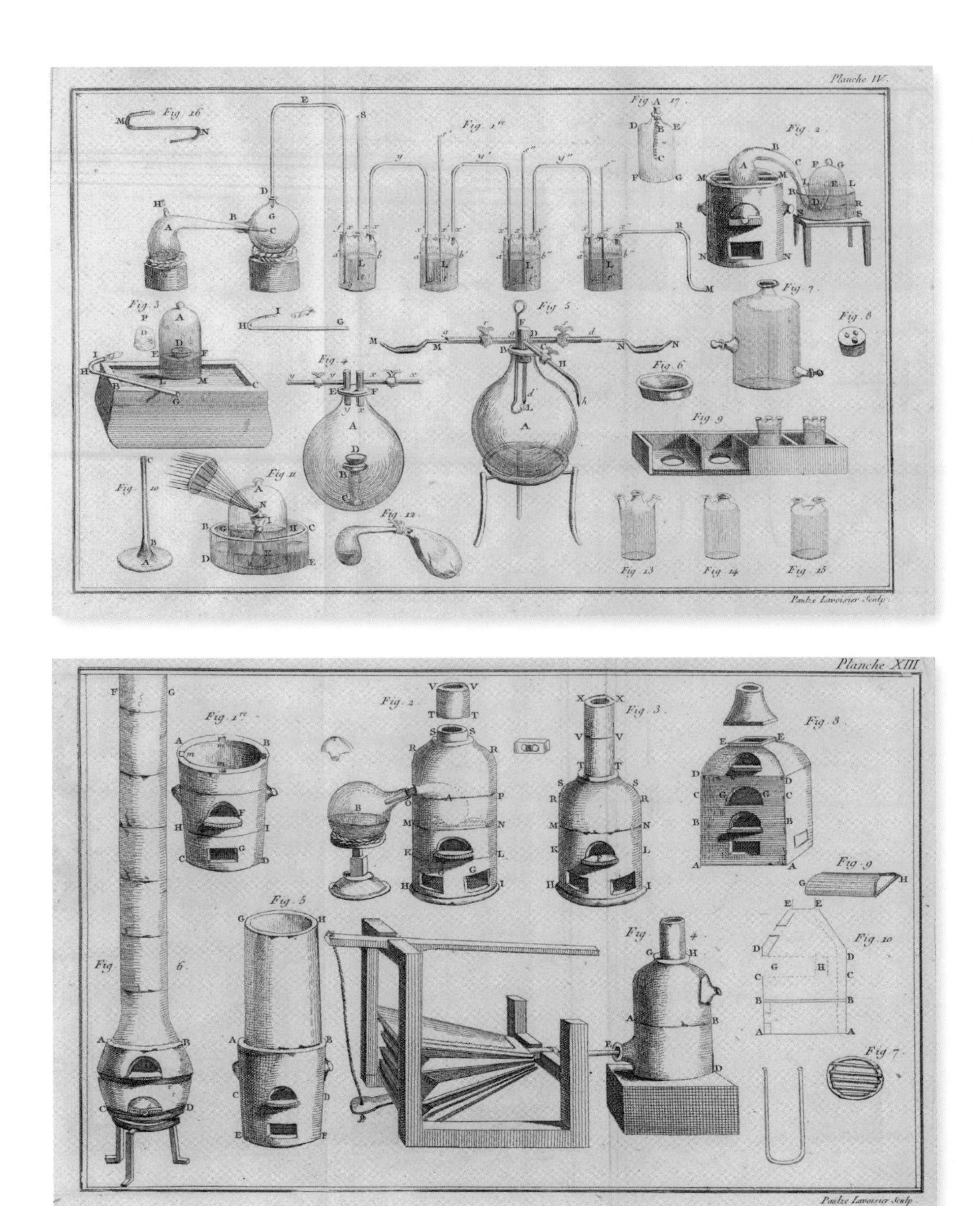

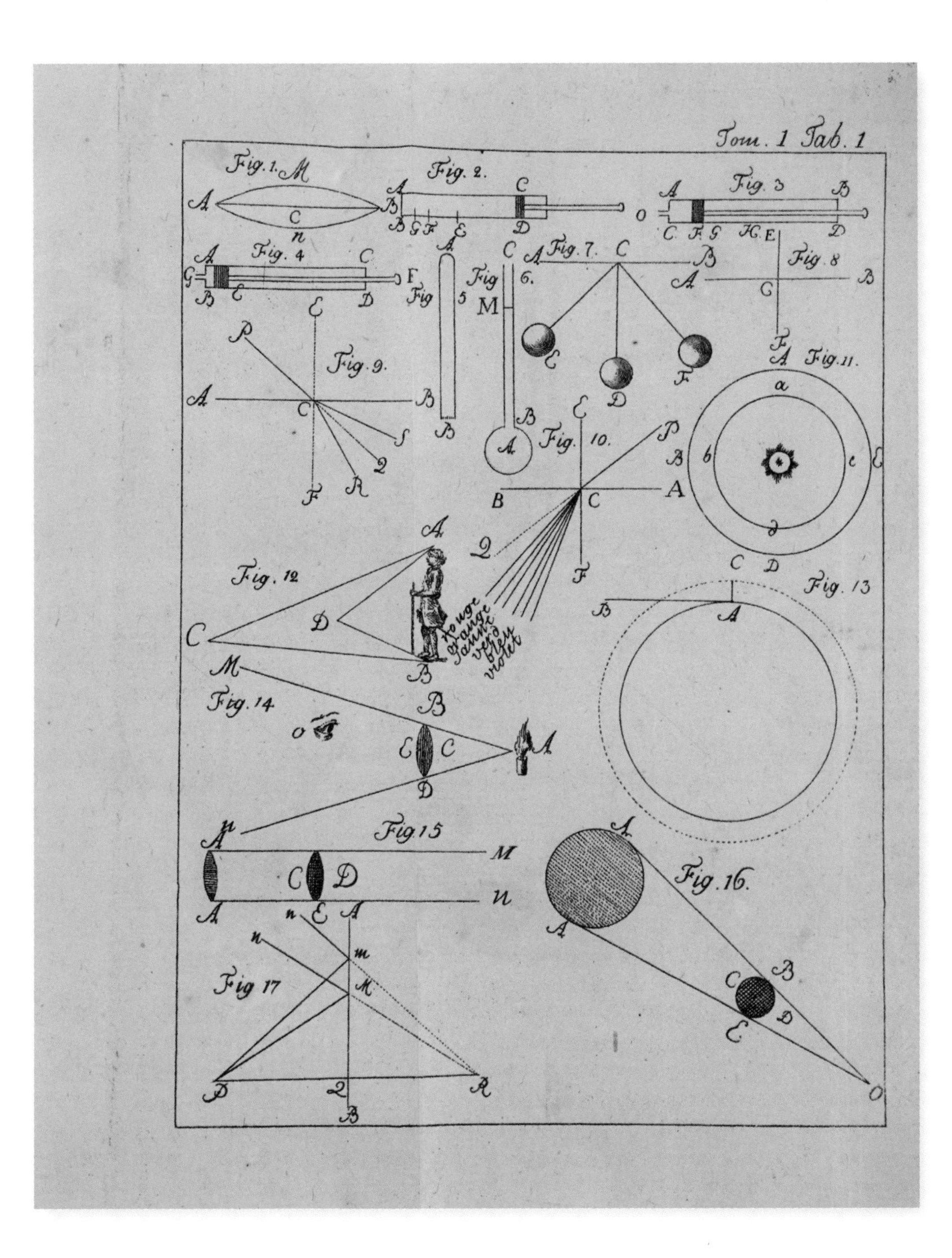

Leonhard Euler
LETTRES À UNE PRINCESSE D'ALLEMAGNE,
BARTHÉLEMY CHIROL, 1775

Illustration covering a wide range of topics from optics to eclipses in this edition of Euler's 1768 title.

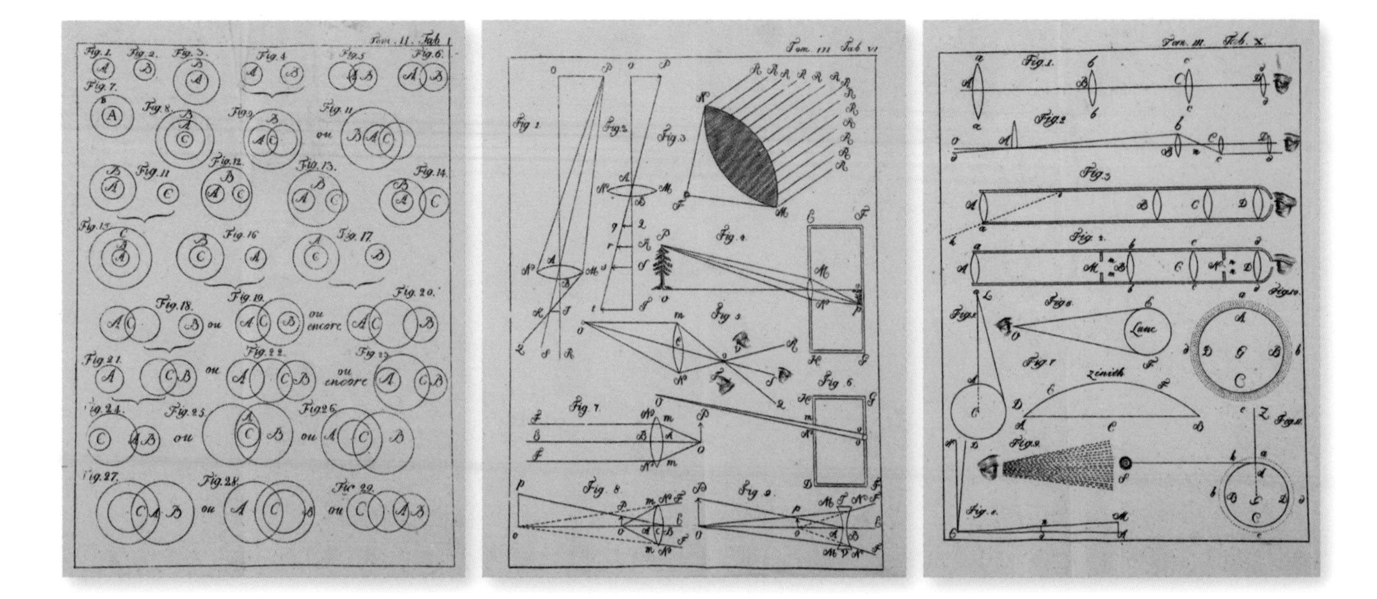

Leonhard Euler
LETTRES À UNE PRINCESSE D'ALLEMAGNE,
BARTHÉLEMY CHIROL, 1775

A further range of diagrams illustrating the range of topics covered in Euler's early popular science work, based on his letters to Friederike Charlotte, princess of Anhalt-Dessau.

mistakes in his list of supposed elements were light, caloric (an imaginary fluid thought to carry heat) and a number of compounds such as lime (calcium oxide).

Although not widely influential, one eighteenth-century publication is of great interest now, as it recognised that women too could have an interest in science. It may seem condescending to have a science book that was explicitly 'for women' as this was – but it was a step unlikely to even have been conceived by writers before this time, who would have assumed their audience was uniformly male. The book in question was *Lettres à une Princesse d'Allemagne sur divers Sujets de Physique et de Philosophie* (Letters to a German Princess, On Diverse Subjects in Physics and Philosophy), published in 1768. The princess in question was Friederike Charlotte, princess of Anhalt-Dessau, one of the nieces of the king of Prussia, and the book pulled together a series of letters written to the princess by Swiss mathematician Leonhard Euler, born in Basel in 1707. The *Lettres* was effectively a popular summary of the latest thinking in science, and when it was translated into English in 1795 by one Henry Hunter, it became a bestseller.

Although certainly patronising, Hunter's introduction to the translation at least showed how opinions were changing: 'The improvement of the female mind; an object of what importance to the world! I rejoice to think I have lived to see female education conducted on a more liberal and enlarged plan. I am old enough to remember the time when well-born young women, even of the north, could spell their own language but indifferently, and some, hardly read it with common decency [...] They are now treated as rational beings, and society is already the better for it.'

For the moment, though, such books, and women who were prepared to face the potential stigma of entering what was universally regarded as a man's world, were few and far between. It would not be until late in the nineteenth century that significant progress would be made in gender equality.

The poetic naturalist

As we move towards the end of the eighteenth century, a familiar family surname crops up, which we will return to in the next chapter: Darwin. In this case, we are dealing with Erasmus Darwin, the grandfather of the better-known Charles, born in Elston, Nottinghamshire in 1731. Erasmus Darwin was a physician, but his interest in natural philosophy went much deeper than medicine, and as a member of the Lunar Society of Birmingham (so called because the group met when there was a full Moon, to make it easier to see their way when returning home) he was part of an elite group of natural philosophers and industrial giants from the English Midlands that also included Matthew Boulton, Joseph Priestley and Josiah Wedgwood.

In terms of pure scientific importance, Darwin's most significant title was *Zoonomia; or the Laws of Organic Life*, from 1794. This was primarily a medical and anatomical text, but its interest now is in the early ideas of evolution that it contains. Darwin suggested that mammals and other warm-blooded animals could all have had a single common ancestor, a simple organism he referred to as a filament. He also prefigured Lamarck's ideas of animals passing on modifications due to their environment to their offspring (see pages 162–5).

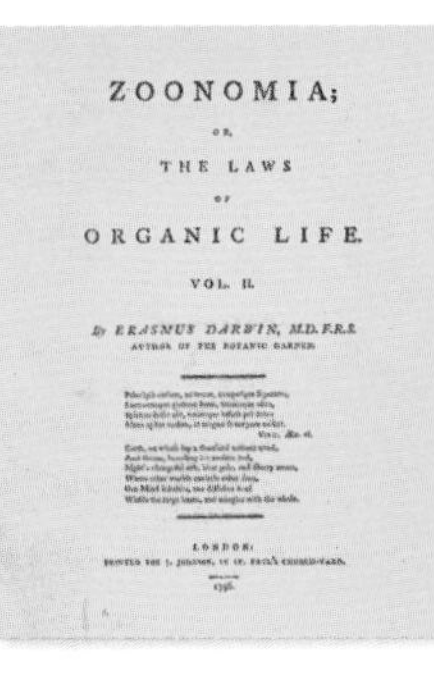

ZOONOMIA;

OR,

THE LAWS

OF

ORGANIC LIFE.

VOL. II.

By ERASMUS DARWIN, M.D. F.R.S.

AUTHOR OF THE BOTANIC GARDEN.

1796.

Erasmus Darwin
ZOONOMIA, JOSEPH JOHNSON, 1796

The title page of *Zoonomia*, two years after original publication.

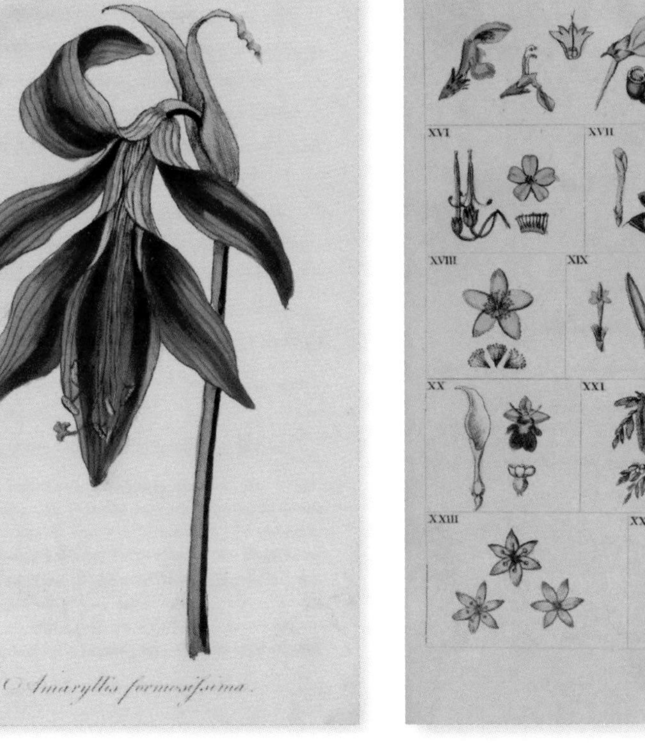

Erasmus Darwin
THE POETICAL WORKS OF ERASMUS DARWIN, JOSEPH JOHNSON, 1806

Plates of illustrations accompanying a coloured edition of the poem 'The Loves of the Plants' in a collection of Darwin's poetical works from 1806. The poem was originally published in 1789 as the second part of *The Botanic Garden*.

However, another of Darwin's works, *The Botanic Garden*, which predated *Zoonomia* by three years, reached a wider audience. We've already seen one scientific poem with a lasting impact – Lucretius's 'De Rerum Natura' (see pages 49–50). *The Botanic Garden* consisted of a pair of poems which were designed to make scientific ideas more approachable to the audience of the period. The first of these, 'The Economy of Vegetation', despite the name, features a significant amount on mining and the inventors of the day, even taking the opportunity to castigate the slave trade and to support the French and American revolutions. The second poem, 'The Loves of the Plants', paints a broader scientific picture, from the creation of the cosmos to the essentials of botany (including the sexual nature of plants, considered rather risqué at the time) and again includes some hints of evolutionary theory.

The dark vision of Malthus

Neatly closing off the eighteenth century, published as it was in 1798, is a book by the English clergyman Thomas Malthus, who was born in the village of Westcott in 1766. *An Essay on the Principle of Population* is in a class of book now sometimes called futurology, which has proved highly influential even though futurology titles are usually wildly incorrect.

Originally published anonymously, Malthus's book foresaw a dire future for humanity, based on a difference in the rate at which population and food production could be expected to grow. The limited scientific content of the book was in its use of mathematical approaches to these growth rates, and its economic assessment of the impact of population changes on wages and inflation. The concept Malthus put forward was that population would double every 25 years, while the rate of production of food would only rise by a number of percentage points each period, meaning that population would spiral far beyond our ability to produce enough food and starvation would set in.

Arguably, the biggest influence of the book was to emphasise limited population data, resulting in the introduction of ten-yearly censuses in Great Britain (which in turn influenced a number of other countries to do so). In the economic part of the book, Malthus argued that growing population would also make labour too readily available, driving down wages, adding poverty to the famine caused by overpopulation.

We can be grateful that the doom-laden predictions of Malthus have not come to pass, primarily because of the benefits reaped from science and technology. Malthus ignored the impact of science and technology in transforming food production, expanding the job market and in making birth control available. Despite this flaw in his theories, Malthus's book is important. It was one of the first science books for a general audience to use statistics, and also one of the first to attempt to bring science into the last remaining area of application, human behaviour. As the nineteenth century dawned, there was new hope, as science and technology transformed everyday life. Like all periods of transition, there were difficulties, and it might have seemed initially that Malthus's predictions were coming true. But in the century when the word 'scientist' was first coined, the best was yet to come.

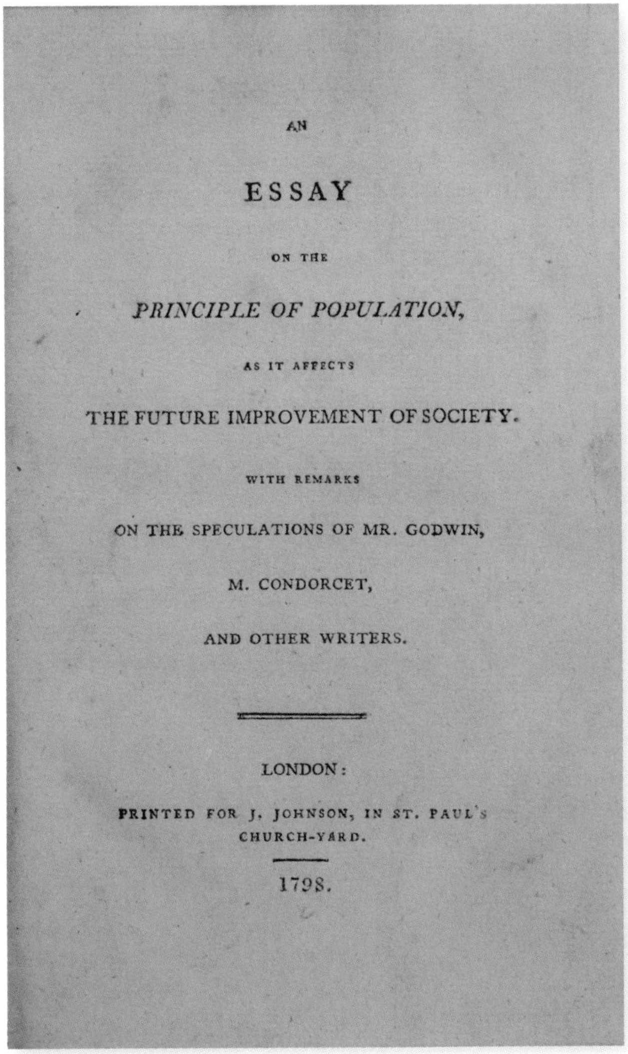

AN

ESSAY

ON THE

PRINCIPLE OF POPULATION,

AS IT AFFECTS

THE FUTURE IMPROVEMENT OF SOCIETY.

WITH REMARKS

ON THE SPECULATIONS OF MR. GODWIN,

M. CONDORCET,

AND OTHER WRITERS.

LONDON:

PRINTED FOR J. JOHNSON, IN ST. PAUL'S CHURCH-YARD.

1798.

Thomas Malthus
AS ESSAY ON THE PRINCIPLE OF POPULATION,
JOSEPH JOHNSON, 1798

The title page of Malthus's gloomy prognostications alongside an 1851 George Cruikshank etching showing an imagined future London as Malthus's vision became reality.

3

MODERN CLASSICAL

VICTORIAN STABILITY

BY THE NINETEENTH CENTURY, science was becoming a more professional business, and science books were an expected and established part of the academic sphere. It was in 1834 that the word 'scientist' was coined (preferred, thankfully, over suggestions of sciencer, scientician and scientman), drawing a parallel with the likes of artist, economist and atheist. Until this period 'natural philosophers' were often wealthy amateurs or more general philosophers, engaging in science as a part-time indulgence.

The development of scientist as a profession was, however, a gradual process. When Michael Faraday joined the Royal Institution in London – an organisation he would eventually head up – in 1813, becoming an assistant to leading light Humphry Davy, his position was more that of a servant than an equal. (In fact, when Faraday accompanied Davy on a European tour, Davy expected him to act as a part-time valet.)

Initially, professional scientists formed a second tier alongside the amateur natural philosophers. Their relationship was not dissimilar to the cricket teams of the nineteenth and early twentieth centuries: largely amateur but including a (working-class) professional to improve their game. More of the most important books we will encounter during this period were written by amateurs than by professionals.

Mr Atom

Despite the prevalence of amateurs, one of the first important science books in the nineteenth century was written by one of the new professional class. John Dalton (like Faraday) had no university education, but became a working scientist through personal drive. Dalton was born in Eaglesfield in the northern English county of Cumberland in 1766. Dalton, also like Faraday, came from a poor family, but even had he been wealthy, he would have been unable to attend university in England as places were still restricted to members of the Church of England, and Dalton was a Quaker.

Dalton was paid to teach, not to undertake research, but to all intents and purposes his scientific work was his main job, with the teaching merely a means to make ends meet. Like most of the scientists of the period, Dalton had wide-ranging interests, from the behaviour of gasses, via meteorology, to the causes of colour-blindness (Dalton was himself colour-blind, and for a while the condition would be known as Daltonism). However, there is no doubt that Dalton's scientific claim to fame was his work on atomic theory, first published in a series of essays in 1802.

William Henry Worthington
JOHN DALTON OF MANCHESTER, ENGRAVING, CA. 1823

Engraving of John Dalton after an oil painting by Joseph Allen.

At the time, atoms were still a subject of debate. In fact, they would not be fully accepted until the twentieth century, when a paper by Einstein gave very strong evidence for their existence. Some scientists accepted that atoms were real; others – probably the majority – felt that they were a useful accounting technique but didn't represent real entities, while a few still felt that atoms were philosophically objectionable. Dalton was strongly in the realist camp and put forward a model of the nature of matter using what was largely Lavoisier's chemistry (see page 133), based on the existence of atoms and of compounds formed by combining them.

Often a new way of looking at what is already known can give insights. Dalton's breakthrough inspiration was that each atom of an element had a distinct mass.

Augustus Pugin and Thomas Rowlandson
LIBRARY OF THE ROYAL INSTITUTION, AQUATINT, 1809

An aquatint of the library of the Royal Institution, Albermarle Street, London.

John Dalton
A NEW SYSTEM OF CHEMICAL PHILOSOPHY,
PART II OF II,
ROBERT BICKERSTAFF, 1810

A plate from part II of Dalton's book showing some of his proposed molecular structures and the boiling points of differing concentrations of compounds.

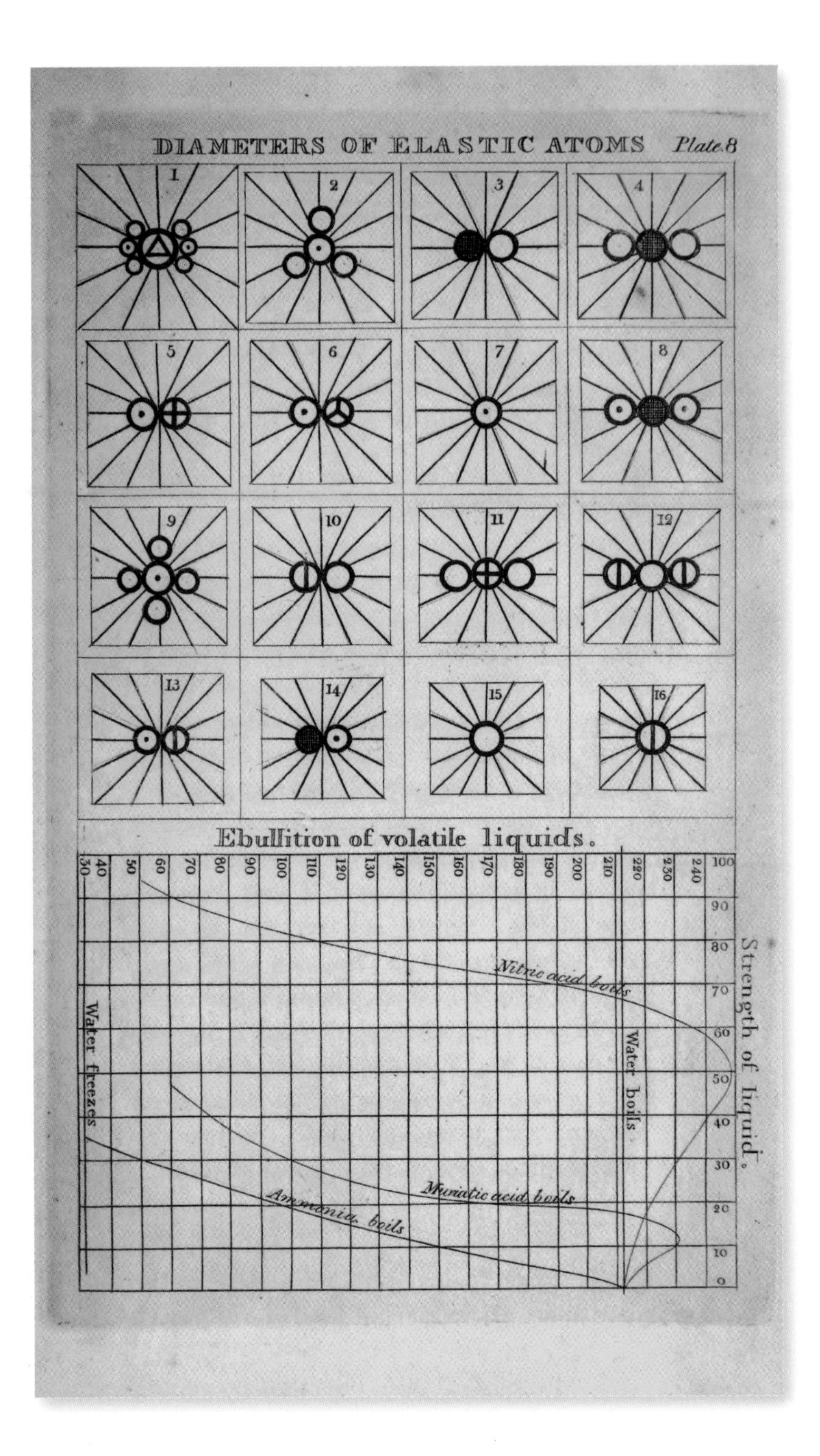

John Dalton

A NEW SYSTEM OF CHEMICAL PHILOSOPHY, PART II OF II, ROBERT BICKERSTAFF, 1810

Plates from part II of Dalton's book showing his symbols for elements, some suggested molecular structures and his ideas on crystalline solids.

John Dalton
TABLE OF ELEMENTS, 1808

A diagram created to accompany Dalton's lectures on atoms and atomic weights showing the relative weights he ascribed to the atoms and his icons for the elements.

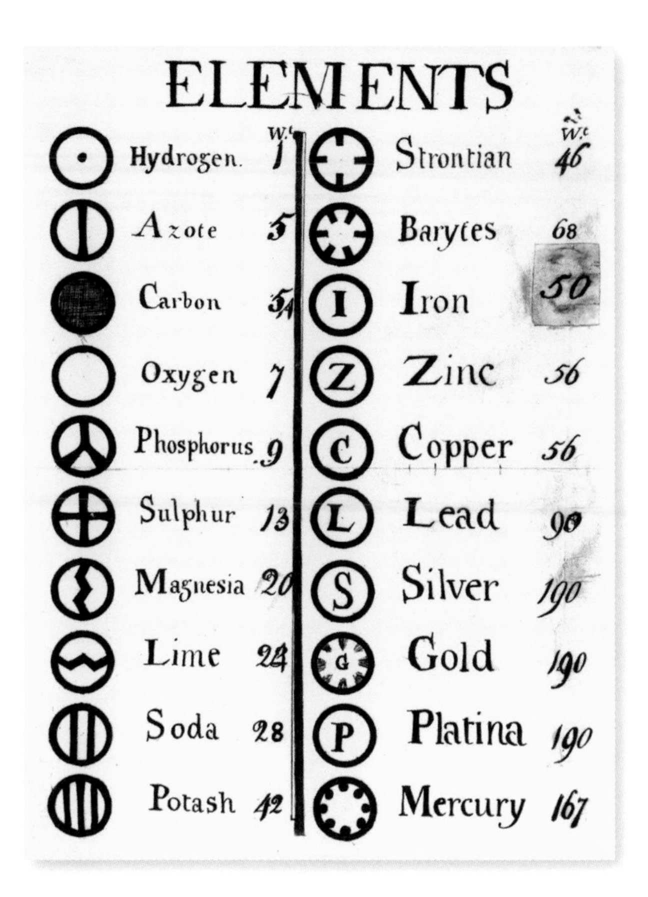

He made the mass of hydrogen (the lightest element) 1 and assigned to the others multiples of this, for example, azote (nitrogen) was 5, oxygen 7 and phosphorus 9. These relative masses were (somewhat confusingly) referred to as atomic weights, the term that is still most commonly used. When two elements reacted to form a compound, the result was that their atoms combined to form molecules with the combined atomic weight of the two. In a flurry of papers from 1802 to the publication of his book *A New System of Chemical Philosophy* in 1808, Dalton circulated and expanded on his theory, providing elegant tables of the elements, each of which he identified with a unique symbol.

Dalton's work had its limitations. It was based on his own experiments, which used poor-quality instruments, even by the standards of the day. As a result, his atomic weights were often incorrect. Furthermore, from his original theory that all atoms had a mass that was a multiple of the mass of hydrogen, he never made the much greater leap to the idea that all atoms were made up of multiples of hydrogen-like components (as is the case). This was probably because Dalton always referred to the ratios of atomic weights as, say, 'nearly 7:1' – he didn't think these ratios were exact, and because he was sure that atoms were spheres of different sizes, his mental model of them did not fit well with the idea that all atoms were constructed from hydrogen-like building blocks.

Like Lavoisier, Dalton failed to spot that some compounds – lime, for example – weren't elements, and he also miscalculated the number of atoms in molecules, usually assuming there was one atom of each element: he thought that water was HO, for example, rather than the now familiar H_2O, and he missed that oxygen came as a two-atom molecule. Dalton also refused to accept the simpler text-based symbols we use today, which were introduced by the Swedish chemist Jöns Jacob Berzelius, and already in use before Berzelius's most significant book, his 1808 *Läroboken i Kemien* (Chemistry Textbook). Instead, Dalton insisted on using his own pretty looking, but hard to remember, icons for each element.

Despite these flaws, we have to remember that Dalton was a pioneer in his field – and so not likely to get everything right – and was working without the support of a university or the high-quality instruments a rich amateur scientist of the period could have obtained. He was also working at a time when some of the elements, such as oxygen and nitrogen, were still relatively newly discovered. His book brought to the world of chemistry a new rigour and method.

The ultimate birdwatcher

Books like Dalton's were still primarily aimed at other natural philosophers, but the nineteenth century also saw the gradual appearance of books that would not only appeal to scientists, or even those with an interest in science, but to a wider audience. A prime example of this is Audubon's *Birds of America*, which at the time was seen much as we might today consider a coffee-table book of images from the Hubble Space Telescope. Although the book's original images were made for a scientific purpose, their primary appeal to the wider public was in their visual beauty – there probably had not been an equivalent book since Hooke's *Micrographia*.

With a remarkable mixed heritage, *Birds of America* was written by a Haiti-born Frenchman who had American citizenship (despite being an illegal immigrant), and it owed its existence to the enthusiasm of a British audience. Published in 1827, it featured 435 prints, each hand-coloured, and contained a remarkable array of birds on its huge 99 x 66-cm (39 x 26-inch) plates. Shortly after his arrival in America in 1803, John James Audubon started to take an interest in the wide range of American birds. Not only did he observe them and paint them, he learnt taxidermy and set up a natural history museum that included a wide range of American native species, both birds and other animals. Audubon's day job was as a merchant, but his enthusiasm for natural history gradually took over more and more of his life.

Unlike many involved in science at the time, Audubon was not independently wealthy. When, for example, his latest venture of a flour mill went under in 1819, Audubon was bankrupted and sent to debtors' prison. Even so, by 1820, he was once again on the road, painting birds, with a grand plan in place of putting together a collection of images of all the birds of North America. By the time he reached Philadelphia in 1824 he had accumulated over 300 images and hoped to make a book from them – but couldn't find a publisher.

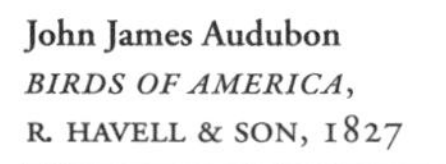

John James Audubon
BIRDS OF AMERICA,
R. HAVELL & SON, 1827

Plate 53 from the original 'double elephant' folio of hand-coloured aquatints, showing painted buntings engraved by Robert Havell junior, who printed the book with his father, Robert Havell senior.

John James Audubon
BIRDS OF AMERICA,
R. HAVELL & SON, 1827

Plates 217 and 386 from the original 'double elephant' folio of hand-coloured aquatints showing a Louisiana heron and a white heron.

On the recommendation of a friend, Audubon travelled to England, and finally found the enthusiastic audience he required (in part inspired by his exotic, backwoodsman persona) and the funding to publish his great work. The whole book was not produced at once. Instead, subscribers would receive five prints at a time in a tin box. Like most books of the time, the single pages would then be bound by the buyers themselves, with a cover added to match their libraries. The book cost a phenomenal amount to print; even at that time, to buy a full set of all 435 prints would have cost around $1,000 – equivalent to around £20,000 or $26,000 in modern terms. A complete *Birds of America* sold at auction in 2010 for £6.5m ($10.3m). Only relatively few copies of the full-sized edition were produced, but a smaller 25.3 x 15.8-cm (10 x 6.25-inch) version had a far wider circulation.

Although the illustrations were issued separately, there was also a five-volume text to accompany them, written by Audubon in collaboration with the Scottish natural historian William MacGillivray. This was published independently to avoid having to provide free copies of the expensive illustrations to the British Crown Libraries, which as the UK national library receives copies of all British publications.

Scientific manufacture

In the modern sense, *Birds of America* did little to advance science. It was purely a matter of data capture, with no theoretical use made of that data. Yet it provided a unique view of the avian natural history of North America, and stood out for the way in which its visual splendour turned a scientifically inspired publication into an object of desire for ordinary citizens. A totally different, but highly influential book, was published just five years later. Here it was not the pictures, but the theory, that was central to its success. The author, Charles Babbage, wanted to bring science to the business of manufacturing.

CHARLES BABBAGE, CA. 1860

A relatively elderly Babbage in an undated photograph.

We now remember Babbage, born in London in 1791, for his work on early mechanical computers. However, his computers were never completed, and he did not write an influential book on the subject. The closest we come to this was more in the form of a paper, the 1842 *Sketch of the Analytical Engine invented by Charles Babbage (with additional notes by Augusta Ada, Countess of Lovelace)*. This was a piece written in French by the Italian general and mathematician Luigi Federico Menabrea, which was translated and doubled in length with notes by Ada King, the Countess of Lovelace, who worked with Babbage on the way his mechanical computer could be programmed.

However, Babbage did not spend all his time on calculating engines. He also considered ways to apply mathematical and organizational principles to the workplace. A much-expanded version of his approach would become a branch of applied mathematics in its own right (called operational research) in the 1940s, when it was first applied to wartime problems and later to business. Babbage made one of the first attempts to bring science to the operations of the business world.

His book, *On the Economy of Machinery and Manufactures* was published in 1832. In researching it, Babbage put time into observing what went on in factories and realised that skilled workers were spending a considerable amount of their time on unskilled

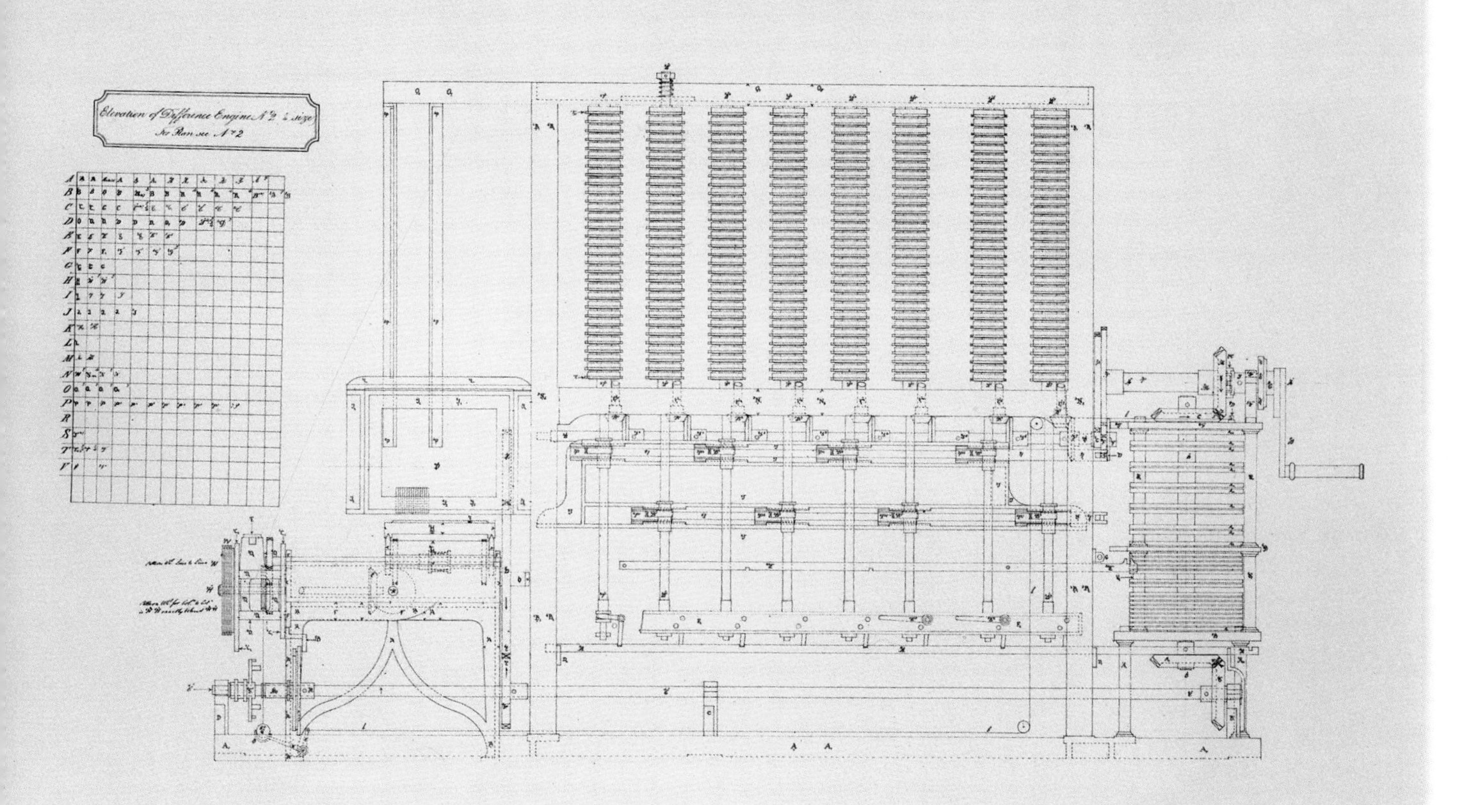

Charles Babbage

DIFFERENCE ENGINE NO. 2,
1847

Technical drawing of Babbage's Difference Engine number 2, his mechanical calculator, which was partly constructed but abandoned as he went on to design his more advanced Analytical Engine.

labour. He suggested that dividing up tasks, so that the skilled workers could concentrate on their skills, would result in greater efficiency. Babbage was also one of the first to look at breaking down costs to understand the profitability of a business and to examine the benefits of profit-sharing.

The dramatic changes in industry of the time were taking place in large part as a result of the introduction of the steam engine. But, initially, the science behind this transformative technology was sketchy at best. The book that would change that was written by a young French engineer, who would sadly be dead of cholera by the age of 36. This was Nicolas Léonard Sadi Carnot, born in Paris in 1796. Carnot effectively began the branch of physics later known as thermodynamics with his 1824 book *Réflexions sur la Puissance Motrice du Feu et sur les Machines Propres a Développer cette Puissance* (Reflections on the Motive Power of Fire and on Machines Fitted to Develop that Power).

Although Carnot was working with a soon-to-be-outmoded model of heat as a fluid called 'caloric' that could pass from body to body, he realised that the efficiency of an engine driven by heat, such as a steam engine, depended on the difference in temperature between the hot part and the cold part of the engine, and so laid the groundwork for one of the central laws of physics: the second law of thermodynamics. At its simplest, this law states that heat moves from a hotter to a colder body, but its more sophisticated form, involving the concept of 'entropy', the measure of disorder in a system, would have implications for everything from information theory to the end of the universe.

From Carnot's viewpoint, though, the important aspect of his work was to find ways to improve the workings of steam engines. The engines of the time were highly

Charles Babbage
ON THE ECONOMY OF MACHINERY AND MANUFACTURES, CHARLES KNIGHT, 1832

The title page of Babbage's book on organizing manufacture alongside James Nasmyth's painting of his own steam hammer, erected in his foundry near Manchester in 1832.

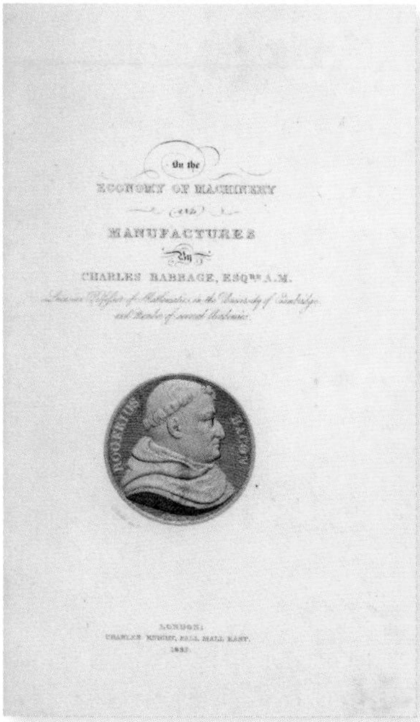
On the
ECONOMY OF MACHINERY
and
MANUFACTURES
By
CHARLES BABBAGE, ESQ^{RE} A.M.
LONDON:
CHARLES KNIGHT, PALL MALL EAST.

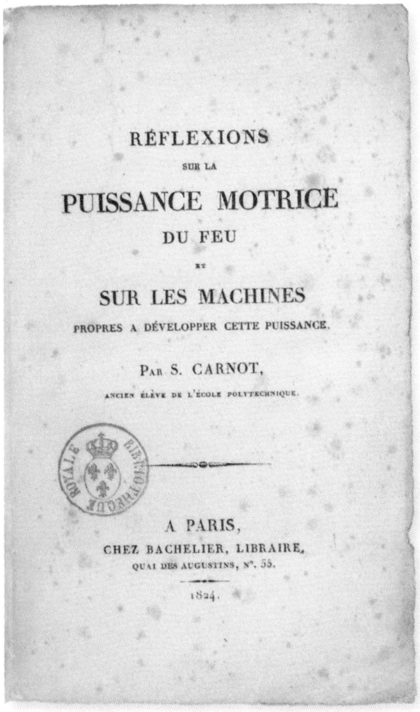

RÉFLEXIONS

SUR LA

PUISSANCE MOTRICE

DU FEU

ET

SUR LES MACHINES

PROPRES A DÉVELOPPER CETTE PUISSANCE.

PAR S. CARNOT,

ANCIEN ÉLÈVE DE L'ÉCOLE POLYTECHNIQUE.

A PARIS,

CHEZ BACHELIER, LIBRAIRE,

QUAI DES AUGUSTINS, N°. 55.

1824.

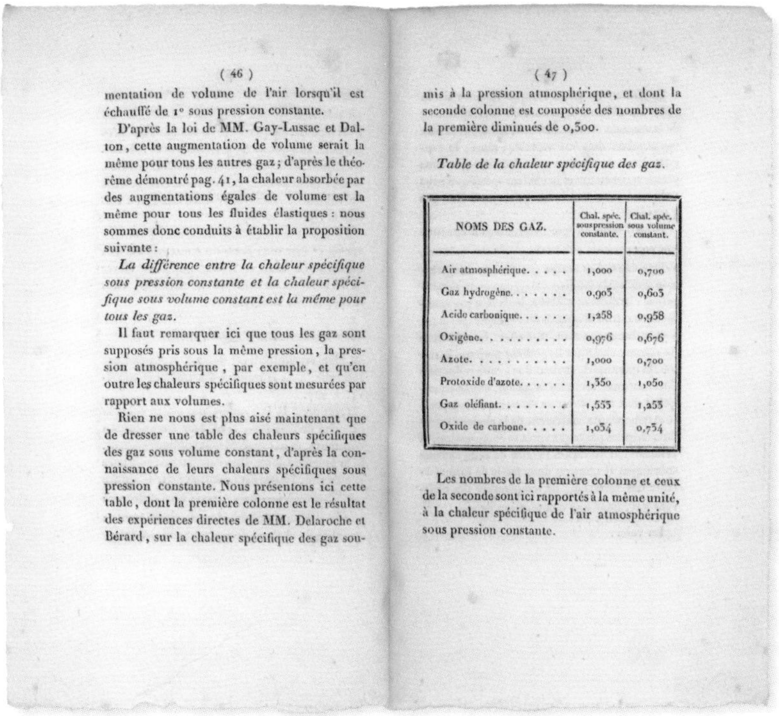

(46)

mentation de volume de l'air lorsqu'il est échauffé de 1° sous pression constante.

D'après la loi de MM. Gay-Lussac et Dalton, cette augmentation de volume serait la même pour tous les autres gaz ; d'après le théorême démontré pag. 41, la chaleur absorbée par des augmentations égales de volume est la même pour tous les fluides élastiques : nous sommes donc conduits à établir la proposition suivante :

La différence entre la chaleur spécifique sous pression constante et la chaleur spécifique sous volume constant est la même pour tous les gaz.

Il faut remarquer ici que tous les gaz sont supposés pris sous la même pression, la pression atmosphérique, par exemple, et qu'en outre les chaleurs spécifiques sont mesurées par rapport aux volumes.

Rien ne nous est plus aisé maintenant que de dresser une table des chaleurs spécifiques des gaz sous volume constant, d'après la connaissance de leurs chaleurs spécifiques sous pression constante. Nous présentons ici cette table, dont la première colonne est le résultat des expériences directes de MM. Delaroche et Bérard, sur la chaleur spécifique des gaz sou-

(47)

mis à la pression atmosphérique, et dont la seconde colonne est composée des nombres de la première diminués de 0,300.

Table de la chaleur spécifique des gaz.

NOMS DES GAZ.	Chal. spéc. sous pression constante.	Chal. spéc. sous volume constant.
Air atmosphérique.	1,000	0,700
Gaz hydrogène.	0,903	0,603
Acide carbonique.	1,258	0,958
Oxigène.	0,976	0,676
Azote.	1,000	0,700
Protoxide d'azote.	1,350	1,050
Gaz oléfiant.	1,553	1,253
Oxide de carbone.	1,034	0,734

Les nombres de la première colonne et ceux de la seconde sont ici rapportés à la même unité, à la chaleur spécifique de l'air atmosphérique sous pression constante.

inefficient, using very little of the energy that was produced in burning their fuel and wasting the rest as heat. *Réflexions* was largely ignored at the time – it would only be in the second half of the nineteenth century that it would really begin to have influence. Unlike Babbage, Carnot did not move in the right circles.

A good friend of Babbage's also published some significant works at around the same time. This was John Herschel, the son of the man who discovered Uranus, William Herschel. John, born in the English town of Slough in 1792, allegedly inspired Babbage's computer work. The story goes that Babbage was helping Herschel painstakingly construct a table of numbers, when he cried out, 'My God, Herschel, how I wish these calculations could be executed by steam!' Herschel wrote a number of important astronomical titles, notably *A Treatise on Astronomy*, and (amusingly, given Babbage's complaint) produced a major update of his father's astronomical catalogues.

However, John Herschel's most significant book was *A Preliminary Discourse on the Study of Natural Philosophy*, which was an attempt to give a more modern synthesis of the scientific method than the earlier work of Bacon and others. Strictly, this wasn't a book in its own right but appeared in 1831 as part of *Lardner's Cabinet Cyclopaedia*. This publication was not, as it sounds, a single encyclopaedia, but rather a curated library of titles, of which Herschel's was volume 14 of 133.

Sadi Carnot
RÉFLEXIONS SUR LA PUISSANCE MOTRICE DU FEU, BACHELIER, 1824

The title and interior pages from Carnot's book that marked the beginnings of thermodynamics, including a table of specific heats of gases. Housed in the Bibliothèque nationale de France.

Shaping the modern world

The visual élan that was so central to the success of *Birds of America* would also be significant in a work that made important advances in scientific theory in the world of geology. In writing *Principles of Geology*, the Scottish geologist Charles Lyell was aiming at an academic audience, and yet the three-volume book proved a hit with the wider public, in part because of its illustrations. Lyell, born into a wealthy family near Kirriemuir in Angus in 1797, trained as a lawyer, but within a few years was occupied full-time by his interest in geology. He cemented his place in the field with the publication of *Principles of Geology*, which came out a volume at a time between 1830 and 1833.

Looking at the topic with modern eyes, geology does not seem like a subject that is likely to inspire much interest in the public. However, at the time, geology was a controversial field of significant public interest, worthy of newspaper headlines. This was because the latest theories, supported and expanded by Lyell, overthrew the long-held ideas on the age of the Earth, based on biblical calculations. The central idea espoused by Lyell was known as 'uniformitarianism', originally developed by the Scottish geologist James Hutton. Up until this time, it was accepted that the Earth had been shaped by a series of catastrophes, such as the biblical flood, forming its current geography with a timescale calculated from the genealogical lists in the book of Genesis, starting from Noah, based on estimated lifespans. This put the date of the creation of the Earth at 4004 BCE.

Charles Lyell
PRINCIPLES OF GEOLOGY, JOHN MURRAY, 1832

The frontispiece to volume two of Lyell's influential work, showing the region around Mount Etna in Sicily. The book was published in three volumes between 1830 and 1833.

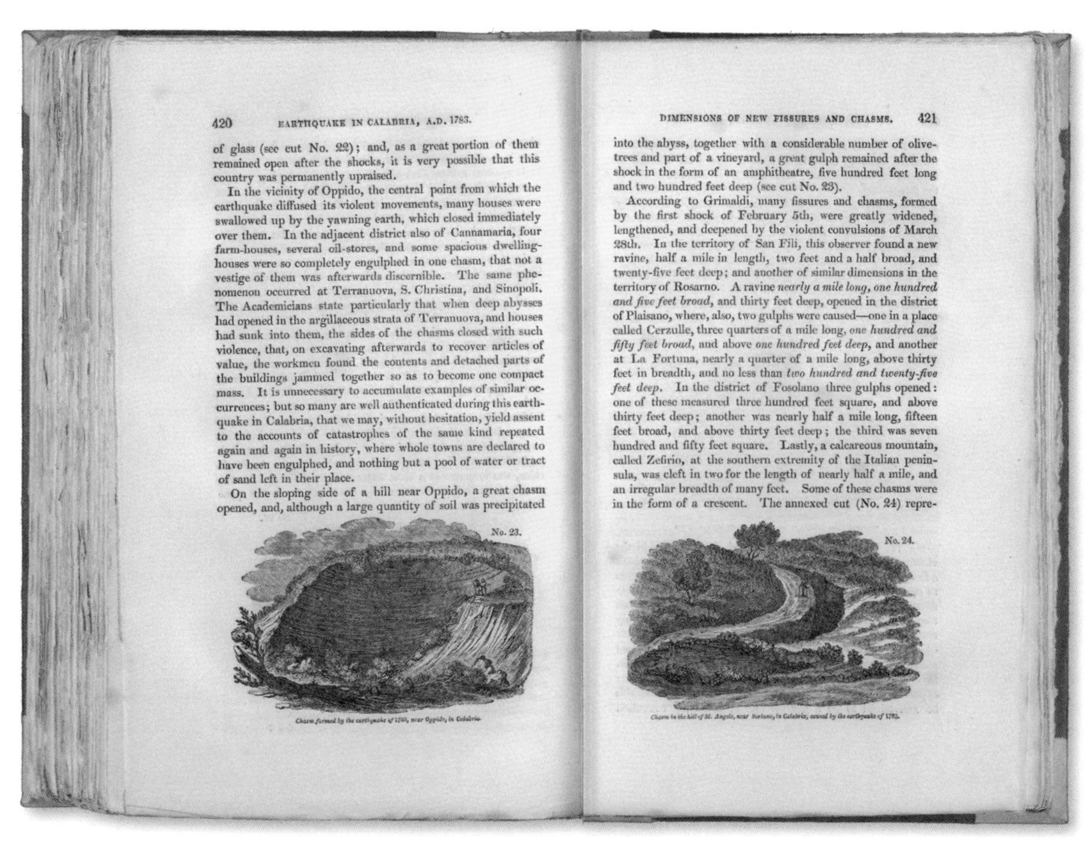

420 EARTHQUAKE IN CALABRIA, A.D. 1783.

of glass (see cut No. 22); and, as a great portion of them remained open after the shocks, it is very possible that this country was permanently upraised.

In the vicinity of Oppido, the central point from which the earthquake diffused its violent movements, many houses were swallowed up by the yawning earth, which closed immediately over them. In the adjacent district also of Cannamaria, four farm-houses, several oil-stores, and some spacious dwelling-houses were so completely engulphed in one chasm, that not a vestige of them was afterwards discernible. The same phenomenon occurred at Terranuova, S. Christina, and Sinopoli. The Academicians state particularly that when deep abysses had opened in the argillaceous strata of Terranuova, and houses had sunk into them, the sides of the chasms closed with such violence, that, on excavating afterwards to recover articles of value, the workmen found the contents and detached parts of the buildings jammed together so as to become one compact mass. It is unnecessary to accumulate examples of similar occurrences; but so many are well authenticated during this earthquake in Calabria, that we may, without hesitation, yield assent to the accounts of catastrophes of the same kind repeated again and again in history, where whole towns are declared to have been engulphed, and nothing but a pool of water or tract of sand left in their place.

On the sloping side of a hill near Oppido, a great chasm opened, and, although a large quantity of soil was precipitated

No. 23.

Chasm formed by the earthquake of 1783, near Oppido, in Calabria.

DIMENSIONS OF NEW FISSURES AND CHASMS. 421

into the abyss, together with a considerable number of olive-trees and part of a vineyard, a great gulph remained after the shock in the form of an amphitheatre, five hundred feet long and two hundred feet deep (see cut No. 23).

According to Grimaldi, many fissures and chasms, formed by the first shock of February 5th, were greatly widened, lengthened, and deepened by the violent convulsions of March 28th. In the territory of San Fili, this observer found a new ravine, half a mile in length, two feet and a half broad, and twenty-five feet deep; and another of similar dimensions in the territory of Rosarno. A ravine *nearly a mile long, one hundred and five feet broad*, and thirty feet deep, opened in the district of Plaisano, where, also, two gulphs were caused—one in a place called Cerzulle, three quarters of a mile long, *one hundred and fifty feet broad*, and above *one hundred feet deep*, and another at La Fortuna, nearly a quarter of a mile long, above thirty feet in breadth, and no less than *two hundred and twenty-five feet deep*. In the district of Fosolano three gulphs opened: one of these measured three hundred feet square, and above thirty feet deep; another was nearly half a mile long, fifteen feet broad, and above thirty feet deep; the third was seven hundred and fifty feet square. Lastly, a calcareous mountain, called Zefirio, at the southern extremity of the Italian peninsula, was cleft in two for the length of nearly half a mile, and an irregular breadth of many feet. Some of these chasms were in the form of a crescent. The annexed cut (No. 24) repre-

No. 24.

Chasm in the hill of St. Angelo, near Soriano, in Calabria, caused by the earthquake of 1783.

Charles Lyell
PRINCIPLES OF GEOLOGY,
JOHN MURRAY, 1830

Interior pages from volume one of three, published between 1830 and 1833. Lyell's description of the impact of the 1783 Calabrian earthquake demonstrates his use of illustrations to engage the reader.

By contrast, uniformitarianism made it clear that the mountains and valleys formed by very gradual, continuous processes, which would have taken millions of years to complete.

Lyell made the details of this theory widely available in *Principles of Geology*, inspiring a young Charles Darwin (see page 166) to think about the implications of these geological timescales for biological species. Not only was Darwin given a copy of the first volume of the book, he was asked by Captain Robert Fitzroy of the HMS *Beagle* – the ship on which he made his famous voyage – to help look for geological formations for Lyell. Later, Darwin and Lyell would become friends.

It is worth noting the full title of Lyell's book: *Principles of Geology: being an attempt to explain the former changes of the Earth's surface, by reference to causes now in operation.* A significant inspiration for the new approach referred to in the title was to take note of the movements of the Earth's surface caused by the likes of volcanoes and earthquakes and consider the long-term implications of such shifts and displacements.

As the discipline of geology advanced, uniformitarianism alone would prove unsatisfactory. One of the main criticisms at the time was evidence of what we now call mass extinctions in the fossil record, which was gradually becoming understood. These mass extinctions, we now know, are indeed the result of sudden, catastrophic changes, but are the exception rather than the rule in shaping the surface of the planet.

Charles Lyell
PRINCIPLES OF GEOLOGY,
JOHN MURRAY, 1833

A range of types of fossilised shell used in helping to comparatively date different geological layers. These are from Lyell's third and final volume.

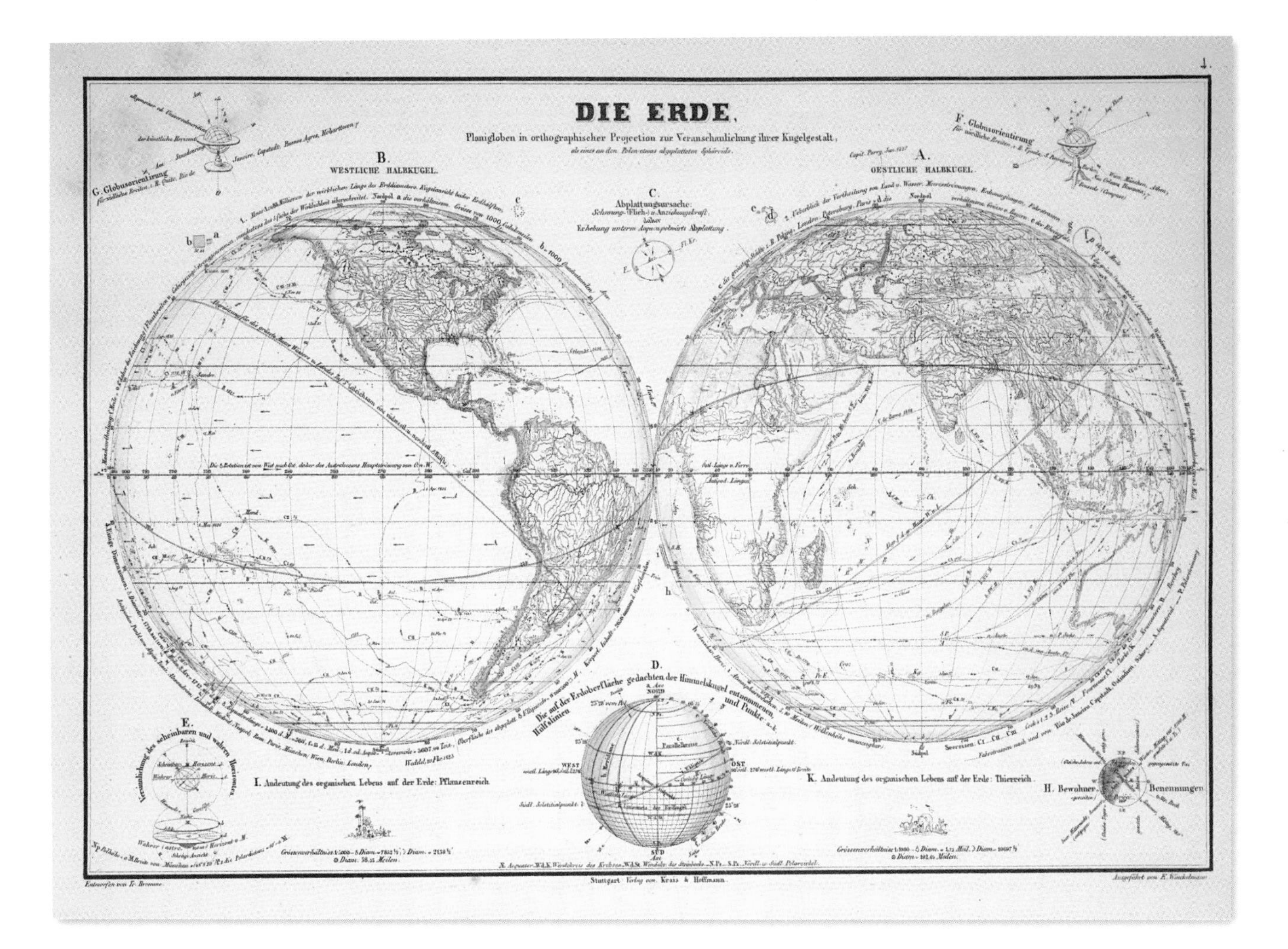

Alexander von Humboldt
KOSMOS, KRAIS & HOFFMANN, 1851

An illustration of the Earth from von Humboldt's popular title.

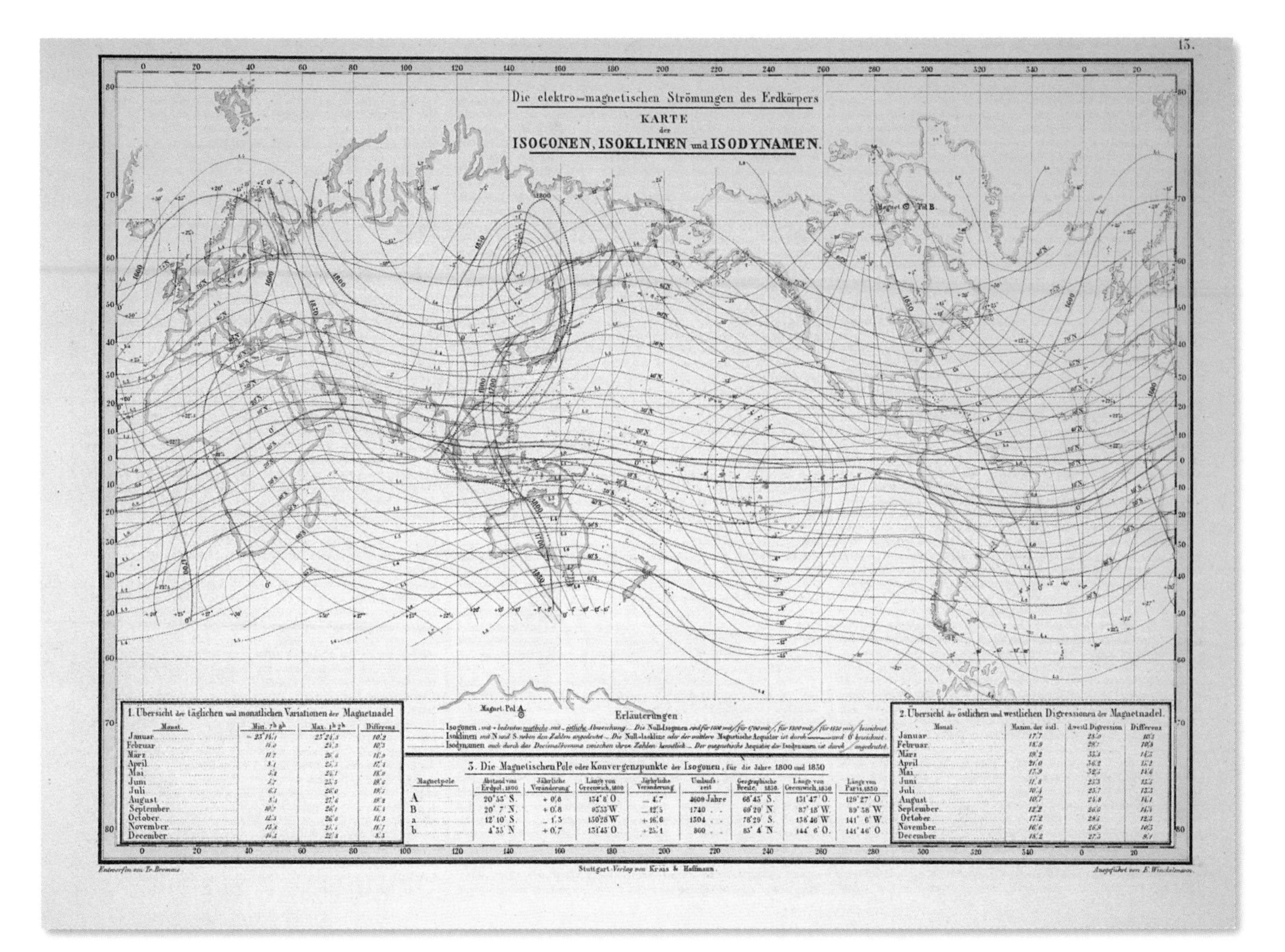
Die elektro-magnetischen Strömungen des Erdkörpers
KARTE
der
ISOGONEN, ISOKLINEN und ISODYNAMEN.
Magnet. Pol A.
Erläuterungen
1. Übersicht der täglichen und monatlichen Variationen der Magnetnadel
3. Die Magnetischen Pole oder Konvergenzpunkte der Isogonen, für die Jahre 1800 und 1850
2. Übersicht der östlichen und westlichen Digressionen der Magnetnadel.
Entworfen von Tr. Bromme
Stuttgart. Verlag von Krais & Hoffmann.

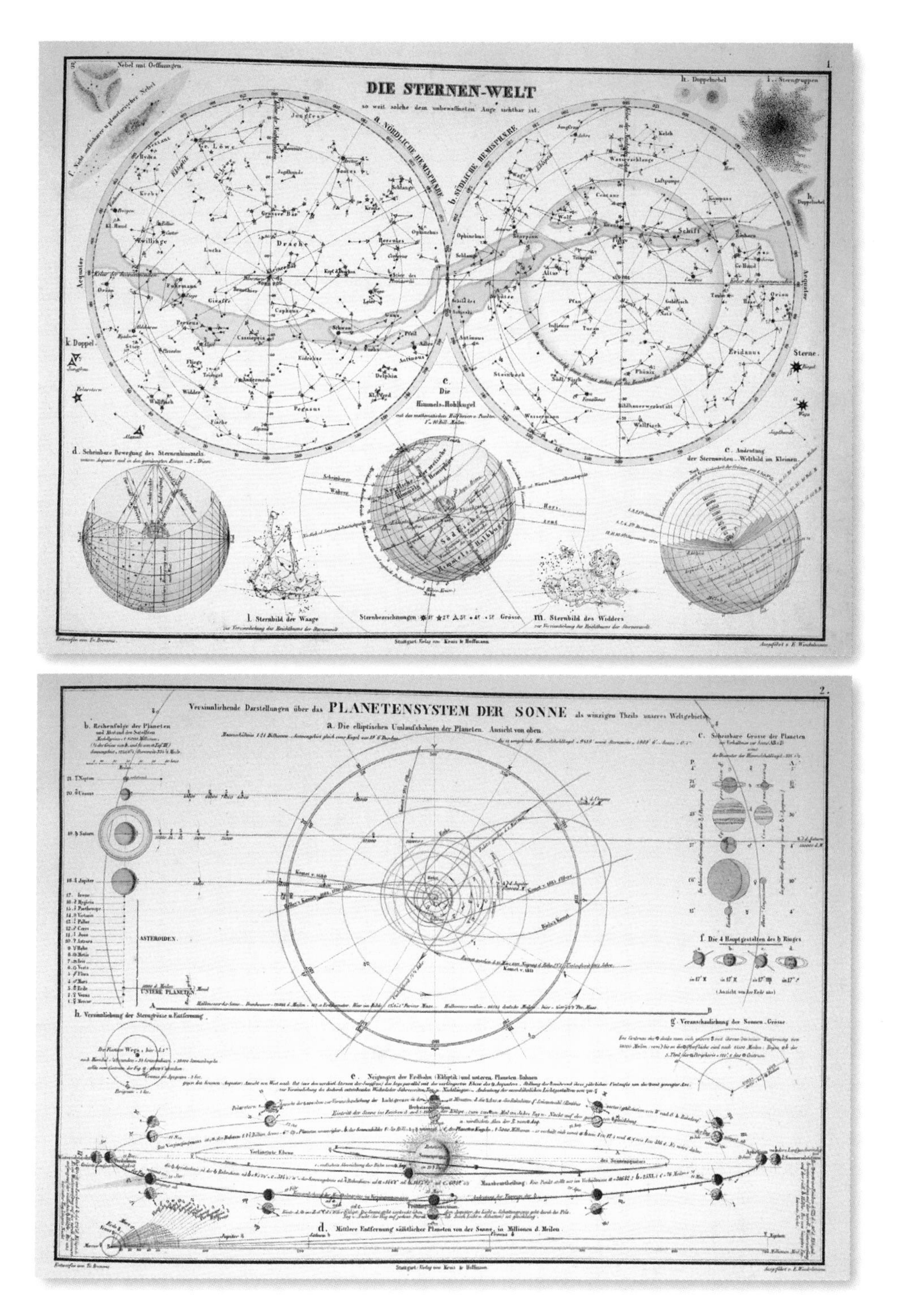

Alexander von Humboldt
KOSMOS, KRAIS & HOFFMANN, 1851

A range of illustrations from von Humboldt's title, demonstrating its mix of geophysical (opposite), meteorological (top) and astronomical (bottom) material.

Alexander von Humboldt
KOSMOS, KRAIS & HOFFMANN, 1851

Another illustration from *Kosmos*, showing world times corresponding to midday in Dresden, reflecting the lack of established time zones, with each city operating on its own time.

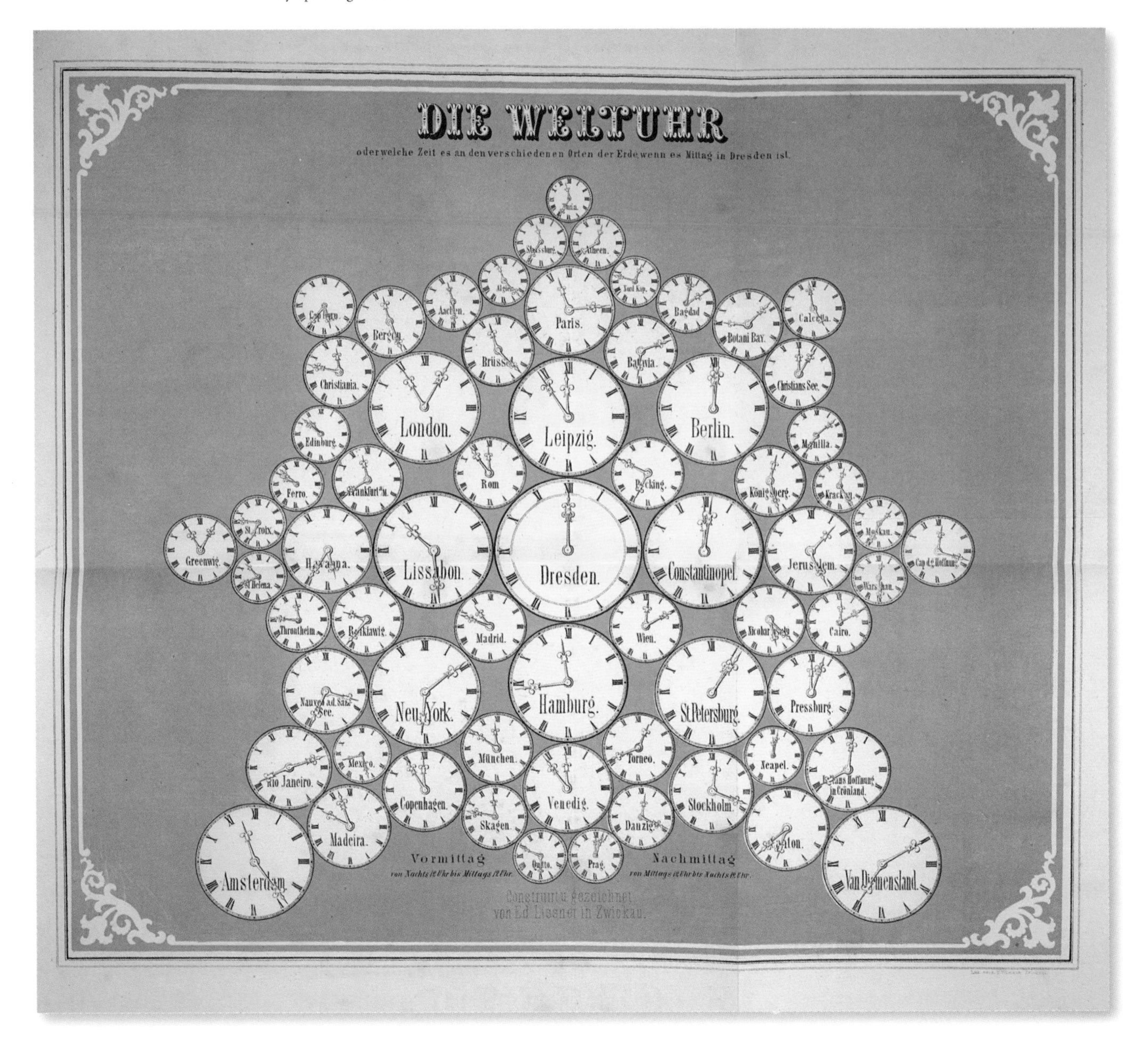

While on the subject of geology and geophysics, it is worth mentioning another book, published a little later, with five volumes coming out between 1845 and 1862. This was *Kosmos* (Cosmos) by the German scientist Alexander von Humboldt, born in Berlin in 1769. Like *Principles of Geology*, *Kosmos* was a hugely popular title in its day, having an appeal far outside the specialists in the area.

Kosmos was based on a series of lectures that Humboldt gave – the closest the period had to the sweeping multi-part science documentaries we now see on television. Although strongly influenced by his expeditions, Humboldt attempted, in *Kosmos*, to provide a holistic picture of the order and structure of the universe, reflected in the Earth. It was Humboldt who, through this book, brought the word 'cosmos' into modern usage – until then it was an obscure Greek word that few would have known. In some ways, Humboldt's book had similarities with Bronowski's *The Ascent of Man* (see page 218), published over a hundred years later. *Kosmos* brought together concepts of the order of nature and of human interpretation and appreciation of beauty. It was both an exploration of the physical universe and of the history of our attempts to understand and appreciate nature.

Humboldt's name is probably most frequently associated with his exploration work. Between 1799 and 1804, Humboldt explored the Americas, bringing back many drawings and accounts of his findings on the flora and fauna (notably an encounter with electric eels) that would find their way into a range of volumes, including *Essai sur la géographie des plantes* (Essay on the Geography of Plants), *Ansichten der Natu*r (Views of Nature), and his popular *Voyage aux Regions Equinoxiales du Noveau Continent* (Travels to the Equinoctial Regions of the New Continent).

Friedrich Georg Weitsch
ALEXANDER VON HUMBOLDT, CA. 1806

A portrait of Alexander von Humboldt, German naturalist and geographer.

The beauty of biology

One of Lyell's most vocal challengers on uniformitarianism was an early expert on fossils, who authored an important biology text. His name was Georges Cuvier, later Baron Cuvier as he was made a life peer by the French state. Born in Montbéliard near the Swiss border in 1769, Cuvier was, scientifically speaking, most successful in his work on palaeontology, which related the limited fossil record known at the time to animals that were still in existence. But his best-known title was a wide-ranging book that tried to categorise all animals into a single structure: *Le Règne Animal* (The Animal Kingdom).

This four-volume work, published in 1817, could be seen as an illustrated equivalent of Linnaeus's *Systema Naturae* with the benefit of nearly 100 years of additional collected information on animals. This enabled Cuvier to produce a more accurate taxonomic structure based on similarities and differences in the animals' anatomy, with the addition of his own experiences from palaeontology. Cuvier got a lot right, considering the crudeness of his method of using visual similarities, though he did have a few stumbling blocks. As well as rejecting uniform geological changes, Cuvier could not accept the principle of evolution, even though in his use of comparative anatomy he made, for example, the connection between the extinct mammoth and the modern elephant. Whatever its conceptual faults, however, *Le Règne Animal* was an impressive effort and a visual delight.

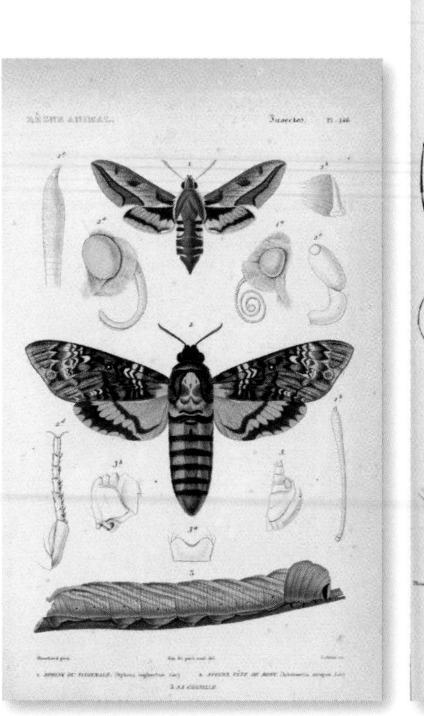

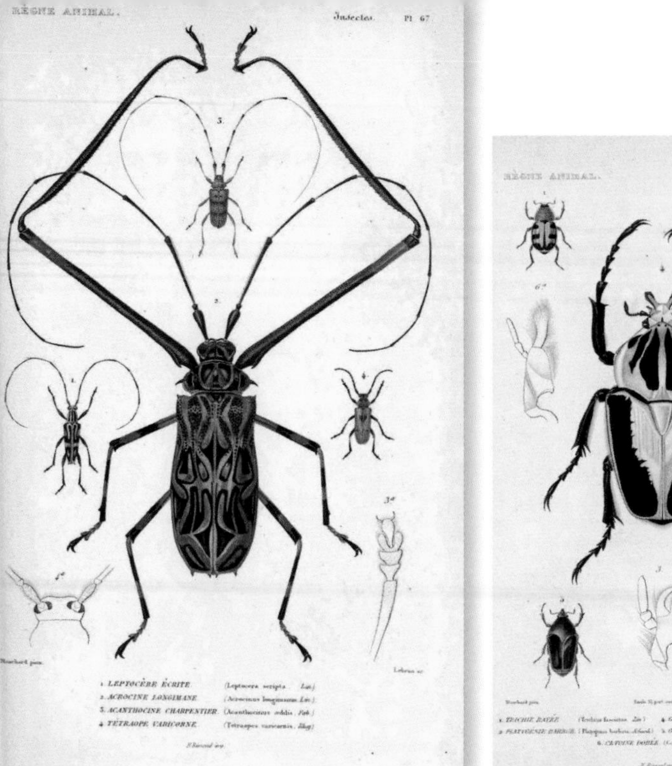

Georges Cuvier
LE RÈGNE ANIMAL,
FORTIN, MASSON ET CIE,
1836–49

Some of Cuvier's elegant illustrations comparing similarities and differences in related species in a later coloured edition, published 19 years after the original publication in 1817.

Cuvier's book was eclipsed in size by an earlier work by another French nobleman, Georges-Louis Leclerc, Comte de Buffon. Buffon's massive *Histoire Naturelle, générale et particulière, avec la description du Cabinet du Roi* (General and Particular Natural History, with a description of the King's Cabinet) ended up 36 volumes long (another eight were later added), written over his working life from 1749 to 1804. This was far more a descriptive work than an attempt to apply a scientific structure to animal classification and was something of a rag bag. While it included elements of physics, chemistry and geology as well as natural history, it excluded, for example, animals other than quadrupeds and birds. This random selection had something in common with the *Cabinet du Roi* of the title – the King's 'cabinet of curiosities' that was an unstructured collection of interesting looking items collected from around the world.

In itself, the *Histoire Naturelle* probably could not be considered a great science book (even at the time it was criticised for a florid writing style and lack of scientific depth, being arguably aimed more to impress the general reader than the natural philosopher), but it would have a significant impact on more important scientific writers such as Cuvier, and it was well illustrated for the time, including nearly 2,000 plates.

Another influential science writer from the period deserving a mention (even though Cuvier seems to have pretty much ignored him) is Jean-Baptiste Lamarck. Born in Bazentin in 1744, the French biologist developed an early (but mistaken) evolutionary theory. Lamarck's idea was that, for example, a giraffe might develop a longer neck during its lifetime as a result of continually stretching to reach for leaves in the trees. This development, he believed, could then be passed on to the animal's offspring. This idea of the inheritance

Georges Cuvier
LE RÈGNE ANIMAL,
DÉTERVILLE, 1817

Comparisons of skulls and fish in the original edition of Cuvier's title.

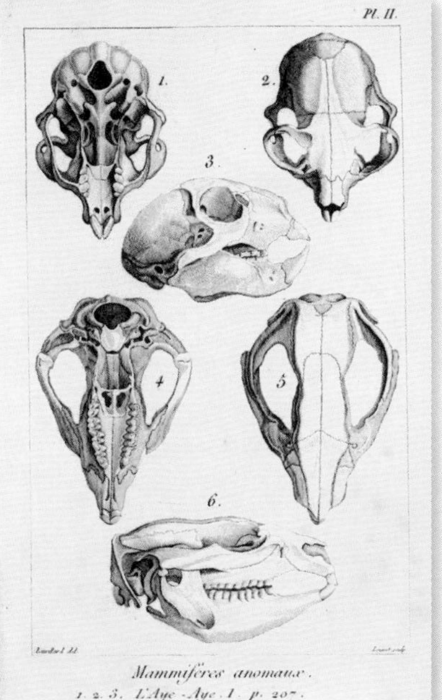

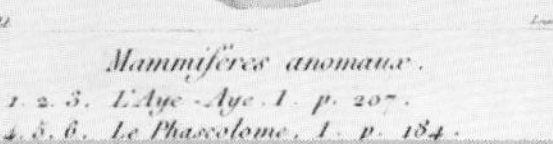

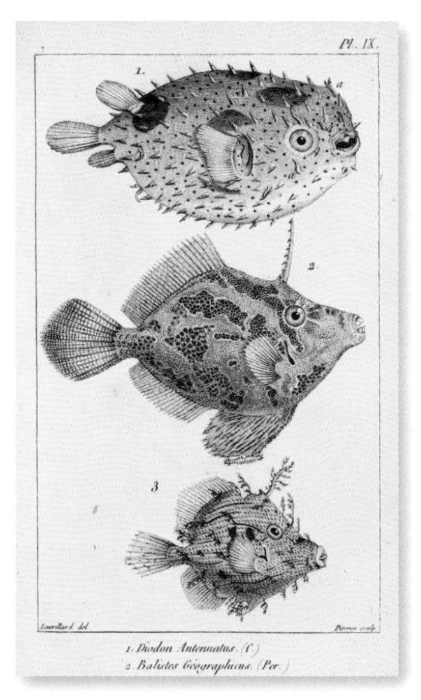

Comte du Buffon
HISTOIRE NATURELLE,
IMPRIMERIE ROYALE,
EIGHTEENTH CENTURY

The title page of the second edition of volume one published in 1750 and an illustration of a rhinoceros from volume four, ca. 1763, housed in the Bibliothèque nationale de France.

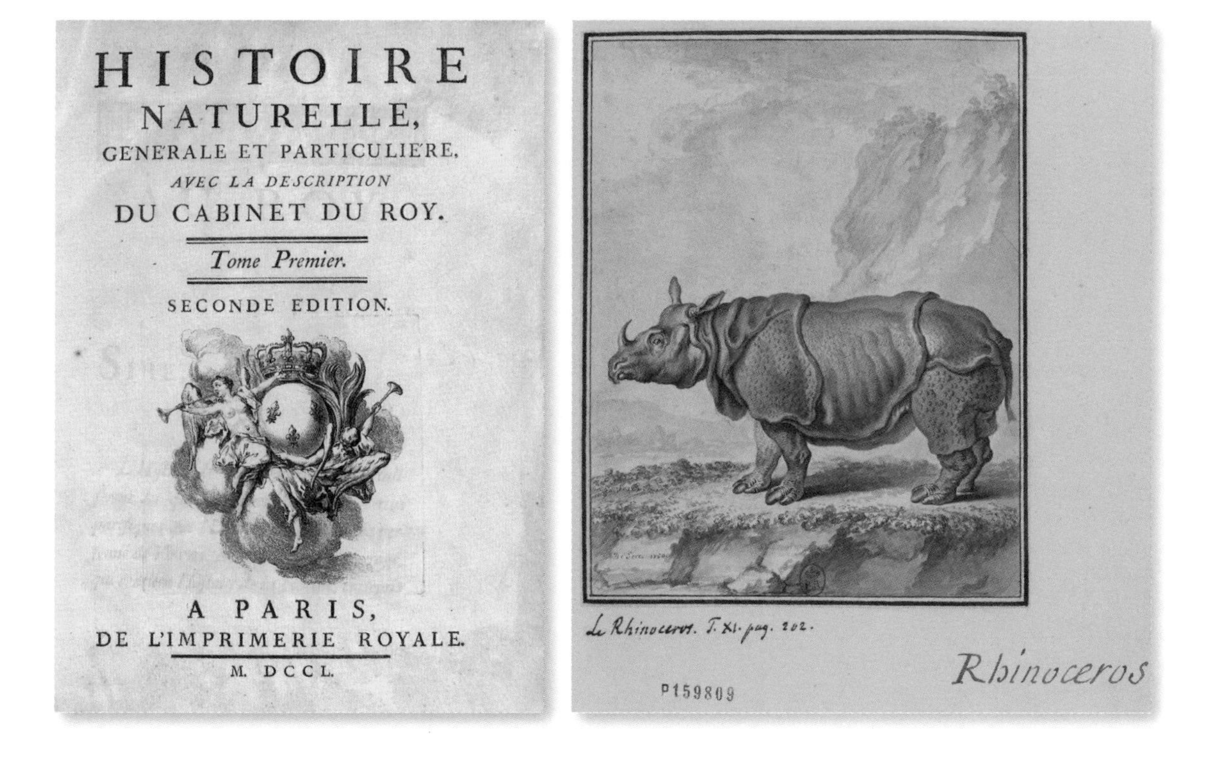

HISTOIRE
NATURELLE,
GÉNÉRALE ET PARTICULIÈRE,
AVEC LA DESCRIPTION
DU CABINET DU ROY.

Tome Premier.

SECONDE EDITION.

A PARIS,
DE L'IMPRIMERIE ROYALE.
M. DCCL.

of acquired characteristics was central to Lamarck's most influential book, *Philosophie Zoologique ou Exposition des considérations relatives à l'histoire naturelle des animaux* (Zoological Philosophy, or Exposition with Regard to the Natural History of Animals), published in 1809.

Lamarck's concept of the inheritance of acquired characteristics would be killed off by the later theory of natural selection and by advances in the understanding of genetics. This meant that for a long time, his work was dismissed. However, modern biology recognises that there are aspects of epigenetics – the way that genes are repeatedly switched on and off by external influences – that can be modified by the environment and passed on to a living thing's descendants. Although, as we have seen, the basis for Lamarck's theory was incorrect, he did at least realise the importance of both adaptation to the environment and survival of the fittest.

In this sense, Buffon and Lamarck were both ahead of Cuvier in their ideas, being more supportive of the ideas of the development of species that would lead to the writing of one of the most influential science books ever produced. Before turning to Charles Darwin, however, it's interesting to look at a book which he himself credited with helping to pave the way for the acceptance of his theory of evolution by natural selection. The title in question was *Vestiges of the Natural History of Creation*. Published in 1844, it was an international bestseller, no doubt helped by the revelation that Prince Albert was said to have read it aloud to Queen Victoria. The book was originally anonymous: it was 40 years later that the author was revealed to be Peebles-born Scottish author and publisher Robert Chambers.

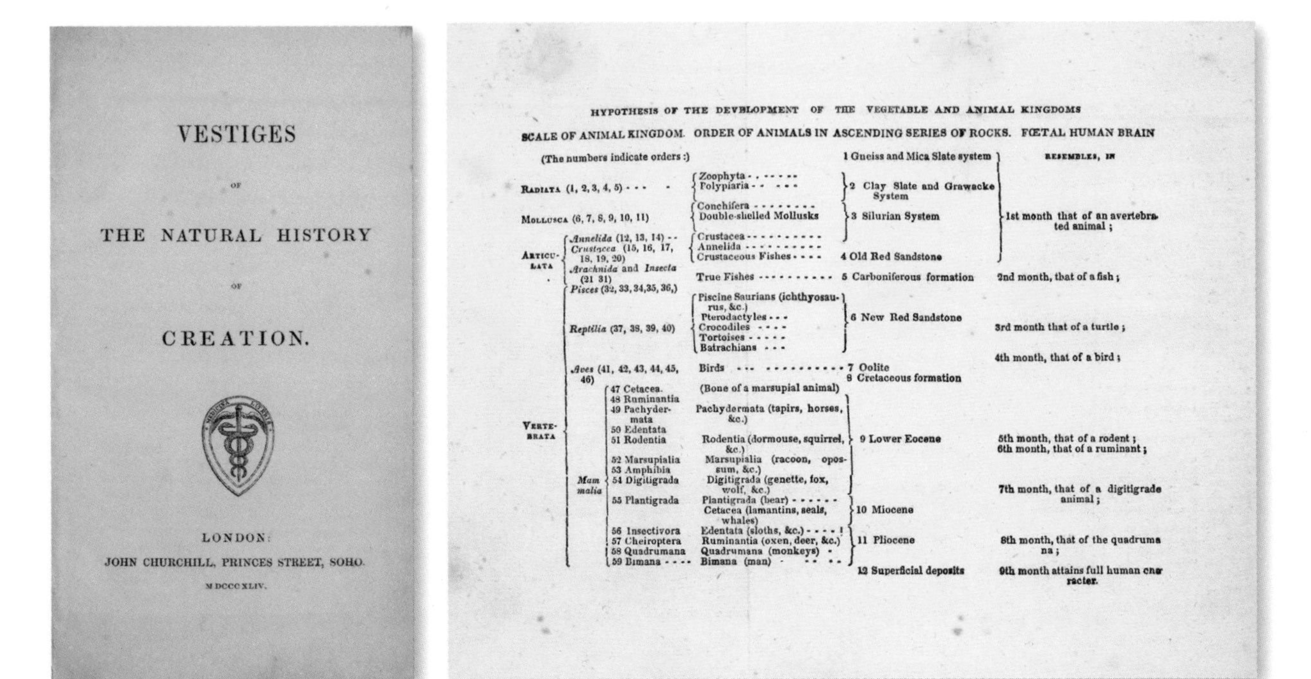

VESTIGES

OF

THE NATURAL HISTORY

OF

CREATION.

LONDON:
JOHN CHURCHILL, PRINCES STREET, SOHO.
MDCCCXLIV.

HYPOTHESIS OF THE DEVELOPMENT OF THE VEGETABLE AND ANIMAL KINGDOMS

SCALE OF ANIMAL KINGDOM. (The numbers indicate orders:)	ORDER OF ANIMALS IN ASCENDING SERIES OF ROCKS.		FŒTAL HUMAN BRAIN RESEMBLES, IN
		1 Gneiss and Mica Slate system	
RADIATA (1, 2, 3, 4, 5)	Zoophyta; Polypiaria	2 Clay Slate and Grawacke System	
MOLLUSCA (6, 7, 8, 9, 10, 11)	Conchifera; Double-shelled Mollusks	3 Silurian System	1st month that of an avertebrated animal;
ARTICULATA: *Annelida* (12, 13, 14); *Crustacea* (15, 16, 17, 18, 19, 20); *Arachnida* and *Insecta* (21 31)	Crustacea; Annelida; Crustaceous Fishes	4 Old Red Sandstone	
VERTEBRATA: *Pisces* (32, 33, 34,35, 36,)	True Fishes	5 Carboniferous formation	2nd month, that of a fish;
Reptilia (37, 38, 39, 40)	Piscine Saurians (ichthyosaurus, &c.); Pterodactyles; Crocodiles; Tortoises; Batrachians	6 New Red Sandstone	3rd month that of a turtle;
Aves (41, 42, 43, 44, 45, 46)	Birds	7 Oolite	4th month, that of a bird;
		8 Cretaceous formation	
Mammalia: 47 Cetacea.	(Bone of a marsupial animal)		
48 Ruminantia; 49 Pachydermata; 50 Edentata; 51 Rodentia; 52 Marsupialia; 53 Amphibia; 54 Digitigrada	Pachydermata (tapirs, horses, &c.); Rodentia (dormouse, squirrel, &c.); Marsupialia (racoon, opossum, &c.); Digitigrada (genette, fox, wolf, &c.)	9 Lower Eocene	5th month, that of a rodent; 6th month, that of a ruminant; 7th month, that of a digitigrade animal;
55 Plantigrada	Plantigrada (bear); Cetacea (lamantins, seals, whales)	10 Miocene	
56 Insectivora; 57 Cheiroptera; 58 Quadrumana; 59 Bimana	Edentata (sloths, &c.); Ruminantia (oxen, deer, &c.); Quadrumana (monkeys); Bimana (man)	11 Pliocene	8th month, that of the quadrumana;
		12 Superficial deposits	9th month attains full human character.

Robert Chambers
VESTIGES OF THE NATURAL HISTORY OF CREATION,
JOHN CHURCHILL, 1844

The title page from the first edition and a table from an 1858 American edition (published by Harper & Brothers) linking animals with rocks thought to date from a similar period.

Vestiges takes a big-picture approach, starting with the idea of stellar evolution to show how everything has developed over time from earlier instances. In its picture of natural and continuous change, *Vestiges* was at odds with creationist views, but was nonetheless respectable as it didn't have any suggestion of an atheist agenda. Some aspects of the science in the book were old-fashioned even by the ideas of the time – Chambers supported both Lamarck's theory of the inheritance of acquired characteristics and the idea of spontaneous generation, for example, where living things were thought to emerge from rotting meat and other substances. But there is no doubt that it had a powerful impact.

Evolving science writing

In 1859, what is probably the best-known science book ever written arrived on the scene. The book was *On the Origin of Species by Means of Natural Selection, or the Preservation of Favoured Races in the Struggle for Life* by Charles Darwin. *Origin of Species* was the book that introduced the concept of evolution by natural selection to the general public. Though, as we have seen, aspects of the concept itself had been debated for a couple of generations, it was Darwin who presented the whole picture for the first time.

This was very much an idea that was in the zeitgeist. With hindsight, evolution by natural selection may seem obvious. Anyone with an understanding of the basics of science, presented with the way that genetic information is passed on from generation to generation would expect that some offspring would be better suited to survive than

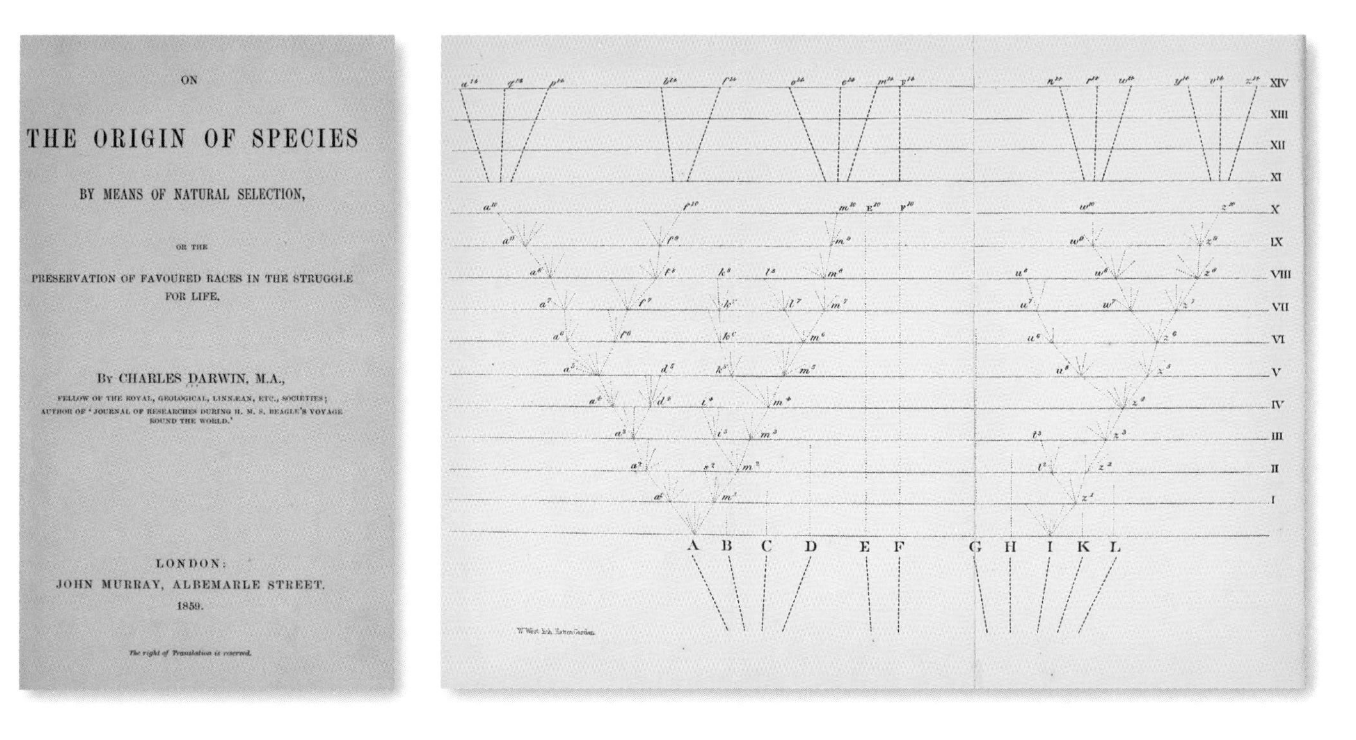

ON

THE ORIGIN OF SPECIES

BY MEANS OF NATURAL SELECTION,

OR THE

PRESERVATION OF FAVOURED RACES IN THE STRUGGLE FOR LIFE.

BY CHARLES DARWIN, M.A.,

FELLOW OF THE ROYAL, GEOLOGICAL, LINNÆAN, ETC., SOCIETIES;
AUTHOR OF 'JOURNAL OF RESEARCHES DURING H. M. S. BEAGLE'S VOYAGE ROUND THE WORLD.'

LONDON:
JOHN MURRAY, ALBEMARLE STREET.
1859.

The right of Translation is reserved.

Charles Darwin
ON THE ORIGIN OF SPECIES, JOHN MURRAY, 1859

The title page of Darwin's famous book with the only illustration in the original, a lithographic diagram by William West of a tree of descent based on degrees of similarity.

FINCHES FROM GALAPAGOS ARCHIPELAGO, 1890

Galapagos finches showing varied beaks from a later publication of Darwin's *Beagle* journals (first published in 1839) alongside a photograph by Julia Margaret Cameron from 1868 of Darwin aged 59.

others, and those with the preferred genetic modifications would be the ones to produce further generations and pass on those adaptations. However, in Darwin's day, that genetic model was not available, making the mental leap harder. As we have seen, though, ideas of evolution were already in the air in Darwin's grandfather's time (see pages 137–8), and by the time Darwin was working on his theory others were hot on his heels, notably another English naturalist, Alfred Russel Wallace, who shared his very similar ideas with Darwin before *Origin of Species* was finished, and would jointly publish papers with Darwin.

Born in Shrewsbury, England, in 1809, Darwin originally trained as a doctor before his interest in natural history pushed his studies aside. Famously, the event that cemented his position in history was being invited to be a naturalist and companion of captain Robert FitzRoy on his HMS *Beagle* voyage. Lasting almost five years, this epic journey exposed Darwin to a huge range of wildlife, from the marsupials of Australia to the finches of the Galapagos Islands. It was Darwin's observation of these birds and the way in which they had markedly diverged on neighbouring islands with different environments, that would prove one of the inspirations for writing *Origin of Species*.

It took Darwin over 20 years from first considering the ideas of evolution (the *Beagle* returned to England in 1836) to publishing *Origin of Species* in 1858. His thinking was advanced by meetings with the likes of Lyell and with those who had studied fossil bones, such as anatomist and palaeontologist Richard Owen. One of the reasons the gestation of the book was so long was that Darwin's work on evolution was something he used to fill in time, rather than being his central activity. After writing accounts of the *Beagle* voyage (notably *Journal and Remarks*, published in 1839, usually referred to as *The Voyage of the Beagle*) and books on the flora and fauna he had encountered, Darwin was distracted by a near-obsession with barnacles. But the arrival of a letter from Wallace that contained an idea very similar to Darwin's own firmly pushed Darwin into completing his masterpiece.

Once published, the book proved popular and, considering how much debate it still generates in some parts of America, was surprisingly lacking in controversy in its early days. With time, debates on the topic (not involving the self-effacing Darwin) became more vocal, most notably after the infamous Oxford 'debate' in 1860. More a discussion than a strict debate, this involved a number of supporters of Darwin, including Robert FitzRoy, but has come to be best remembered for the verbal sparring between a highly vocal Darwin enthusiast, Thomas Huxley, and the bishop of Winchester, Samuel Wilberforce. Wilberforce asked Huxley if he was happy to have a monkey or ape as a (great) grandparent (the exact wording is uncertain as there was no word-for-word account), a jibe Huxley effortlessly put down.

Surprisingly, there seems to have been less negative reaction still to Darwin's follow-up title, published in 1871, *The Descent of Man, and Selection in Relation to Sex*. This focused on the human implications of evolutionary theory and brought in an additional selection mechanism to natural selection in the form of sexual selection. The lack of negative reaction seems to have been because, despite the popularity of the book, the implications of evolution for the origins of humanity had been argued thoroughly in the previous decade and more. The idea that humans had not been created as they were, with no antecedent species, had lost the power to shock. Only the aspect of sexual selection – the idea that

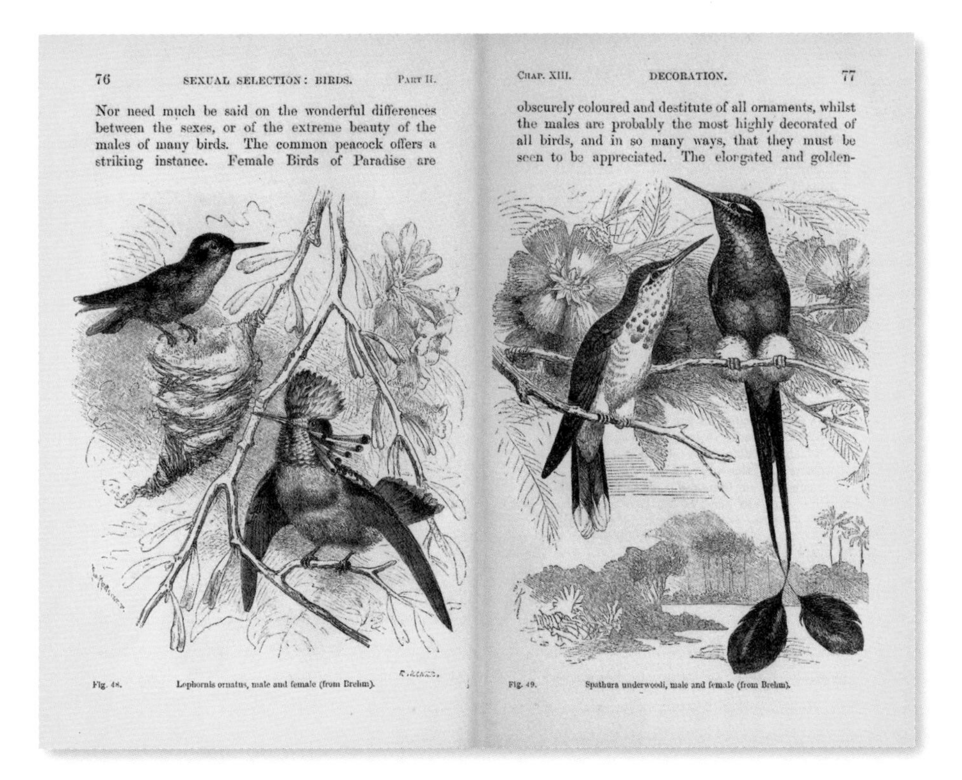
76 SEXUAL SELECTION: BIRDS. PART II.

Nor need much be said on the wonderful differences between the sexes, or of the extreme beauty of the males of many birds. The common peacock offers a striking instance. Female Birds of Paradise are

Fig. 48. Lophornis ornatus, male and female (from Brehm).

CHAP. XIII. DECORATION. 77

obscurely coloured and destitute of all ornaments, whilst the males are probably the most highly decorated of all birds, and in so many ways, that they must be seen to be appreciated. The elongated and golden-

Fig. 49. Spathura underwoodi, male and female (from Brehm).

Charles Darwin
THE DESCENT OF MAN,
JOHN MURRAY, 1871

Illustrated pages from the first edition of volume two of two, showing the deployment of colouration and extreme feather structures in sexual selection of birds.

animals could keep changes that would be disadvantageous under natural selection because they made mating more likely, such as the peacock's tail – was a novelty, and was too technical to worry the general reader.

Darwin's last notable title, *The Autobiography of Charles Darwin*, was published in 1887, five years after his death. It was originally written for his children as *Recollections of the Development of my Mind and Character* and is surprisingly unstuffy for a piece of Victorian writing. Given the way that Darwin tends to be hero-worshipped (or demonised) today, *The Autobiography* is wonderfully self-deprecating in describing his path from enthusiastic slaughterer of game to amateur geologist and self-taught naturalist. The book doesn't cover the *Beagle* voyage, but rather the way that Darwin tried to apply the same kind of scientific rigour that Lyell applied to geology and use Francis Bacon's approach of collecting facts without hypothesis, before going further to his ideas on natural selection and evolution.

Genetics and the scientific paper

One thing Darwin definitely got wrong was his idea of how evolution took place on the microscopic level, assuming that it involved some kind of blending of the natures of an animal's parents. He would have benefited hugely if he had been able to read *Versuche über Pflanzenhybriden* (Experiments on Plant Hybridisation) by Gregor Mendel. Strictly speaking, this was an academic paper rather than a book, only running to around 50 pages, though it was subsequently published in book form with additional material.

Gregor Mendel
'VERSUCHE ÜBER PFLANZENHYBRIDEN', *VERHANDLUNGEN DES NATURFORSCHENDEN VEREINES IN BRÜNN*, 1866

The 1866 copy of the Proceedings of the Natural History Society of Brünn, which carried Mendel's largely overlooked paper.

In effect, Mendel's work was the missing piece of the jigsaw that Darwin never found. Gregor Mendel was an Austrian friar, born in 1822 in Heinzendorf bei Odrau (now in the Czech Republic). By cross-breeding pea plants with different traits, such as height and flower colour, Mendel discovered that there were factors that we would now call genes that determined the way these traits were inherited in future generations. This was the mechanism necessary for Darwin's evolutionary theory to work.

Unfortunately for Darwin and his colleagues, Mendel's work, published in 1866 in the proceedings of an obscure natural history society, had no impact until it was rediscovered in the early twentieth century. Since then, *Versuche über Pflanzenhybriden* has been criticised as the results are simply too close to Mendel's expectations – he may have shaded the results to match his theory. However, whether or not this was the case, Mendel's text deserves a special place in the history of science writing.

Versuche über Pflanzenhybriden reflects the change in the way that scientists communicated with each other, which accelerated in the nineteenth century. The idea of scientific papers published in journals grew out of letters exchanged between scientists. In an age of email and social media, it's easy to forget how important letter writing was in past times. A small illustration of this is that the Post Office in Scotland decided to place a postbox on the entrance to the scientist James Clerk Maxwell's country estate for his use, so large was the flow of correspondence between Maxwell and his colleagues.

With the foundation of scientific societies, such as the Accademia dei Lincei in Rome in 1603 and the Royal Society in London in 1660, these institutions became hubs for the

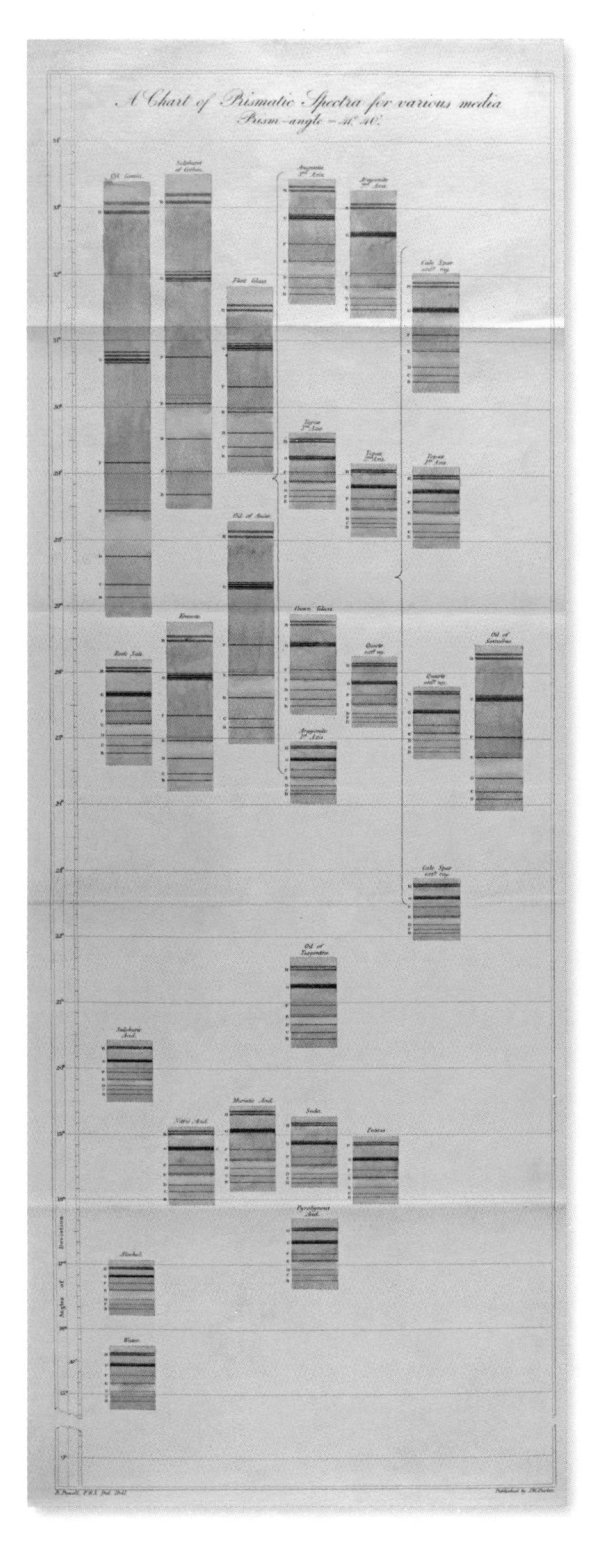

Baden Powell
A GENERAL AND ELEMENTARY VIEW OF THE UNDULATORY THEORY, JOHN W. PARKER, 1841

Spectral charts (left) from Powell's experiments which were first published in *Philosophical Transactions* before being collected together in a book on the mathematician's work on optics.

William Herschel
'ASTRONOMICAL OBSERVATIONS RELATING TO THE SIDEREAL PART OF THE HEAVENS', *PHILOSOPHICAL TRANSACTIONS OF THE ROYAL SOCIETY OF LONDON*, VOL. 104, 1814

Image above of nebulae accompanying an article by Herschel in *Philosophical Transactions*.

exchange of scientific information. This was often in the form of letters, funnelled through the society's secretary, but increasingly became focused on their publications. The Royal Society started the world's first scientific journal, *Philosophical Transactions* (still in print today), in 1665, bringing with it the idea of peer review, whereby other natural philosophers would provide appropriate criticism of the content before publication.

Such journals, published both by academic institutions and private publishing companies, flourished and multiplied. Getting a paper published in a journal was a far quicker process than completing a book and getting that to press. So successful were journals that, with time, it became increasingly difficult to keep on top of the sheer volume of publications. Some attempt was made through abstracts. The idea was to write a very short summary of a paper which would be abstracted into a separate publication, providing a top-level summary of current developments, enabling someone interested in reading more on a particular topic to then request the full paper. But it still was all too easy for a paper in an obscure publication to be overlooked, as Mendel discovered.

As the rate of scientific discovery took off, it became natural to share ideas via journals, making it less necessary for scientists to go to the effort of writing books. The peer-to-peer communication of the academic scientific book would never entirely disappear, but it would become far less essential for scientists' work to be recognised.

From logic to anatomy

Writing at the same time as Darwin were two leading mathematicians, George Boole and John Venn, whose books were highly influential in their field. They are now largely remembered for a symbolic approach to logic that is central to the workings of computers and to simple diagrams that illustrate logical connections. Boole, born in Lincoln in 1815, was the first into the fray with a book called *An Investigation of the Laws of Thought on Which are Founded the Mathematical Theories of Logic and Probabilities*, published in 1854. Since ancient Greek times, logic had been a matter of words, but Boole brought to it the symbolic tools of mathematics. The result, 'Boolean algebra', is at the heart of the way that gates – the fundamental logical components of computers – work, and is in action every time we type a search term into a computer.

As for John Venn, born in Hull in 1834, his first important book was *The Logic of Chance* from 1866, which was a major work in the development of the ideas of probability. Venn's mathematical approach to the probability of something happening was based on the number of times such an event would occur if you had a very large number of trials. This was probably Venn's more significant contribution to mathematics, although he is better remembered for Venn diagrams, which first appeared in his 1881 book *Symbolic Logic*. These familiar diagrams use shapes to illustrate relationships in Boolean algebra. Typically, they feature two or more overlapping circles within a space. Depending on where an item is placed within the space, the circles it occupies show the logical combination of properties assigned to it.

Just as Venn diagrams and Boolean searches are familiar terms today, so too is the title of what remains probably the best-known medical textbook of all time, Gray's *Anatomy* –

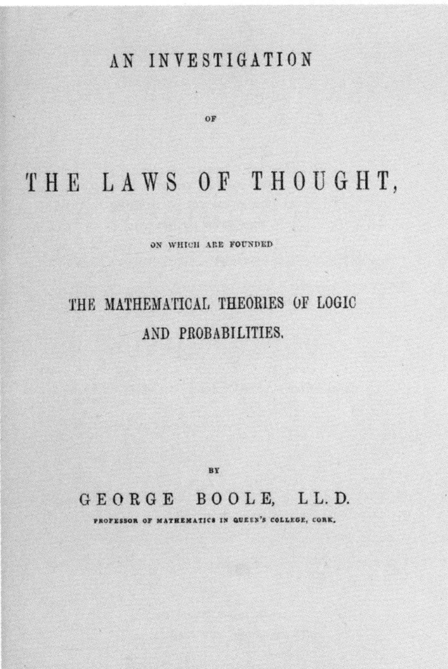

AN INVESTIGATION

OF

THE LAWS OF THOUGHT,

ON WHICH ARE FOUNDED

THE MATHEMATICAL THEORIES OF LOGIC AND PROBABILITIES.

BY

GEORGE BOOLE, LL.D.

PROFESSOR OF MATHEMATICS IN QUEEN'S COLLEGE, CORK.

DOVER PUBLICATIONS, INC.

George Boole
AN INVESTIGATION OF THE LAWS OF THOUGHT,
DOVER PUBLICATIONS, 1854

The title page of the book in which Boole introduced Boolean algebra – a notation for the mathematical representation of logic.

THE

LOGIC OF CHANCE

AN ESSAY

ON THE FOUNDATIONS AND PROVINCE OF THE THEORY OF PROBABILITY,

WITH ESPECIAL REFERENCE TO ITS LOGICAL BEARINGS AND ITS APPLICATION TO

MORAL AND SOCIAL SCIENCE, AND TO STATISTICS,

BY

JOHN VENN, Sc.D., F.R.S.,

FELLOW AND LECTURER IN THE MORAL SCIENCES, GONVILLE AND CAIUS COLLEGE, CAMBRIDGE.

LATE EXAMINER IN LOGIC AND MORAL PHILOSOPHY IN THE UNIVERSITY OF LONDON.

"So careful of the type she seems
So careless of the single life."

THIRD EDITION, RE-WRITTEN AND ENLARGED.

London:
MACMILLAN AND CO.
AND NEW YORK
1888

[All Rights reserved.]

118 *Randomness and its scientific treatment.* [CHAP. V.

character of the rows of figures displayed by the incommensurable or irrational ratios in question.

As it may interest the reader to see an actual specimen of such a path I append one representing the arrangement of the eight digits from 0 to 7 in the value of π. The data are taken from Mr Shanks' astonishing performance in the calculation of this constant to 707 places of figures (*Proc. of R. S.*, XXI. p. 319). Of these, after omitting 8 and 9, there remain 568; the diagram represents the course traced out by following the direction of these as the clue to our path. Many of the steps have of course been taken in opposite directions twice or oftener. The result seems to me to furnish a very fair graphical indication of randomness. I have compared it with corresponding paths furnished by rows of figures taken from logarithmic tables, and in other ways, and find the results to be much the same.

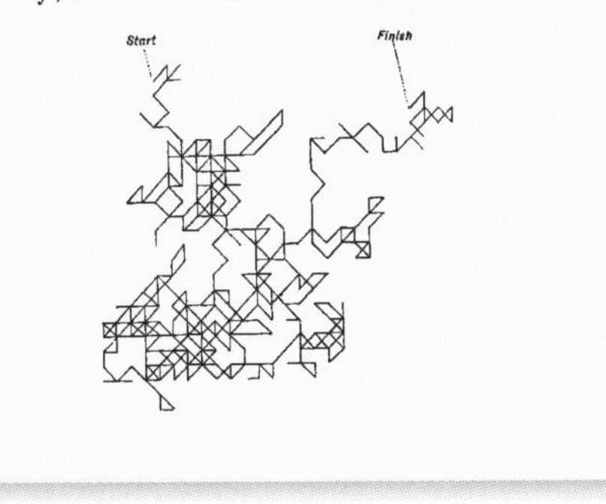

John Venn
THE LOGIC OF CHANCE,
MACMILLAN AND CO., 1888

The title page and an illustration of a 'drunkard's walk' or random walk from the third edition of Venn's important book on probability, first published in 1866.

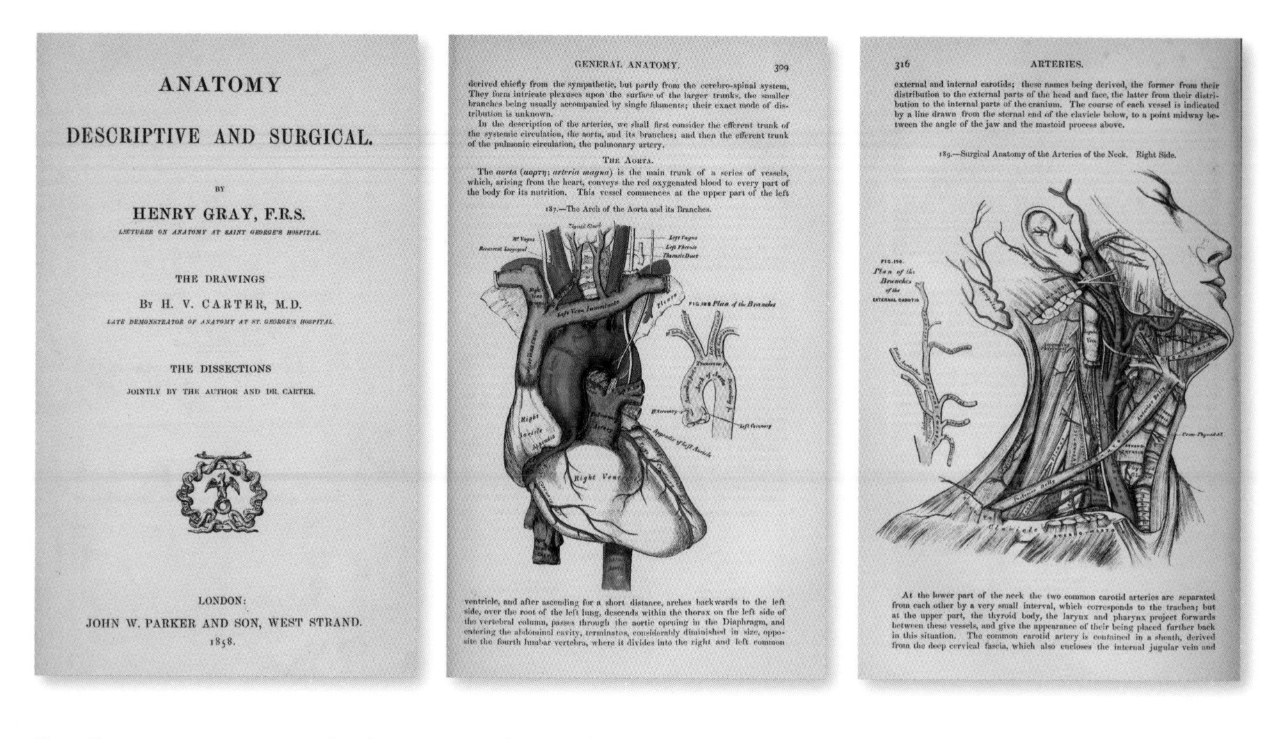

ANATOMY

DESCRIPTIVE AND SURGICAL.

BY

HENRY GRAY, F.R.S.

LECTURER ON ANATOMY AT SAINT GEORGE'S HOSPITAL.

THE DRAWINGS

BY H. V. CARTER, M.D.

LATE DEMONSTRATOR OF ANATOMY AT ST. GEORGE'S HOSPITAL.

THE DISSECTIONS

JOINTLY BY THE AUTHOR AND DR. CARTER.

LONDON:

JOHN W. PARKER AND SON, WEST STRAND.

1858.

GENERAL ANATOMY. 309

derived chiefly from the sympathetic, but partly from the cerebro-spinal system. They form intricate plexuses upon the surface of the larger trunks, the smaller branches being usually accompanied by single filaments; their exact mode of distribution is unknown.

In the description of the arteries, we shall first consider the efferent trunk of the systemic circulation, the aorta, and its branches; and then the efferent trunk of the pulmonic circulation, the pulmonary artery.

THE AORTA.

The *aorta* (*αορτη*; *arteria magna*) is the main trunk of a series of vessels, which, arising from the heart, conveys the red oxygenated blood to every part of the body for its nutrition. This vessel commences at the upper part of the left

187.—The Arch of the Aorta and its Branches.

ventricle, and after ascending for a short distance, arches backwards to the left side, over the root of the left lung, descends within the thorax on the left side of the vertebral column, passes through the aortic opening in the Diaphragm, and entering the abdominal cavity, terminates, considerably diminished in size, opposite the fourth lumbar vertebra, where it divides into the right and left common

316 ARTERIES.

external and internal carotids; these names being derived, the former from their distribution to the external parts of the head and face, the latter from their distribution to the internal parts of the cranium. The course of each vessel is indicated by a line drawn from the sternal end of the clavicle below, to a point midway between the angle of the jaw and the mastoid process above.

189.—Surgical Anatomy of the Arteries of the Neck. Right Side.

At the lower part of the neck the two common carotid arteries are separated from each other by a very small interval, which corresponds to the trachea; but at the upper part, the thyroid body, the larynx and pharynx project forwards between these vessels, and give the appearance of their being placed further back in this situation. The common carotid artery is contained in a sheath, derived from the deep cervical fascia, which also encloses the internal jugular vein and

Henry Gray
ANATOMY, JOHN W. PARKER & SON, 1858

The title page and coloured anatomical illustrations from Gray's famous book.

thanks in part to the popular US television drama of the same name. First published in 1858, updated versions of Gray's *Anatomy* are still in print.

Henry Gray was an English surgeon, born in London in 1827. Historically, surgeons had been considered very much second-rate to physicians, members of a profession that began as little more than barbers with a small amount of ad-hoc practical experience of human anatomy. However, just as professional science was beginning to take off, so the idea that the practice of surgery could be given more of a scientific basis was also coming into vogue. Working with his accomplished illustrator, also a surgeon, Henry Vandyke Carter, Gray worked through a series of dissections of human bodies in the process of writing the book. Sadly, Gray lived only three years from the first publication of the book, contracting smallpox while treating his own nephew and dying aged just 34. As a result, the vast majority of the extant copies of Gray's *Anatomy* owe as much to other authors as they do to Gray. However, his name will always be linked to this definitive medical title.

This was an important period for medical work. Just three years after Gray's *Anatomy* first appeared, another book was published, marking the work of a man whose observations saved the life of many women in childbirth. This was *Die Ätiologie, der Begriff und die Prophylaxis des Kindbettfiebers* (The Causes, Description and Preventative Treatment of Childbed Fever) by the Hungarian physician Ignaz Semmelweis, born in Buda in 1818. In Europe at the time, the number of women dying of what we now know were infections of the reproductive tract had shot up. In the worst cases around four in ten women would not survive the period around childbirth.

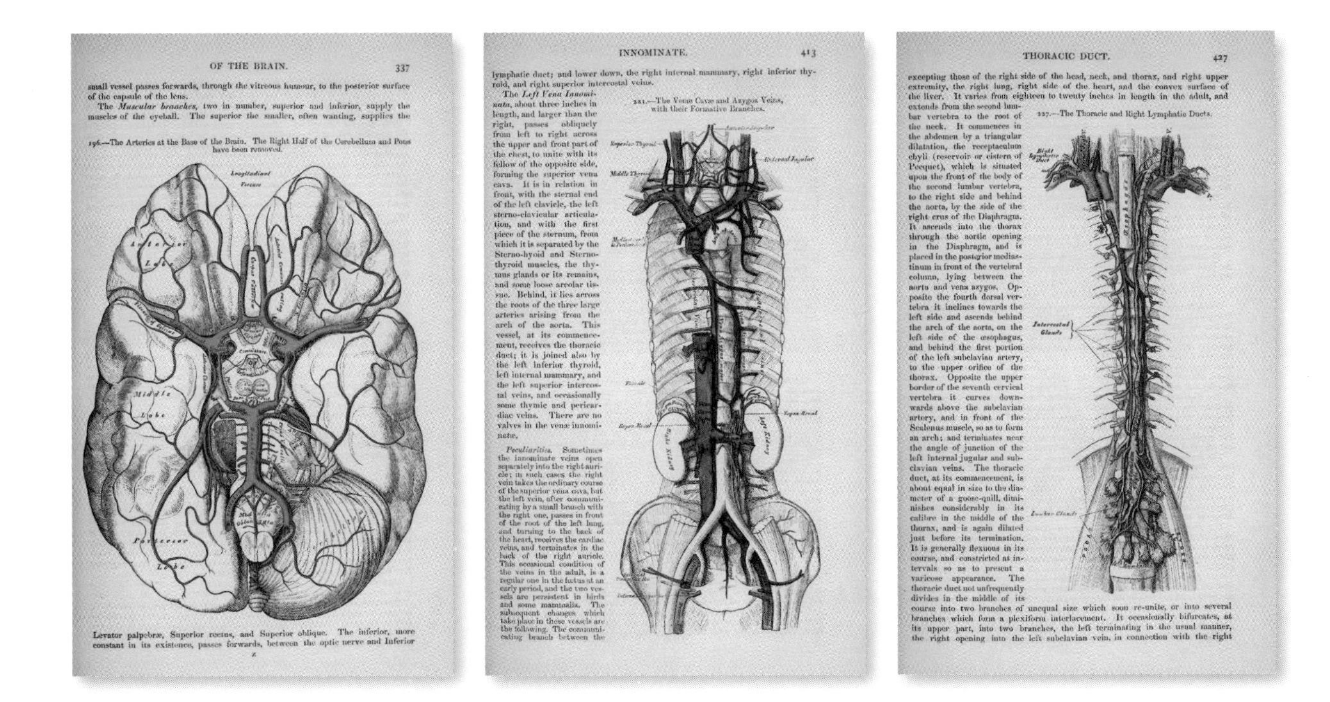

OF THE BRAIN. 337

small vessel passes forwards, through the vitreous humour, to the posterior surface of the capsule of the lens.

The *Muscular branches*, two in number, superior and inferior, supply the muscles of the eyeball. The superior the smaller, often wanting, supplies the

196.—The Arteries at the Base of the Brain. The Right Half of the Cerebellum and Pons have been removed.

Levator palpebræ, Superior rectus, and Superior oblique. The inferior, more constant in its existence, passes forwards, between the optic nerve and Inferior

z

INNOMINATE. 413

lymphatic duct; and lower down, the right internal mammary, right inferior thyroid, and right superior intercostal veins.

The *Left Vena Innominata*, about three inches in length, and larger than the right, passes obliquely from left to right across the upper and front part of the chest, to unite with its fellow of the opposite side, forming the superior vena cava. It is in relation in front, with the sternal end of the left clavicle, the left sterno-clavicular articulation, and with the first piece of the sternum, from which it is separated by the Sterno-hyoid and Sterno-thyroid muscles, the thymus glands or its remains, and some loose areolar tissue. Behind, it lies across the roots of the three large arteries arising from the arch of the aorta. This vessel, at its commencement, receives the thoracic duct; it is joined also by the left inferior thyroid, left internal mammary, and the left superior intercostal veins, and occasionally some thymic and pericardiac veins. There are no valves in the venæ innominatæ.

Peculiarities. Sometimes the innominate veins open separately into the right auricle; in such cases the right vein takes the ordinary course of the superior vena cava, but the left vein, after communicating by a small branch with the right one, passes in front of the root of the left lung, and turning to the back of the heart, receives the cardiac veins, and terminates in the back of the right auricle. This occasional condition of the veins in the adult, is a regular one in the fœtus at an early period, and the two vessels are persistent in birds and some mammalia. The subsequent changes which take place in these vessels are the following. The communicating branch between the

221.—The Venæ Cavæ and Azygos Veins, with their Formative Branches.

THORACIC DUCT. 427

excepting those of the right side of the head, neck, and thorax, and right upper extremity, the right lung, right side of the heart, and the convex surface of the liver. It varies from eighteen to twenty inches in length in the adult, and extends from the second lumbar vertebra to the root of the neck. It commences in the abdomen by a triangular dilatation, the receptaculum chyli (reservoir or cistern of Pecquet), which is situated upon the front of the body of the second lumbar vertebra, to the right side and behind the aorta, by the side of the right crus of the Diaphragm. It ascends into the thorax through the aortic opening in the Diaphragm, and is placed in the posterior mediastinum in front of the vertebral column, lying between the aorta and vena azygos. Opposite the fourth dorsal vertebra it inclines towards the left side and ascends behind the arch of the aorta, on the left side of the œsophagus, and behind the first portion of the left subclavian artery, to the upper orifice of the thorax. Opposite the upper border of the seventh cervical vertebra it curves downwards above the subclavian artery, and in front of the Scalenus muscle, so as to form an arch; and terminates near the angle of junction of the left internal jugular and subclavian veins. The thoracic duct, at its commencement, is about equal in size to the diameter of a goose-quill, diminishes considerably in its calibre in the middle of the thorax, and is again dilated just before its termination. It is generally flexuous in its course, and constricted at intervals so as to present a varicose appearance. The thoracic duct not unfrequently divides in the middle of its course into two branches of unequal size which soon re-unite, or into several branches which form a plexiform interlacement. It occasionally bifurcates, at its upper part, into two branches, the left terminating in the usual manner, the right opening into the left subclavian vein, in connection with the right

227.—The Thoracic and Right Lymphatic Ducts.

Semmelweis believed that the extremely high levels of mortality were due to the doctors, who were examining women without handwashing, and often went from woman to woman in a maternity ward. He recommended handwashing between examinations using antiseptic, which when applied drastically reduced the deaths. This wasn't an entirely new idea – Scottish doctor Alexander Gordon had written a treatise with the same conclusion as early as 1795, for example – but it was Semmelweis who systematically investigated the problem, trying out different approaches at Vienna's General Hospital from around 1847.

By the time Semmelweis published his book in 1861, he had excellent evidence that the increase in fatalities had coincided with the introduction of regular internal examinations and that doctors washing their hands with a chlorinated solution made childbirth much safer. Sadly, Semmelweis was mocked by many in the medical profession and his ideas were initially largely rejected in continental Europe. He suffered from increasing mental health issues, dying of an infected wound in an institution in 1865. It would be decades before Semmelweis's ideas were fully adopted, though thankfully many physicians felt there was little harm to be had from the handwashing, and as a result, infection levels fell rapidly from the 1860s.

The book of the lecture

We started this chapter with a mention of one of the nineteenth century's greatest scientists, Michael Faraday. As we have seen, Faraday came from a poor family; born in London 1791, he started his working life as a bookbinder's apprentice and had a limited formal education. Perhaps because of this, although a number of books bear his name, the majority are either collections of his scientific papers or write-ups of his popular lectures from the Royal Institution. One of these, particularly worthy of note as a landmark example of early popular science, was *The Chemical History of the Candle*.

Published in 1861, this consisted of transcripts of six lectures given by Faraday in 1848, one of his contributions to the annual series of Christmas lectures for children which Faraday instituted and are still run today. Faraday's idea was to use a simple candle as a starting point to look at what is happening in the flame, the results of burning, the nature of the atmosphere that allows it to burn and so forth. Faraday even includes in the book a number of experiments to try at home – the lecture series, as the Christmas lectures always have, would have featured a wide range of exciting demonstrations. Although *The Chemical History of the Candle* was not initially conceived as a book, the approach taken in its lightness of touch and consideration of a non-technical audience would set the scene for the development of more approachable popular science titles in the twentieth century.

It's worth mentioning in parallel with Faraday's work a series of three titles by an Irish physicist who was also mainly linked with the Royal Institution: John Tyndall. Born in Leighlinbridge in 1820, Tyndall is now best remembered for his explanation of why the

Leighton Bros
FARADAY AT THE ROYAL INSTITUTION, CA. 1855

A tinted lithograph after Alexander Blaikley, showing Michael Faraday giving one of the Royal Institution Christmas lectures to an audience including Prince Albert and his son.

Michael Faraday
THE CHEMICAL HISTORY OF THE CANDLE, HARPER & BROTHERS, 1861

Pages from Faraday's 'book of the lecture series', showing the reduction in pressure as steam condenses and the collection of carbon dioxide from exhalation.

74 CONTRACTION OF STEAM WHEN CONDENSED.

take place, I will take this tin flask, which is now full of steam, and close the top. We shall see what takes place when we cause this water or steam to return back to the fluid state by pouring some cold water on the outside. [The lecturer poured the cold water over the vessel, when it immediately collapsed.] You see what

Fig. 12.

has happened. If I had closed the stopper, and still kept the heat applied to it, it would have burst the vessel; yet, when the steam returns to the state of water, the vessel collapses, there

170 PRODUCTS OF RESPIRATION.

and by means of a pipe I get my mouth over it so that I can inhale the air. By putting it

Fig. 33.

over water, in the way that you see, I am able to draw up this air (supposing the cork to be quite tight), take it into my lungs, and throw it back into the jar: we can then examine it, and see the result. You observe, I first take up the air, and then throw it back, as is evident from the ascent and descent of the water; and now, by putting a taper into the air, you will see the state in which it is by the light being extinguished. Even one inspiration, you see, has completely spoiled this air, so that it

John Tyndall
SOUND, LONGMANS, GREEN, AND CO., 1869

First published in 1867, Tyndall's *Sound*, like Faraday's books, was based on his series of lectures at the Royal Institution.

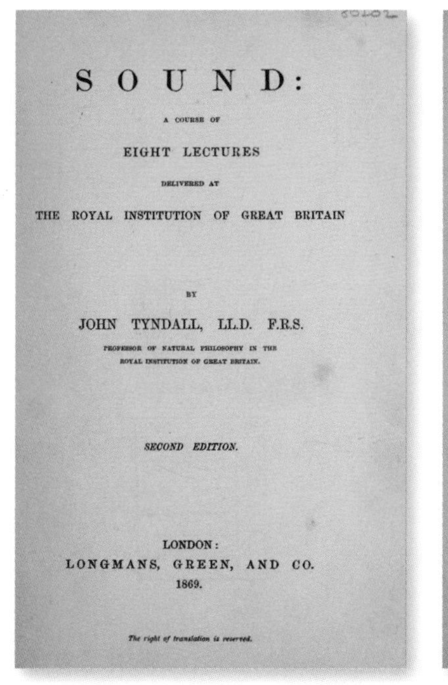

SOUND:

A COURSE OF

EIGHT LECTURES

DELIVERED AT

THE ROYAL INSTITUTION OF GREAT BRITAIN

BY

JOHN TYNDALL, LL.D. F.R.S.

PROFESSOR OF NATURAL PHILOSOPHY IN THE ROYAL INSTITUTION OF GREAT BRITAIN.

SECOND EDITION.

LONDON:
LONGMANS, GREEN, AND CO.
1869.

The right of translation is reserved.

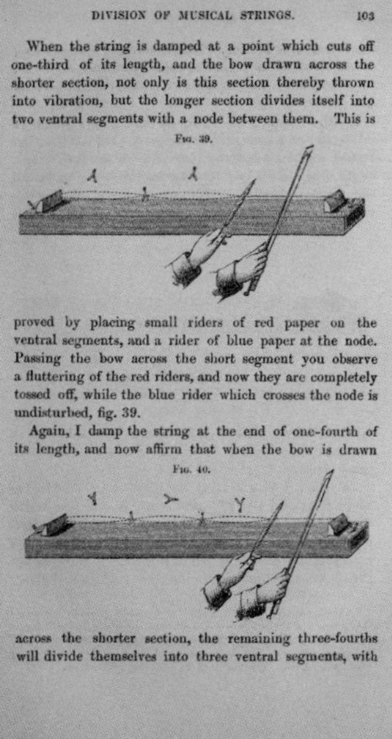

DIVISION OF MUSICAL STRINGS. 103

When the string is damped at a point which cuts off one-third of its length, and the bow drawn across the shorter section, not only is this section thereby thrown into vibration, but the longer section divides itself into two ventral segments with a node between them. This is

Fig. 39.

proved by placing small riders of red paper on the ventral segments, and a rider of blue paper at the node. Passing the bow across the short segment you observe a fluttering of the red riders, and now they are completely tossed off, while the blue rider which crosses the node is undisturbed, fig. 39.

Again, I damp the string at the end of one-fourth of its length, and now affirm that when the bow is drawn

Fig. 40.

across the shorter section, the remaining three-fourths will divide themselves into three ventral segments, with

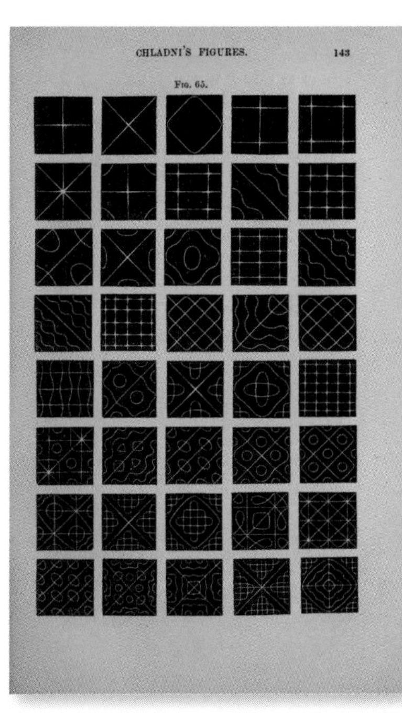

CHLADNI'S FIGURES. 143

Fig. 65.

sky is blue and for discovering the principle that would be used in fibre optics. But while Professor of Physics at the Royal Institution, Tyndall, like Faraday, was a great believer in the popularisation of science.

In three 'tutorial' books, collecting material from lectures, Tyndall covered some of the most significant basic physics topics of the day: heat, light and sound. His aim was to make these topics approachable. As he says in the introduction to *Sound: delivered in eight lectures* from 1867: 'In the following pages I have tried to render the science of Acoustics interesting to all intelligent persons, including those who do not possess any special scientific culture. . . . There is a growing desire for scientific culture throughout the civilised world. The feeling is natural, and, under the circumstances, inevitable. For a power which influences so mightily the intellectual and material action of the age, could not fail to arrest attention and challenge imagination.' *Sound* was followed by *Heat: a mode of motion* in 1868 and *Six lectures on Light* in 1873. Although aimed at the general public, in each case, Tyndall tried as much as was possible to bring out the latest scientific thinking.

The master of electromagnetism

James Clerk Maxwell
TREATISE ON ELECTRICITY AND MAGNETISM,
CLARENDON PRESS, 1873

A range of illustrations from Maxwell's definitive title on electromagnetism, including the lines of force introduced by Michael Faraday.

Michael Faraday in many ways acted as a precursor to one of the greatest physicists of the nineteenth century that few have heard of. Born in Edinburgh in 1831, James Clark Maxwell became a professor at the age of 25 and went on to have an impressive career, working both in universities and during breaks from academia in his home in Scotland. This culminated with his becoming the Cavendish Professor of Experimental Physics in Cambridge, where he set up the world-famous Cavendish Laboratory before his untimely death at the age of 48.

In his day, Maxwell was best-known for his work on statistical mechanics, explaining the actions of gasses through the statistical behaviour of molecules, as well as for work on the perception of colour, which led to him producing the colour model still used in colour televisions and computer screens today. His best-known book at the time was his *Theory of Heat*, first published in 1871. This is a textbook, but a relatively approachable one, spanning the gap between the academic and wider spheres, as the subtitle ('Adapted for the use of artisans and students in public and science schools') made clear.

However, Maxwell's greatest work, and the most significant for future generations, was on electromagnetism. It was Maxwell who identified light as an electromagnetic wave, and he who, by developing first mechanical and then mathematical models of electricity and magnetism, was able to pull the two concepts together to produce a uniform whole, describing the action of electromagnetism by mathematics that would later be simplified down to four short equations. For physicists who have grown up with it, Maxwell's mathematical modelling method was the natural approach to take – but for his contemporaries, even great physicists like his Scottish colleague Lord Kelvin, Maxwell's abstracted theories were hard to understand, which was why it would not be until after his death that their greatness was truly appreciated.

For this reason, it is the *Treatise on Electricity and Magnetism*, published in 1873, two years after *Theory of Heat*, that is now regarded as by far Maxwell's most influential piece

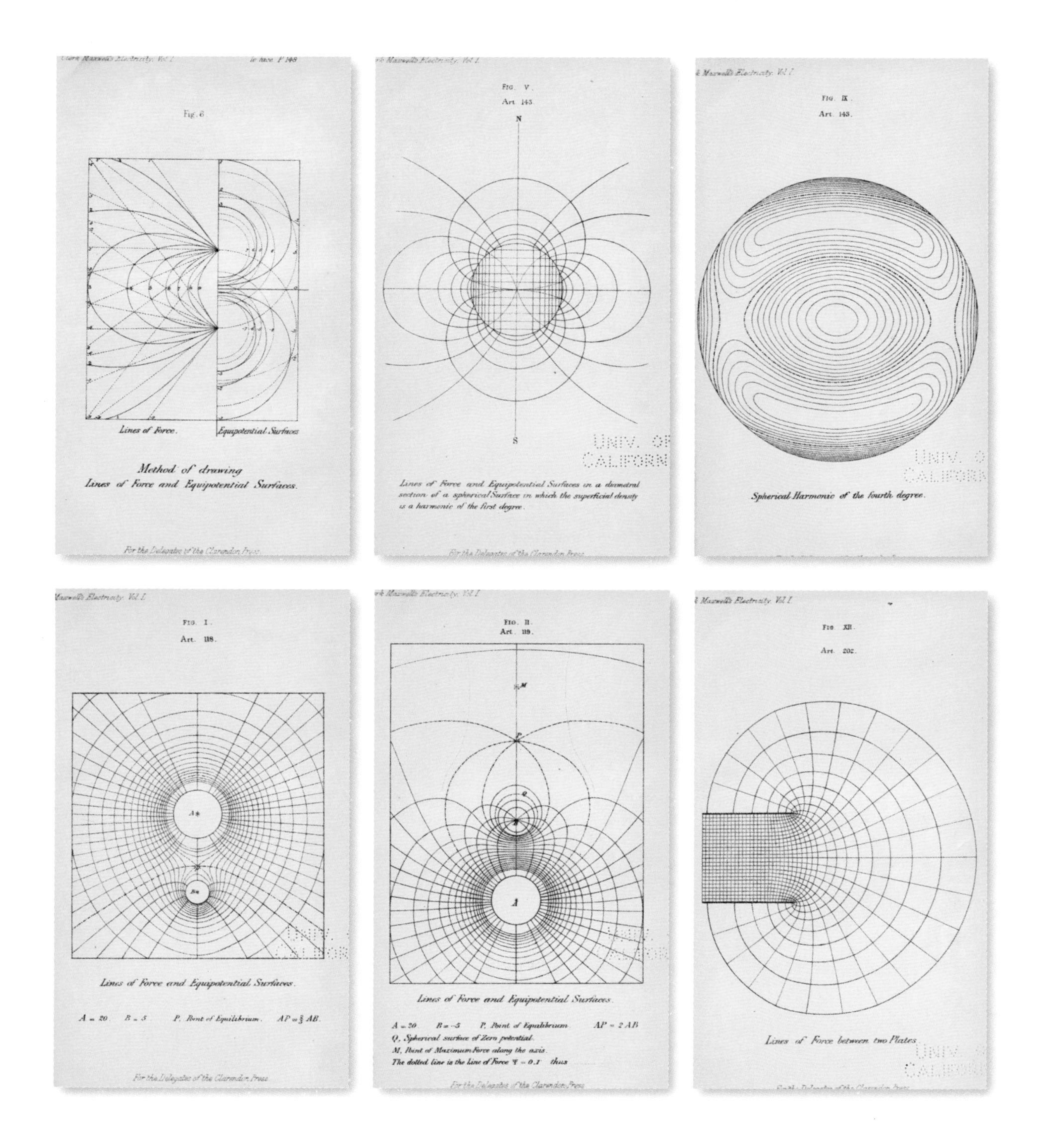
Fig. 6
Lines of Force.
Equipotential Surfaces
Method of drawing
Lines of Force and Equipotential Surfaces.
Fig. V.
Art. 143
N
S
Lines of Force and Equipotential Surfaces in a diametral section of a spherical Surface in which the superficial density is a harmonic of the first degree.
Fig. IX.
Art. 143.
Spherical Harmonic of the fourth degree.
Fig. I.
Art. 118.
Lines of Force and Equipotential Surfaces.
Fig. II.
Art. 119.
Lines of Force and Equipotential Surfaces.
Q, Spherical surface of Zero potential.
M, Point of Maximum Force along the axis.
Fig. XII.
Art. 202.
Lines of Force between two Plates.
For the Delegates of the Clarendon Press

of writing. It would continue to be used as a textbook well into the twentieth century, and was one of the reasons that Einstein said of Maxwell, 'There would be no modern physics without Maxwell's electromagnetic equations; I owe more to Maxwell than to anyone.' Echoing Einstein, American physicist Richard Feynman commented, 'From a long view of the history of mankind – seen from, say, ten thousand years from now – there can be little doubt that the most significant event of the nineteenth century will be judged as Maxwell's discovery of the laws of electrodynamics.' And the *Treatise* was Maxwell's definitive word on the subject.

At the borders of arts and sciences

With Maxwell as the zenith of nineteenth-century science writing, making mathematics central to physics in a way that had never been the case before, it's timely to take a look at a less earth-shattering title that has nonetheless had lasting, if quirky appeal. This is *Flatland: a romance of many dimensions* by the unimaginatively named Edwin Abbott Abbott. Abbott, born in London in 1838, spent the majority of his career as a school headmaster. An ordained priest, he mostly wrote on the English language, but the slim volume that is *Flatland*, published in 1884, takes us on a journey into a two-dimensional world.

Although much of *Flatland* is a satire on social classes and mores, its descriptions of the interactions of two-dimensional shapes and the implications of living in a two-dimensional world, especially when a three-dimensional object appears in it, help the reader to get a feel for the mathematician's approach to dimensions. Abbott even put a toe into the fourth dimension long before Einstein would use it in his work on relativity. To the modern reader, Abbott's writing style is on the dull side, but the ideas contained in *Flatland* are still fresh.

Just as physics and maths books were taking steps forward towards the end of the nineteenth century, biology was also moving on in the hands of a biologist who took the work of Linnaeus and Cuvier, mixed in Darwin and brought natural history up to date. This was German scientist Ernst Haeckel, born in Potsdam in 1834. Although Haeckel was very much a contemporary of Maxwell in birth, his much longer life allowed him to publish into the twentieth century.

In some ways, Haeckel typified that insult ascribed to physicist Ernest Rutherford that all science is either physics or stamp collecting. His great contribution was primarily one of cataloguing, adding thousands of species to the Linnaean structure and coining a whole range of new terms from 'ecology' to 'phylum'. Unlike his predecessors, however, Haeckel had Darwin's work to build on, so he was able to construct one of the earliest family trees (literally in the form of a tree and branches) across animal species. Probably Haeckel's best-known contribution to biological theory, which is now disregarded, was recapitulation theory. This was the idea that embryos pass through the different forms of their ancestral predecessor species before taking on the current form before birth. Nevertheless, Haeckel was hugely influential.

As well as being a scientist and philosopher, Haeckel was an impressively skilled artist, and his best-known work demonstrates not so much the breadth of his discoveries, but rather the effectiveness of his illustrations. *Kunstformen der Natur* (Artforms in Nature)

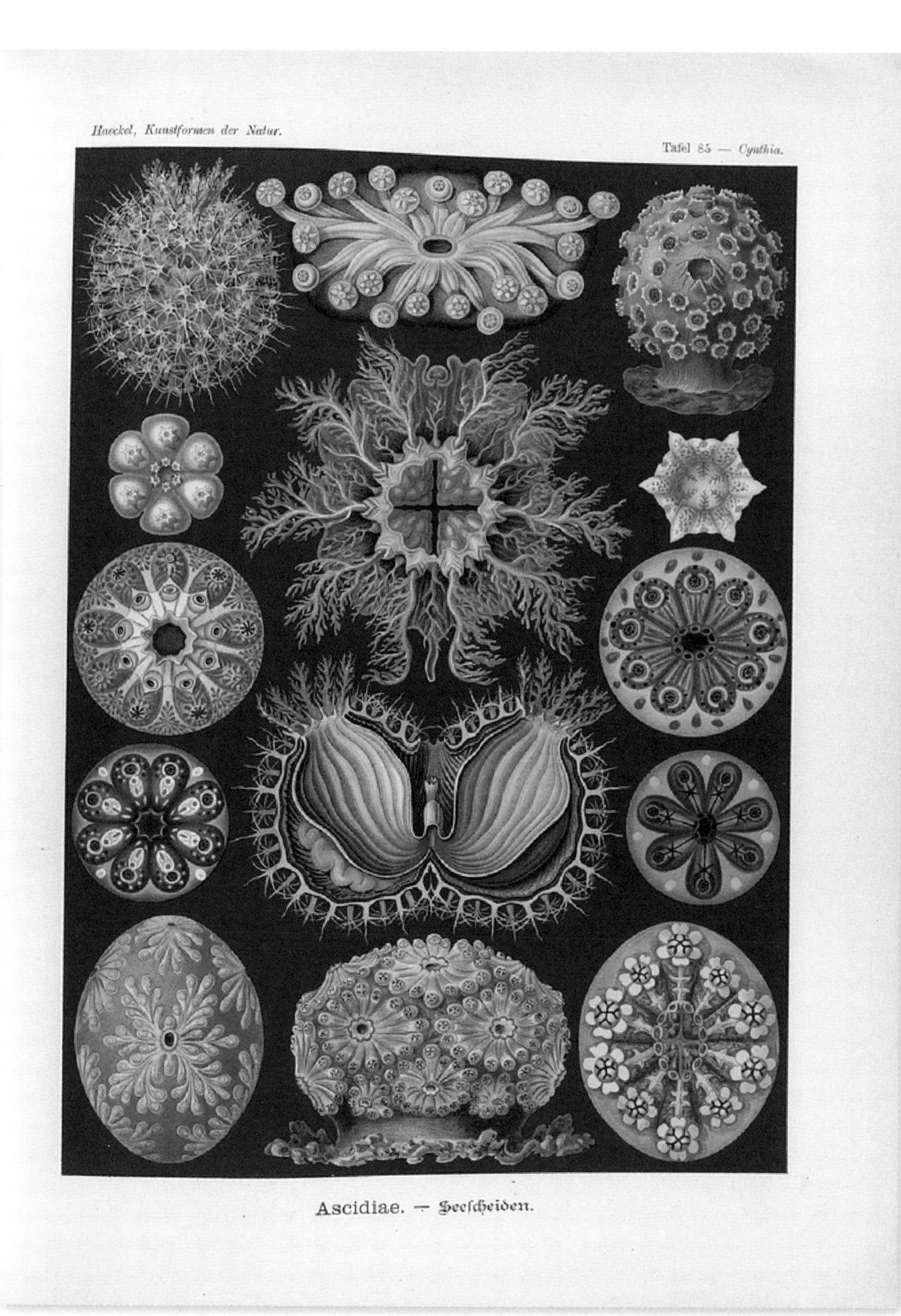

Ernst Haeckel
KUNSTFORMEN DER NATUR, BIBLIOGRAPHISCHEN INSTITUTS, 1899–1904

One of the beautiful illustrations of the marine invertebrate class Ascidiacea, known as sea squirts, from Haeckel's collection of prints.

Ernst Haeckel
KUNSTFORMEN DER NATUR,
BIBLIOGRAPHISCHEN INSTITUTS,
1899–1904

Images of organisms of the order Desmidiales, a type of green algae from which land plants developed (left), and Acantharea, radiolarian protozoa with hard mineral skeletons (right).

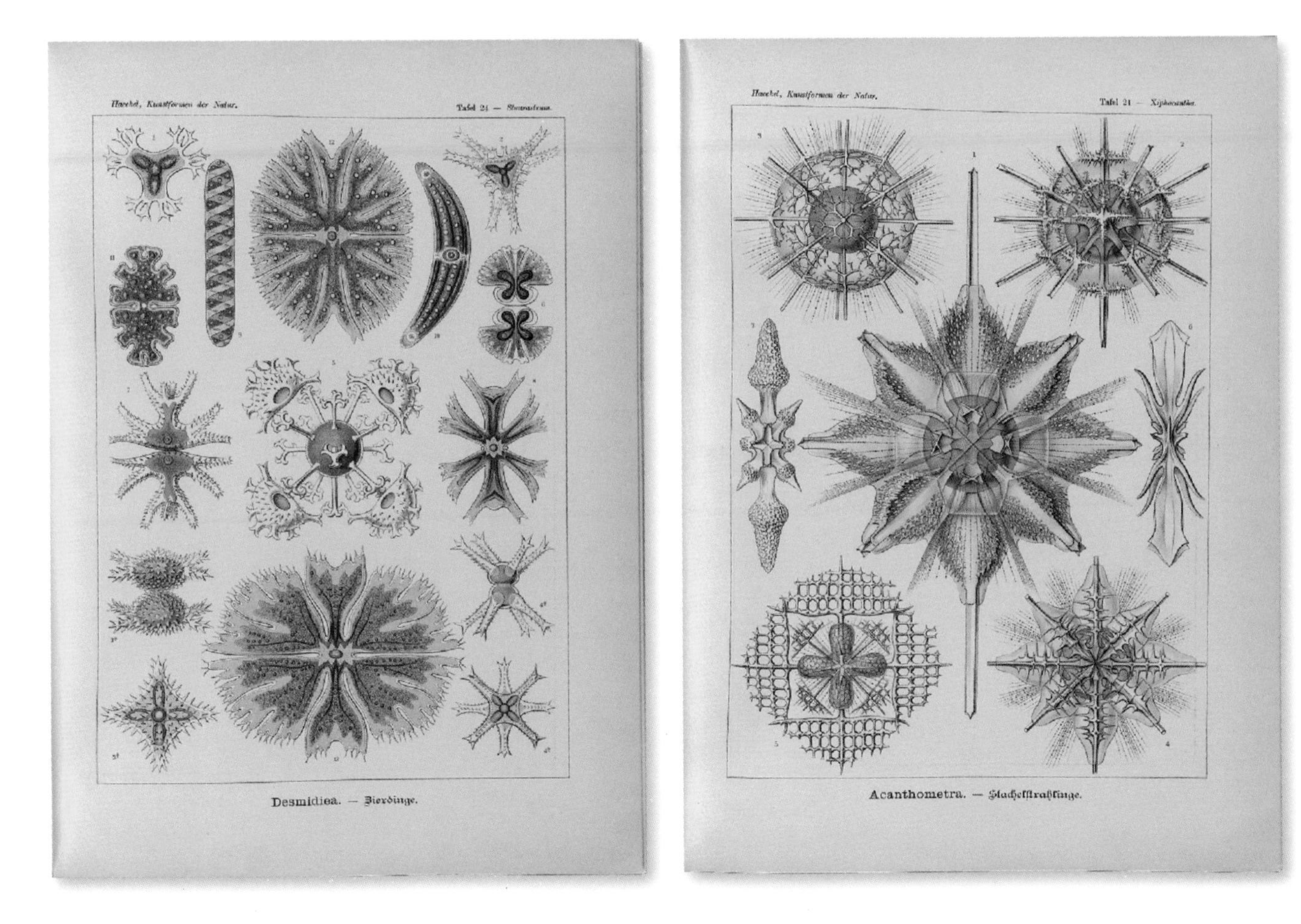

Ernst Haeckel
KUNSTFORMEN DER NATUR,
BIBLIOGRAPHISCHEN INSTITUTS,
1899–1904

More images of Radiolaria (left), showing their jewel-like structures and of Blastoidea, extinct sea animals similar to sea urchins (right).

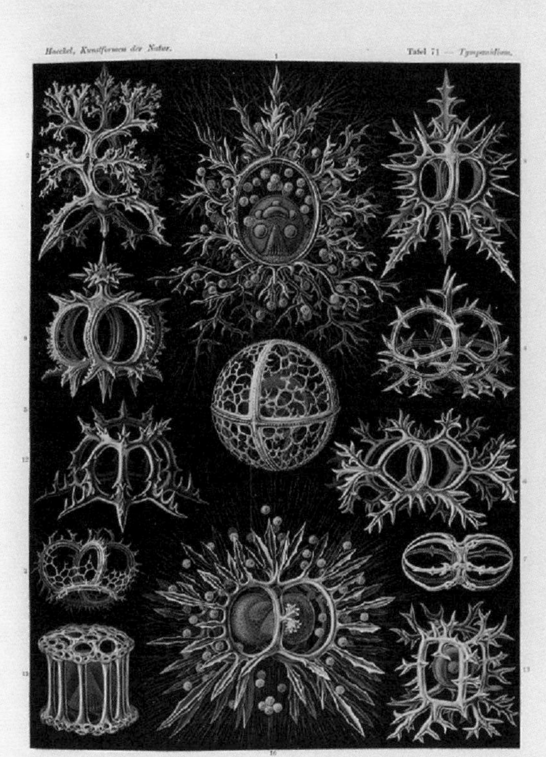

JEAN-HENRI FABRE, CA. 1900

A portrait by a French photographer of Fabre at work.

started off as sets of prints published from 1899, ending up in book form in 1904. It contained just 100 images, devised and structured to put across his ideas on the relationships between organisms and the importance of structural symmetry in species. Many of the prints were coloured, and the images, which typically showed a collection of related organisms, function as much as works of art as they do as exemplars of natural history. A number of artists and fabric designers of the period were strongly influenced by the book.

It's interesting to contrast Haeckel's highly visual approach with that of the French entomologist Jean-Henri Fabre, whose books on insects were also highly popular, not so much for the pictures as for his engaging writing style. Fabre, born in Saint-Léons in 1823, taught himself entomology and, without anthropomorphism, managed to give his descriptions of insects and their lives a kind of biographical style that made them highly approachable to a wider audience. He wrote a wide range of books, but his best-known was *Souvenirs Entomologiques* (Entomological Memories), first published in 1879 and continuing to be updated for two decades.

The last two decades of the nineteenth century, tipping into the twentieth century, also saw another example of illustrations from nature being used both scientifically and as works of art – but the big change here was that the illustrations were not reproductions of sketches and paintings, but photographs. The photographer in question was the remarkable Englishman, Eadweard Muybridge.

Born Edward Muggeridge in Kingston-upon-Thames in 1830, Muybridge spent most of his working life in America, first in San Francisco and later in Philadelphia. He started

Eadweard Muybridge
ANIMAL LOCOMOTION,
UNIVERSITY OF
PENNSYLVANIA, 1887

Boxers and a bucking mule from Muybridge's stop-motion images; when the boxers were shown animated at the Royal Institution, the Prince of Wales was reportedly 'delighted'.

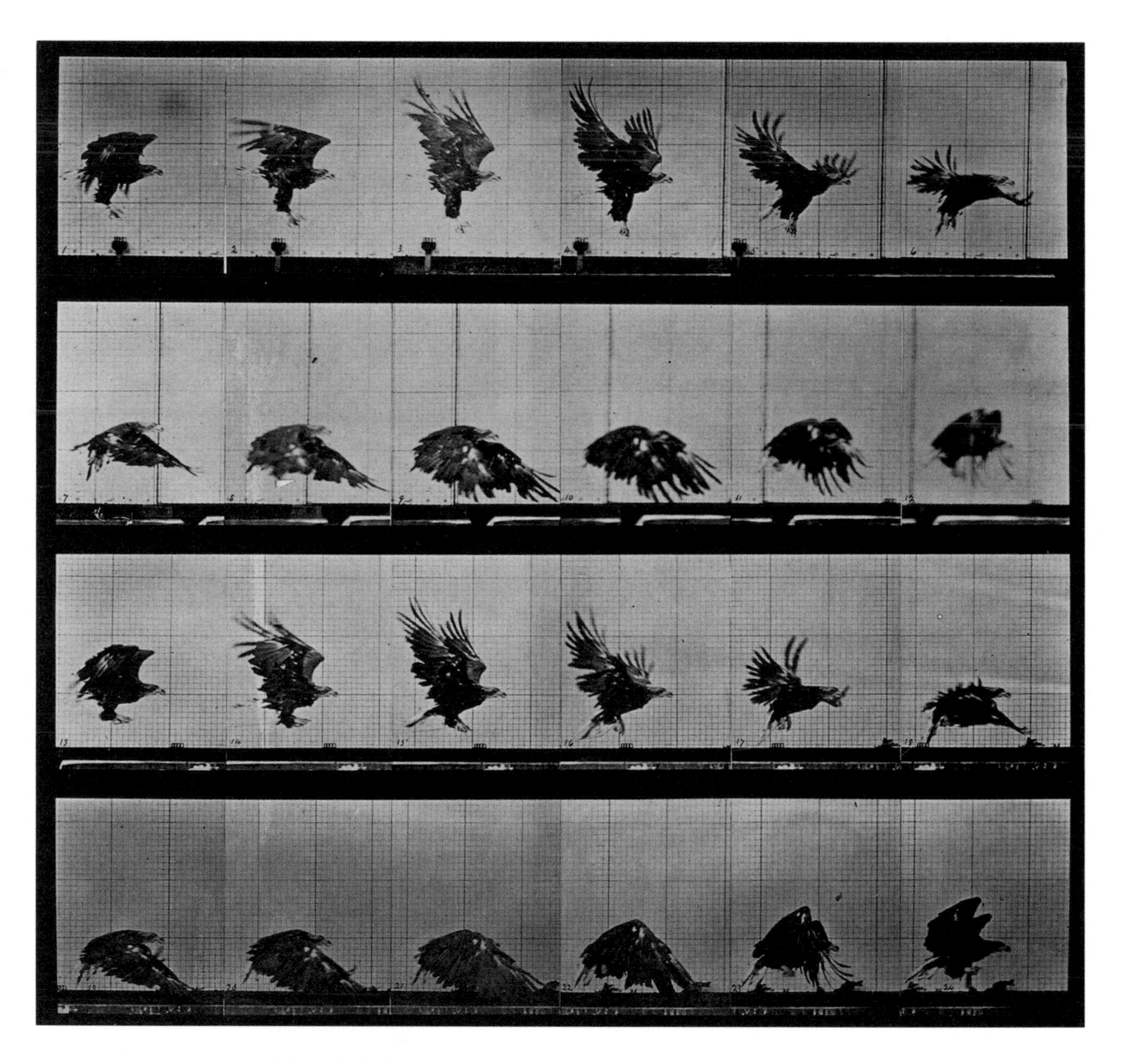

Eadweard Muybridge
ANIMAL LOCOMOTION,
UNIVERSITY OF
PENNSYLVANIA, 1887

Eagle in flight, from a range of sequences taken by Muybridge at Philadelphia Zoological Gardens using a 5-centimetre (2-inch) grid against a white cloth backdrop.

out as a landscape photographer, acquired notoriety when he murdered his wife's lover (the jury found him not guilty as they considered this an appropriate response) and came to photographic and scientific fame when he devised a mechanism for taking a series of photographs in rapid succession – using a battery of cameras – which enabled him to analyse the motion of first horses and then other animals. In book form, Muybridge's images were presented as a series of shots forming a now-iconic look in the history of photography, but he also put his series of photographs onto spinning discs, which were projected to provide very short sequences of moving pictures, amongst the first movies ever seen.

Muybridge produced a vast number of photographic series of everything from wild animals to the beating heart of a turtle and of humans carrying out various tasks (often naked to show their musculature). Rather like Haeckel, he started off by selling sets of prints, leading up to the production of what was, in some ways, his masterpiece, *Animal Locomotion*. This was not a book, but rather a collection of photograph prints with a catalogue and descriptive introductory text. It comprised 781 48 x 61-cm (19 x 24-inch) photographs, each of which could have as many as 24 images within it. The full collection sold mostly to institutions at a hefty $600, while there were also sets of 100 prints in a leather portfolio available for $100.

The reason *Animal Locomotion* was not published as a book was that in 1887, the technology to reproduce photographs in a book was in its infancy. However, Muybridge's work had already found its way into book form, in a manner that put his whole career at risk. In 1882, while in London on a successful tour with his moving pictures, he had been due to give a talk at the Royal Society when he was summoned to account for himself and shown a book called *The Horse in Motion* by a J. D. B. Stillman. This contained engravings based on his photographs – but Muybridge himself was not mentioned. The book had been put out on behalf of Muybridge's former sponsor, railroad magnate and founder of Stanford University, Leland Stanford. The way that the book did not consider Muybridge worthy of mention seemed to indicate to the Royal Society's committee that Muybridge had been nothing more than a technician, rather than the inventor of this technology. Muybridge sued Stanford with little success, but luckily the Stillman book did not sell well, and when Muybridge went on to produce vastly superior images from his work at the University of Pennsylvania, his expertise was recognised.

By the end of the century, technology had advanced, and Muybridge was able to publish two books that brought his works to a much wider public. These were *Animals in Motion* and *The Human Figure in Motion*, published in the UK in 1899 and 1901 respectively. The photographic quality might not have been as great as that in *Animal Locomotion*, but the book form brought the images and text covering the technology and the scientific value of the images to a much wider audience. As the *Human Figure* book contained a considerable amount of nudity, one review, in *The Graphic* in December 1901, thought it wise to warn, 'It will be understood that the volume, not being intended virginibus puerisque [literally: for virgins and for boys] (unless they be full-fledged students) should not be left on the drawing room table.'

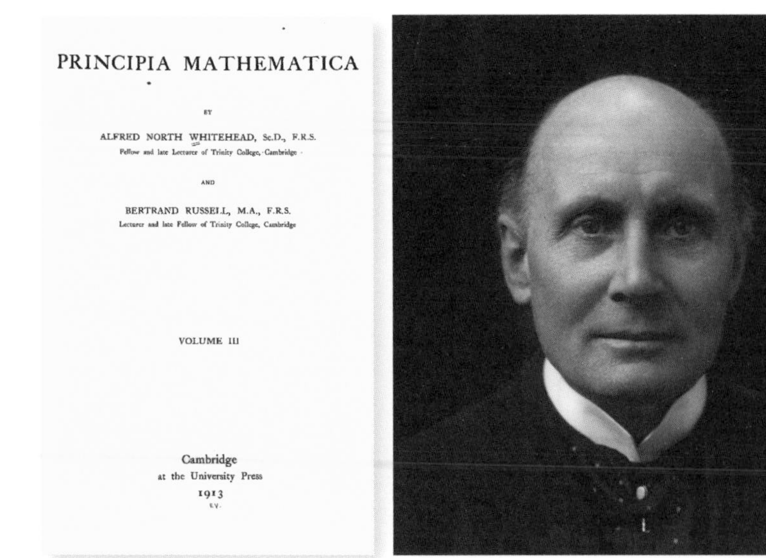

PRINCIPIA MATHEMATICA

BY

ALFRED NORTH WHITEHEAD, Sc.D., F.R.S.
Fellow and late Lecturer of Trinity College, Cambridge

AND

BERTRAND RUSSELL, M.A., F.R.S.
Lecturer and late Fellow of Trinity College, Cambridge

VOLUME III

Cambridge
at the University Press
1913

Alfred North Whitehead and Bertrand Russell
PRINCIPA MATHEMATICA, CAMBRIDGE UNIVERSITY PRESS, 1913

The title page of the third volume of Whitehead and Russell's mathematical tour de force, alongisde portraits of Whitehead ca. 1925 (left) and Russell in 1927 from *Vanity Fair* magazine (right).

Uncovering the foundations of mathematics and continents

We have seen through the nineteenth century a gradual move from science books written purely as a means of communication between natural philosophers to those serving as more of a bridge between the science community and the general public. Although the public may not have appreciated the biological message in *Kunstformen* or the technical wizardry of Muybridge's photography, there was no doubt that they could enjoy the artistry. This did not mean, though, that books that could only be of interest to a particular field had entirely disappeared.

There were, of course, still textbooks, from those for school-age students up to postgraduate works. However, the textbook has always been a very specialist form. And though the majority of scientists exchanged information with their peers through the form of a scientific paper published in a journal, sometimes they would set off on an endeavour of sufficient depth that it required a book to get their ideas across.

Such works included *Grundlagen der Geometrie* (Foundations of Geometry) from 1899 by the German mathematician David Hilbert, which set modern starting points for Euclidean geometry, and, more notably, the three-volume *Principia Mathematica* (Principles of Mathematics) by mathematician Alfred North Whitehead and philosopher Bertrand Russell, published between 1910 and 1913. The latter book, despite its Latin title (no doubt a nod to Newton), was written in English – or more realistically in mathematics, as the purpose was to build the structure of mathematics from its most basic form, starting from a set of simple assumptions (axioms) in symbolic logic and constructing as much of the hierarchy of mathematics from there as was possible.

Perhaps the best-known aspect of this weighty book was the appearance, on page 379 of the original edition, of the words 'From this proposition it will follow, when

DIE WISSENSCHAFT
EINZELDARSTELLUNGEN AUS DER NATUR-
WISSENSCHAFT UND DER TECHNIK · BD. 66
A. WEGENER
DIE ENTSTEHUNG DER
KONTINENTE UND OZEANE
FRIEDR. VIEWEG & SOHN BRAUNSCHWEIG

Alfred Wegener
DIE ENTSTEHUNG DER KONTINENTE UND OZEANE, FRIEDRICH VIEWEG & SOHN, 1920

The cover of the second edition of Wegener's (literally) groundbreaking title setting out the concept of continental drift; his 'tectonic plate' theory was not accepted in his lifetime. The first edition was published in 1915.

arithmetical addition has been defined, that 1 + 1 = 2.' It took 379 pages to get that far. Given this, by the end of the third volume the authors, not surprisingly, had to admit defeat, realising they could only cover a tiny fraction of the mathematics of the time. Ironically, by 1931 the Austrian mathematician Kurt Gödel had proved that any system of mathematics must either be inconsistent or incomplete – such a book could never in fact be written for the whole of mathematics. Nonetheless, *Principia Mathematica* was a landmark book in the philosophy of mathematics.

Alfred Wegener
DIE ENTSTEHUNG DER KONTINENTE UND OZEANE, FRIEDRICH VIEWEG & SOHN, 1920

Images, as they were orientated, from the second edition of Wegner's book. They show his concept of how the continents had originally been positioned against each other and drifted apart.

Published two years after the final volume of *Principia Mathematica*, a title by German geophysicist and meteorologist Alfred Wegener is particularly interesting for being so far ahead of its time. Wegener, born in Berlin in 1880, set out a theory in his *Die Entstehung der Kontinente und Ozeane* (The Origin of Continents and Oceans) that would not be widely accepted until at least 20 years after his death in 1930. In a way, it's not surprising. The central idea of *Die Entstehung der Kontinente und Ozeane*, first published in 1915, was a remarkable one: that the apparently solid surface of the Earth in fact consisted of great plates of rock that were moving with respect to each other.

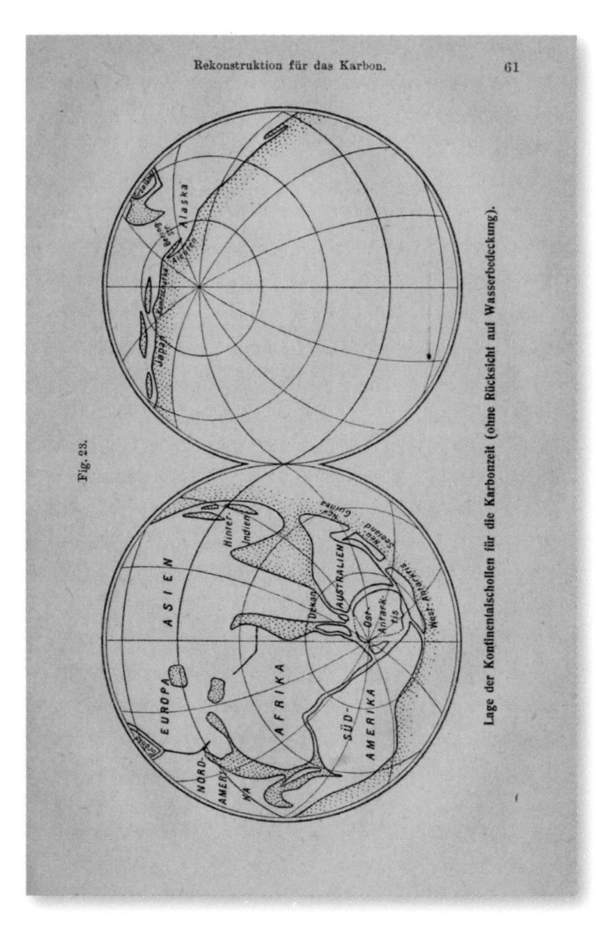

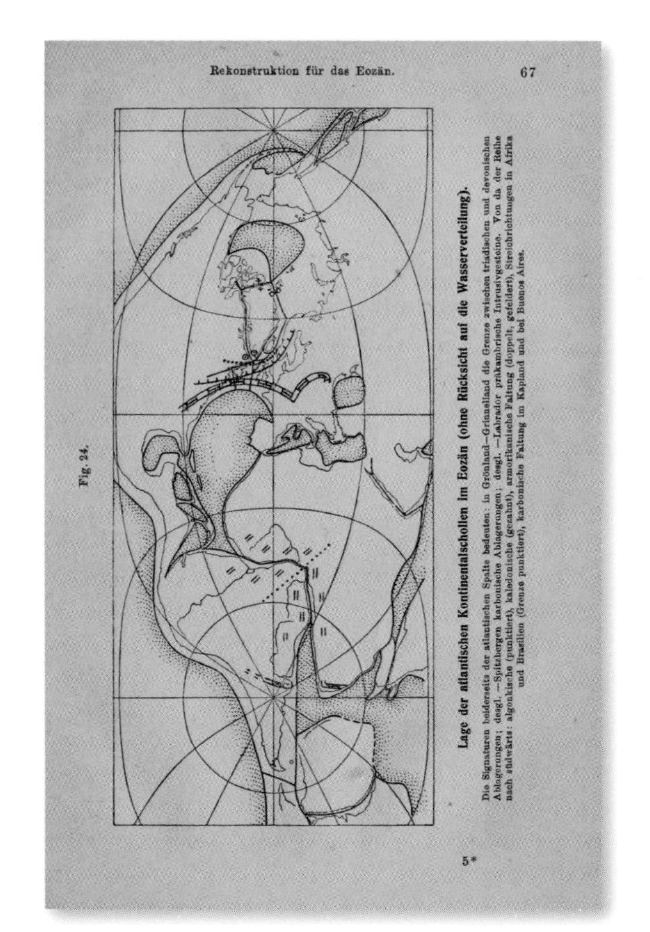

The starting point of Wegener's great idea was the way that different landmasses, such as the Americas set alongside Africa and Europe, seemed to fit together like a jigsaw puzzle that had been split apart. He also noted similarities in the fossil record between continents now separated by ocean, as if these landmasses had once been contiguous. Wegener proposed that the continents very gradually moved relative to each other, so that over geological time, new continental structures could form and break up.

There were many reasons why Wegener's theory was not accepted during his lifetime. Wegener was better known at the time for his work in meteorology and Greenland expeditions than geology – he died as supplies ran out under intense Arctic weather on the last of his expeditions. And it didn't help that he could not come up with a convincing mechanism for the movements of the Earth's surface on such a vast scale. He also overestimated the rate of continental drift by a factor of 100, which not surprisingly resulted in a considerable degree of scepticism. Even so, his book would come to be regarded as a posthumous masterpiece when it was realised how well his theory matched a growing understanding of the way the Earth operated.

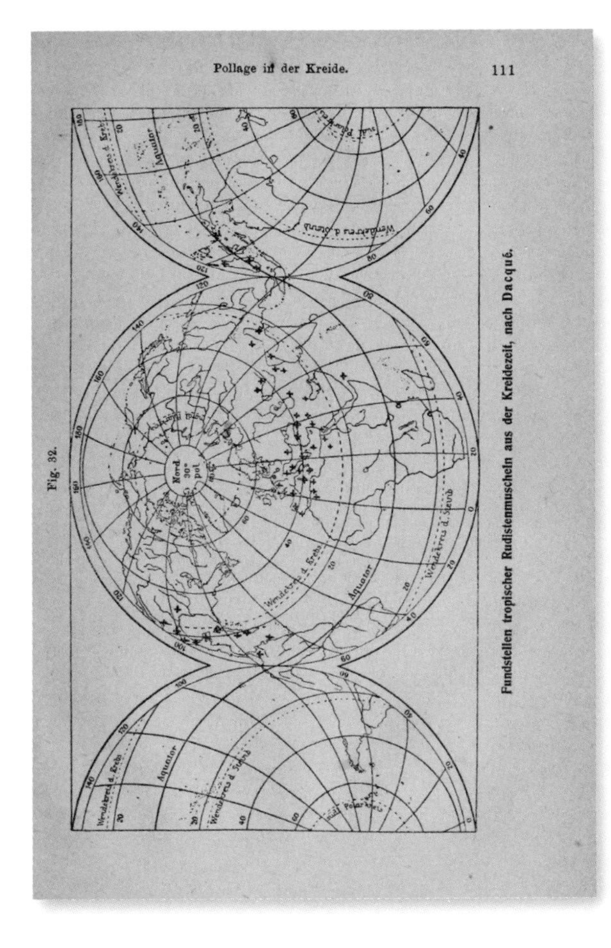

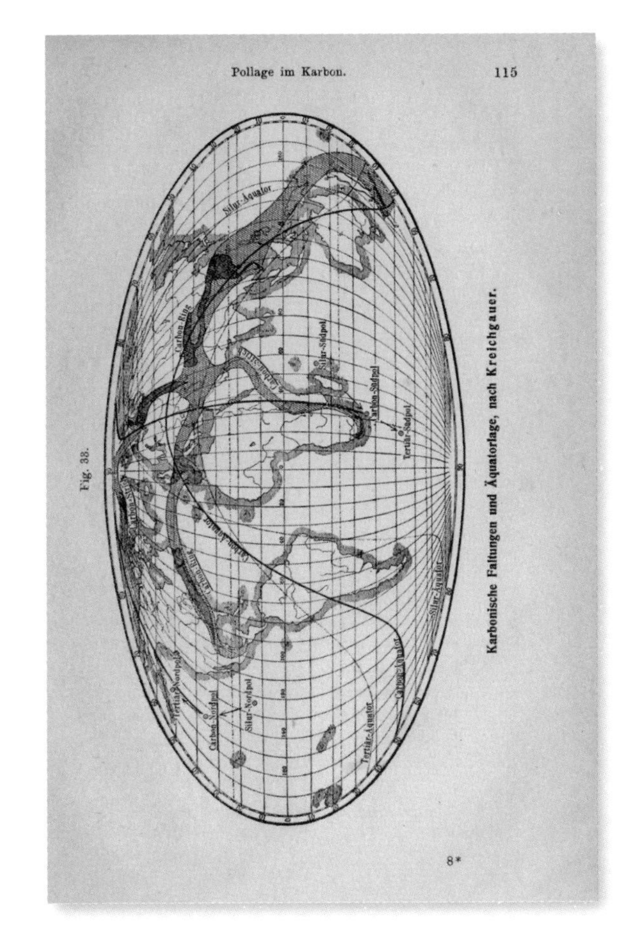

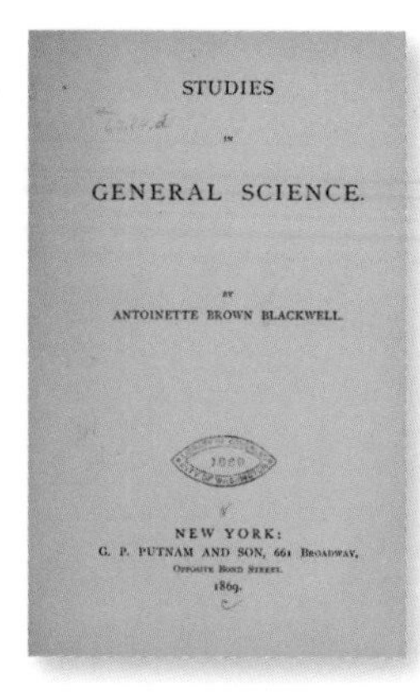
STUDIES

IN

GENERAL SCIENCE.

BY

ANTOINETTE BROWN BLACKWELL.

NEW YORK:

G. P. PUTNAM AND SON, 661 BROADWAY,

OPPOSITE BOND STREET.

1869.

Antoinette Brown Blackwell
STUDIES IN GENERAL SCIENCE, G. P. PUTNAM AND SON, 1869

The title page of Brown Blackwell's book, which received a positive response from Darwin who (somehow) assumed the author was male.

The absence of women

Before concluding this chapter, we need to revisit the place of women in science and science writing. A slow revolution was under way during the nineteenth century, but it's notable that there were still very few books by women on science written during this period. We don't have to look further than the words of one of the heroes of the nineteenth century, Charles Darwin, to see why. In his books, notably *The Descent of Man*, Darwin's opinions aligned with the widely accepted view of the period that women were intellectually inferior (despite Darwin himself getting considerable assistance in his work from female members of the family). He wrote, 'If two lists were made of the most eminent men and women in poetry, painting, sculpture, music . . . history, science and philosophy . . . the two lists would not bear comparison.'

This view was countered by American author Antoinette Brown Blackwell. She had sent a copy of her first book, *Studies in General Science*, to Darwin and received a very positive response, but noticed that Darwin had assumed she was male. She would respond by writing a book called *The Sexes Throughout Nature* in 1875, which directly challenged the idea of female inferiority. Darwin's opinion of this is not recorded.

There has been some effort to rescue Darwin's reputation by noting the considerable correspondence he had with women who he seemed to treat as intellectual equals, but this rather sidesteps his damning correspondence with another American woman, Caroline Kennard. She wrote to Darwin in 1881, shocked to discover that Darwin was being cited as a source of scientific principles demonstrating the inferiority of women. Mrs Kennard asked him, 'If a mistake has been made, the great weight of your opinion and authority should be righted.' She was clearly expecting Darwin to support the concept of equality. Unfortunately, he replied, 'I certainly think that women though generally superior to men [in] moral qualities are inferior intellectually, and there seems to me to be a great difficulty from the laws of inheritance, (if I understand those laws rightly) in their becoming intellectual equals of man.' Darwin's response seemed to have been based on the idea that men had been forced to evolve more due to their traditional roles, while women had been unable to keep up, despite inheriting some positive traits from the male line.

Darwin also argued that for women to be equal they would have to become breadwinners, potentially damaging their children and the happiness of their households. Mrs Kennard wrote back an angry response, pointing out that in the majority of (lower-class) households women *did* work, but were prevented from doing anything other than menial jobs, and any perceived inferiority was due to their environment and restraints on what they were allowed to do, not their ability.

Nonetheless, despite the attitudes of Darwin and others, towards the end of the nineteenth century, science was gradually changing, and barriers were being broken down. Women were being admitted to study sciences at an increasing number of universities around the world and would begin to contribute to science writing. It was a slow process, and many of the great scientists of the time proved conservative. For example, when James Clerk Maxwell opened the Cavendish Laboratory for experimental physics at Cambridge in 1874, he was reluctant to allow women into the building.

Ironically, Maxwell's wife Katherine had been actively involved in some of her husband's experimental work, but it seems that having women as undergraduates was one step too far for Maxwell. Later in the 1870s he relented and allowed women into the laboratories, though his assistant William Garnett noted, 'At last [Maxwell] gave permission to admit women during the Long Vacation, when he was in Scotland, and I had a class who were determined to go through a complete course of electrical measurements during the few weeks when the laboratory was open to them.'

By the start of the twentieth century, the first female professional scientists became established. Famously, Marie Curie (see page 194) became the first woman to win a Nobel Prize in a science subject in 1903 with the Physics prize, and she went on to take the Chemistry prize in 1911. (Less impressively, it's worth noting that by 2018 there have been only two other female winners of the Nobel Prize in Physics: Maria Goeppert-Mayer in 1963, and Donna Strickland in 2018.) Progress was slow, however, and if anything, science publishing would prove even slower to give opportunities to women writers, with only a handful of notable publications appearing before the 1970s.

Despite this glacial advance in gender equality, the twentieth century would bring both a revolution in science and in the way that science books were written.

DOMESTIC CHEMISTRY CLASS, 1907

Female students at work in a laboratory at Battersea Polytechnic, London.

4

POST-CLASSICAL

THE WORLD TURNED UPSIDE DOWN

In many areas of science, notably physics and biology, the close of the nineteenth century marks the end of what's usually called the 'classical period'. Physics before the twentieth century could be seen as a constructive process, building from the Renaissance and particularly making use of the work of Galileo and Newton. But in the twentieth century, the terrible twins of relativity and quantum theory changed the outlook on everything.

Relativity showed that concepts that were previously thought of as fixed and independent – notably time and space – were interwoven and impossible to separate. Apparently common-sense ideas such as two events being simultaneous no longer had any meaning. And gravity moved on from Newton's mysterious action at a distance to become a more logical but mind-bending warping of space and time. Equally, quantum physics showed that light could behave both as a wave and as a particle, that quantum particles had no location when not interacting with their surroundings, and that probability was essential for the operation of the universe. Reality could no longer be considered the clockwork mechanism of Newton's vision.

At the same time, biology was transformed from being little more than a process of cataloguing species, anatomies and behaviours to a complete science. The process started with evolution, but the driving force in the twentieth century was genetics – a gradual understanding of the mechanisms of the genome that would lead to the discovery of the twin helix structure of DNA – and the incorporation of more and more chemistry into biology. Molecular biology, the understanding of biological process at the level of individual molecules and the remarkable molecular machines found in cells, has become a major component of both general biological studies and medicine.

These changes were, of course, represented in the content of science books of the period, but as we saw in the previous chapter, this was also the time when science books written for the public began to dominate.

The nature of reality

Some of the earliest twentieth-century physics titles were written by two of the best-known scientists in history – Albert Einstein and Marie Curie. Curie, born Maria Skłodowska in Warsaw, Poland, in 1867, was not only one of the very few women to be a science Nobel laureate, but also one of the even rarer individuals to win two Nobel Prizes: Physics in 1903 and Chemistry in 1911. The latter achievement came the year after publishing her book *Traité de Radioactivité* (Treatise on Radioactivity).

The whole subject of radioactivity had moved on quickly since the phenomenon was discovered by Henri Becquerel (who shared the 1903 Nobel Prize with Curie and her husband) in the 1890s. The progress can be seen in the difference between Curie's doctoral thesis on the subject from 1903, which ran to just 142 pages, and her *Traité*, containing nearly 1,000 pages in two volumes. As an attempt to pull together all that was known on the topic, it was widely read in the specialist field, but made little impact outside it. Things would be very different when it came to anything written by the second of these big names of early twentieth century science, Albert Einstein.

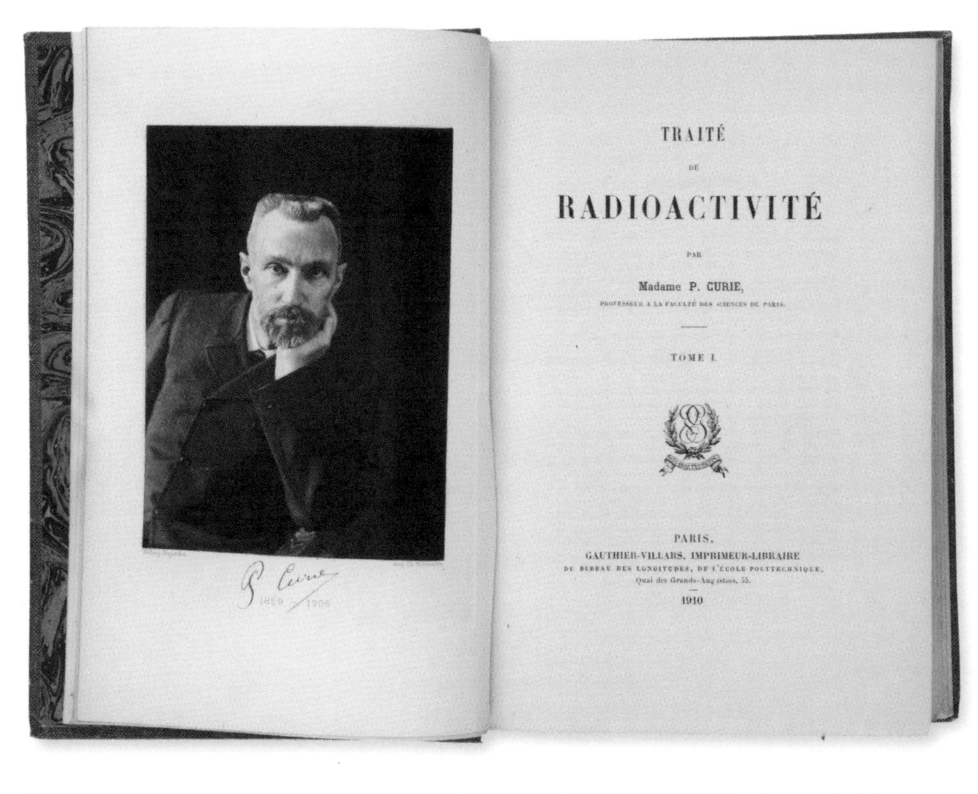

TRAITÉ

DE

RADIOACTIVITÉ

PAR

Madame P. CURIE,

PROFESSEUR À LA FACULTÉ DES SCIENCES DE PARIS.

TOME I.

PARIS,

GAUTHIER-VILLARS, IMPRIMEUR-LIBRAIRE

DU BUREAU DES LONGITUDES, DE L'ÉCOLE POLYTECHNIQUE,

Quai des Grands-Augustins, 55.

1910

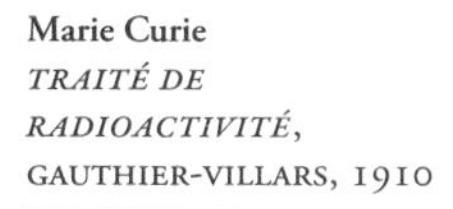

Marie Curie
TRAITÉ DE RADIOACTIVITÉ,
GAUTHIER-VILLARS, 1910

The frontispiece and title page of volume 1 (of 2) of Marie Curie's book, rather oddly featuring a picture of her husband Pierre.

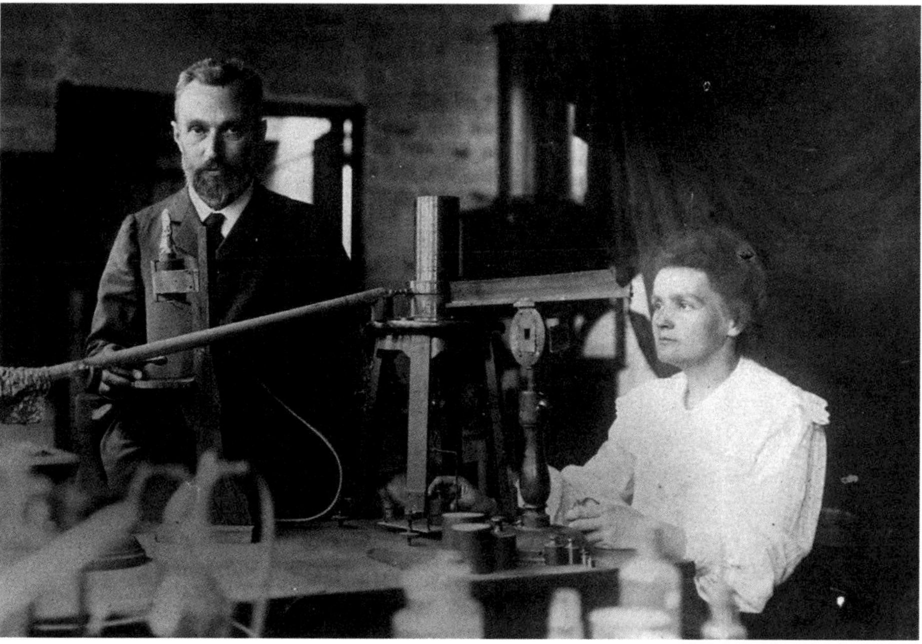

MARIE AND PIERRE CURIE, 1904

The Curies using equipment in their laboratory on Rue Cuvier in Paris.

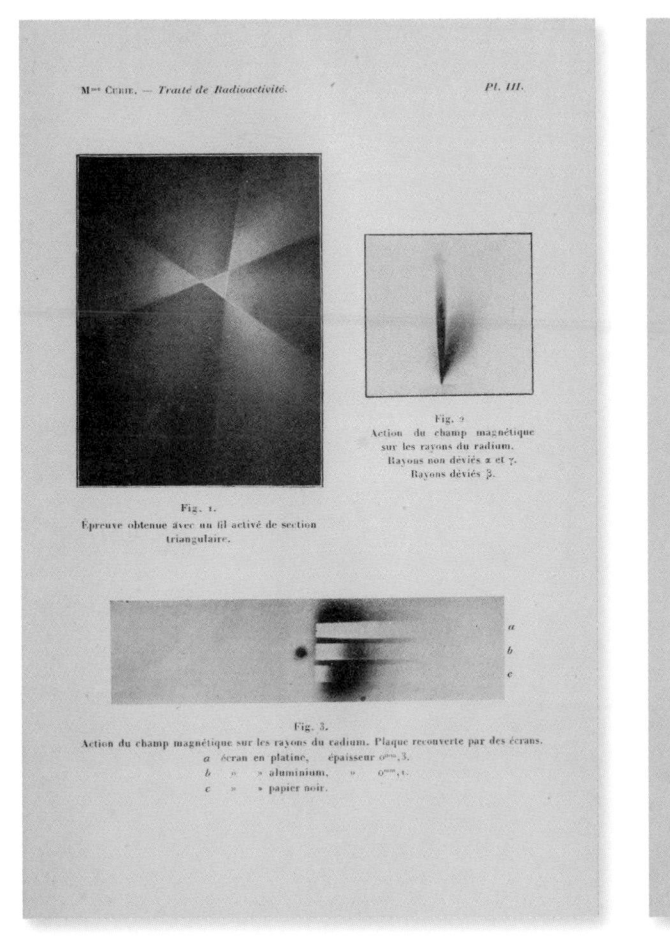

Fig. 1.
Épreuve obtenue avec un fil activé de section triangulaire.

Fig. 2
Action du champ magnétique sur les rayons du radium.
Rayons non déviés α et γ.
Rayons déviés β.

Fig. 3.
Action du champ magnétique sur les rayons du radium. Plaque recouverte par des écrans.
a écran en platine, épaisseur 0mm,3.
b » » aluminium, » 0mm,1.
c » » papier noir.

Mme Curie. — *Traité de Radioactivité.* *Pl. IV.*

Fig. 1.
Dispositif de Becquerel pour isoler des rayons β simples.

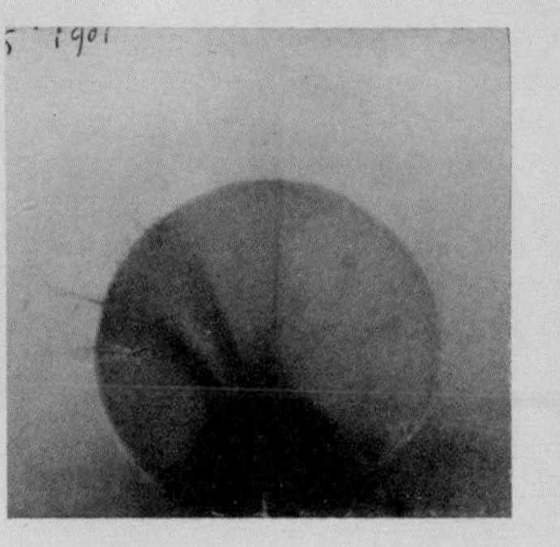

Fig. 2.
Épreuve obtenue avec le dispositif de la figure 1.

Fig. 3.
Épreuve relative à la mesure du rapport $\frac{e}{m}$ par les rayons β du radium (Kaufman).

Fig. 4.
Déviation magnétique des rayons de l'uranium.

Fig. 5.
Passage de rayons β simples au travers d'une couche de paraffine (épaisseur 2mm).

Marie Curie
TRAITÉ DE RADIOACTIVITÉ,
GAUTHIER-VILLARS, 1910

Plates from volume 1 (of 2) showing the action of magnetic fields on the 'radium rays' – alpha particles produced by radium decaying to radon.

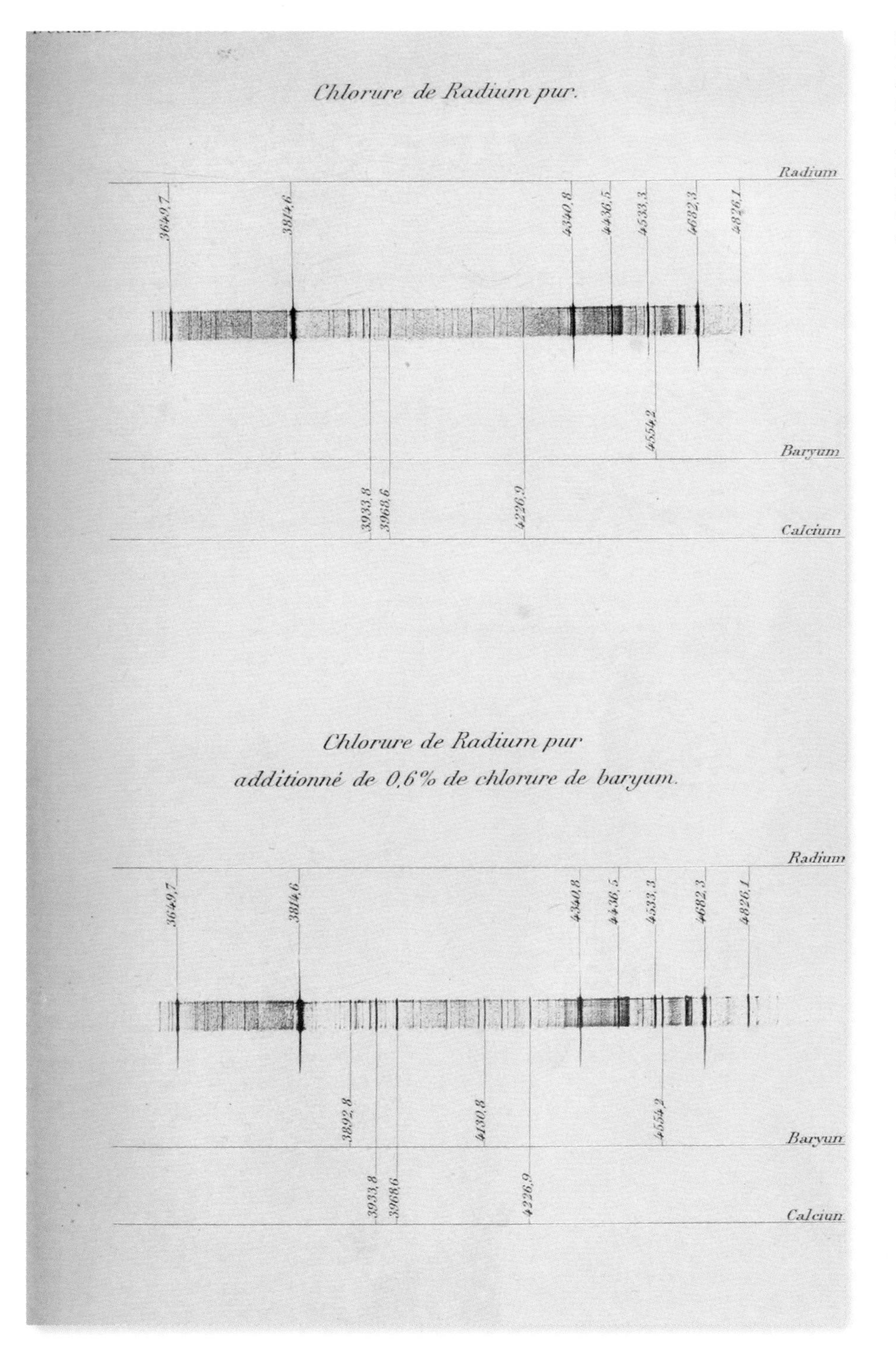

Marie Curie
TRAITÉ DE RADIOACTIVITÉ,
GAUTHIER-VILLARS, 1910

A plate from volume 2 showing spectra from radium chloride, and radium chloride with a small barium chloride impurity.

Some modern scientists who are very visible in the media are primarily communicators who achieved very little in their field before becoming well-known, but Einstein's fame was entirely justified. Born in the German city of Ulm in 1879, Einstein was undistinguished academically until the year of 1905. At the time, he was working in the Swiss patent office in Bern, having failed to get an academic post. But in that one year he published four major papers, one of which would win him the Nobel Prize in Physics in 1921.

In that first burst of output Einstein established the size of molecules (as a result providing evidence for the existence of atoms), helped lay the foundations of quantum physics, described the linkage of space and time in the special theory of relativity and showed that $E=mc^2$. During the following 12 years he worked on his masterpiece, the general theory of relativity, showing how matter warped space and time to produce gravity, and also predicted the existence of gravitational waves and developed the theory that led to the invention of the laser.

Einstein wrote a number of textbooks, but by this stage his scientific breakthroughs were communicated to other scientists via academic papers. However, he also produced a number of books for the general reader, most notably *Über die Spezielle und die Allgemeine Relativitätstheorie, Gemeinverständlich* (On the Special and General Theories of Relativity, A Popular Exposition), published in 1917 and translated into English in 1920.

There was an astonishing public appetite for Einstein's theories. His ideas made front-page headline news in national newspapers around the world; when he turned up

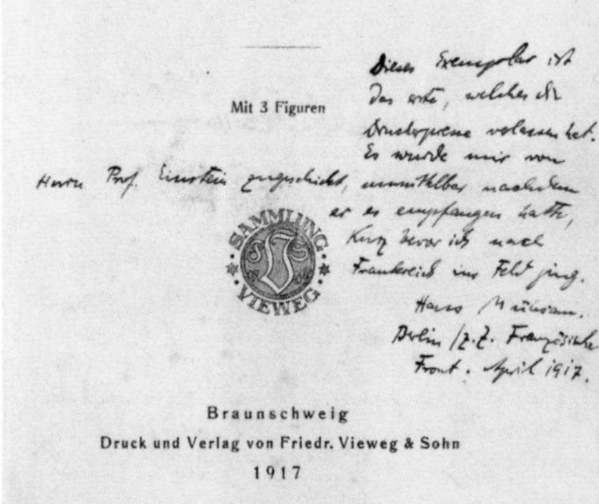

Über die spezielle und die
allgemeine Relativitätstheorie
(Gemeinverständlich)

Von
A. EINSTEIN

Mit 3 Figuren

Braunschweig
Druck und Verlag von Friedr. Vieweg & Sohn
1917

RELATIVITY
THE SPECIAL AND THE GENERAL THEORY

BY PROFESSOR ALBERT EINSTEIN, PH.D., LL.D.

Translated by Professor Robert W. Lawson, M.Sc.

EINSTEIN'S own explanation of his famous discovery—

THE Einstein law has been accepted by astronomers and physicists as an epoch-making discovery. Up to the present Newton's law of gravitation has been universally accepted, but the new theory goes farther, and, apart from supplying the laws of Newtonian mechanics when certain approximations are made, it enables us to predict the exact course of all motions resulting from gravitation. In this book, which is a popular exposition written for the average reader, Professor Einstein explains his famous theory which has so excited the scientific world. This volume is intended primarily for those readers who, though interested in the trend of modern theory, are not conversant with the mathematical analysis used in theoretical physics. The author's aim has been to give an exact insight into the theory of relativity, and to present the main ideas in the clearest and simplest form. He has succeeded admirably, and those who desire an authoritative and understandable explanation of the Einstein theory will find it between the covers of this book.

HENRY HOLT AND COMPANY

Albert Einstein
ÜBER DIE SPEZIELLE UND DIE ALLGEMEINE RELATIVITÄTSTEHORIE, (RELATIVITY: THE SPECIAL AND THE GENERALY THEORY)

(Left) Einstein sent the very first copy of this book, published in 1917 by Friedrich Viewig & Sohn, to his friend, the physician Hans Mühsam. On the right is the 1920 first edition English translation published by Henry Holt and Company.

in a city to give a lecture it would sell out with the rapidity of a modern rock concert. The comments of the leading British science journal *Nature* give a flavour of how the book was received:

> **A popular exposition of the doctrine of Relativity and what it implies: for this the world has been crying since the astronomers announced that the stars had proved it true. Here is an excellent translation of Einstein's own book; we hasten to it to know the whole truth and nothing but the truth. The reviewer on this occasion should be the man in the street, the man who, with thousands, has been asking, 'What is Relativity?' 'What is the matter with Euclid and with Newton?' 'What is this message from the stars?'**

ALBERT EINSTEIN, 1921

Einstein photographed on his arrival at New York on his first trip to the United States.

The *Nature* reviewer also warns that Einstein 'must needs speak largely in parables' as the subject was too technical for the general reader. His mention of Euclid was not random. Einstein opens his book by reminding the reader of their likely study of Euclid at school and builds up to relativity in a manner that, though worded in a friendly fashion, is more like the systematic approach of a textbook than a modern title intended for the public. There is no historical or personal context, despite the book being written by the man who came up with the theory, and it's not until 26 pages in that we get a clear example of the implications of the theory, using railway trains and lightning flashes. Nonetheless, the book was remarkably popular – in this respect, it prefigured Stephen Hawking's *A Brief History of Time* in being more frequently bought than thoroughly read.

A far more approachable and successful exposition of relativity and other aspects of the modern physics of the time would come from one of Einstein's greatest champions, English physicist Arthur Eddington. Born in Kendal in 1882, Eddington was an astrophysicist, most significantly developing our understanding of the structure of stars. However, to the British public he was the star science communicator of the day. Eddington had been a vocal supporter of Einstein and led an expedition in 1919 to observe the solar eclipse, providing evidence to support Einstein's general theory of relativity.

Shortly after the publication of Einstein's paper on general relativity, Eddington obtained some notoriety for his tongue-in-cheek response to a reporter's question. Asked if it were true that only three people in the world understood the theory, Eddington is said to have replied, 'Who is the third?' From his books, this seems likely to have been an attempt at humour rather than a criticism of Einstein's theories.

Eddington wrote a number of titles for the public, despite academics who popularised science being frowned on by their peers. His most successful work was *The Nature of the Physical World* from 1928, which made use of material from radio broadcasts and lectures he had given. Part of the appeal of Eddington's writing – and a lesson for scientists who came after him that still is frequently not observed – is that he realised the importance of giving the reader context. Rather than simply present scientific theories and observations, he considered the philosophical and even theological implications – important for the culture of the time. He was also happy to make literary references and to use humour in getting his message across. Eddington was one of the first modern scientists who realised how much better Galileo's approach to writing science had been than that of Newton.

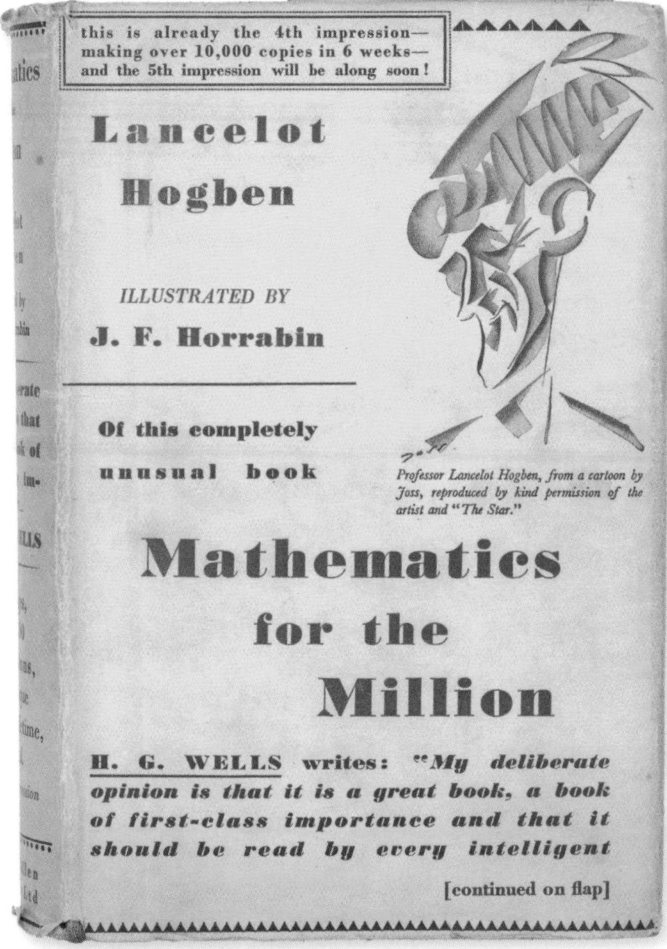

this is already the 4th impression—
making over 10,000 copies in 6 weeks—
and the 5th impression will be along soon!
Lancelot
Hogben
ILLUSTRATED BY
J. F. Horrabin
Of this completely
unusual book
Professor Lancelot Hogben, from a cartoon by Joss, reproduced by kind permission of the artist and "The Star."
Mathematics
for the
Million
H. G. WELLS writes: "My deliberate
opinion is that it is a great book, a book
of first-class importance and that it
should be read by every intelligent
[continued on flap]

Lancelot Hogben
MATHEMATICS FOR THE MILLION, GEORGE ALLEN & UNWIN LTD, 1936

Mathematics for the Million was an instant hit, with six printings in 1936 alone; this is the fourth printing.

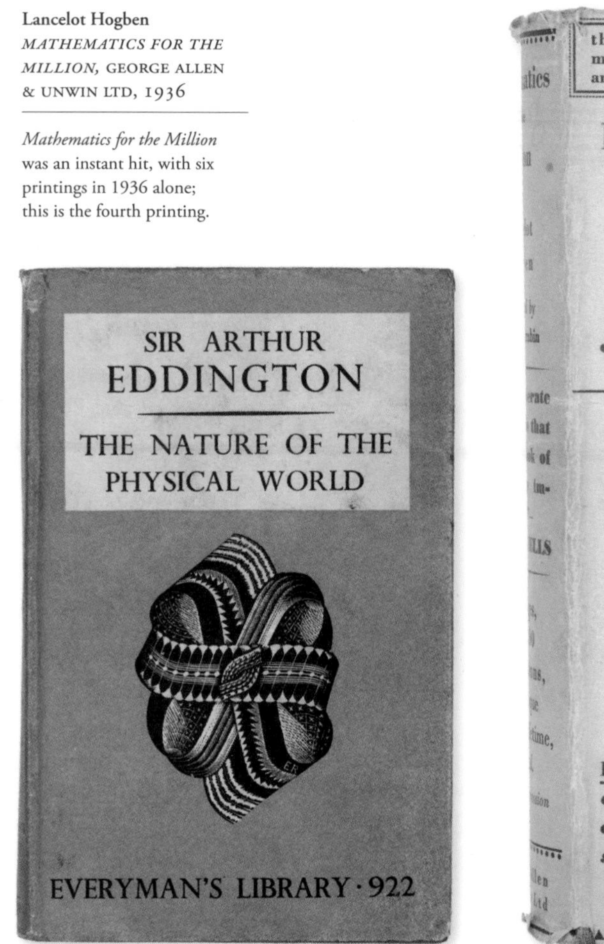

Sir Arthur Eddington
THE NATURE OF THE PHYSICAL WORLD, EVERYMAN'S LIBRARY, 1935

Arthur Eddington's bestseller was first published in 1928 by Cambridge University Press.

The immodest mathematician

Eddington kept maths to a minimum in his books for the public, but a title that came out while Eddington was still active grappled with the hard-to-handle topic of mathematics full on. As the book's title, *Mathematics for the Million*, makes clear, its English author, Lancelot Hogben, made no secret of his aspirations to reach the masses. Born in Portsmouth in 1895, Hogben was a zoologist who took a particular interest in statistics. His aim was to bring the person in the street up to the level of mathematics of a secondary-school, or high-school, graduate specialising in the subject.

Published in 1936, *Mathematics for the Million* is one of the strangest maths books ever written. Unusually for a science book, we know exactly how it came into being. Hogben was having lunch in the fashionable Simpson's in the Strand restaurant in London with American publisher William Warder Norton. Over lunch, Norton confided to

Hogben that there ought to be 'a big market for a book that could do for mathematics what [H.G.] Wells had done for world history in his *Outline*.' (Norton was referring to Wells's hugely successful popular history title *The Outline of History*.)

Norton wasn't asking Hogben to write this new book, but hoped that he could persuade the philosopher Bertrand Russell to do so. Hogben played down Russell, suggesting that he would find it difficult to write down to the level of the public. But, Hogben was able to reveal, he himself had 'already written the book [Norton] wanted'. He had done so during a lengthy stay in hospital to while away the time, but had not tried to get the book published because he was a candidate to become a Fellow of the Royal Society and 'its hierarchy frowned formidably on what they regarded as science popularisation'. This is somewhat ironic, given that today the Society awards an annual prize for science books written for the public. But by the time he met with Norton, Hogben had gained his fellowship and felt safe.

In a distinguished, sometimes downright over-wordy style, Hogben makes use of the way that maths has been employed in history to introduce basic principles. So, for example, he brings in geometry by showing the reader how ancient Egyptian architects might have used it, while trigonometry is demonstrated by showing how it was used to navigate by the stars. Although aimed at the general reader, his book is no mere entertainment. It is full of exercises and is clearly intended as a self-teach manual. Perhaps its greatest advance was its historical context, and the fact that Hogben realised it was unnecessary to, say, plough through the whole of Euclid (as many schoolchildren would continue to do long after this book was published). He simply gave the essentials of geometry and moved on.

Hogben was a respected scientist, though his book was on a subject well removed from his day-to-day activities. Strangely, considering that his mathematical speciality was statistics, this branch of maths only gets a limited exposure in *Mathematics for the Million*, probably because it did not feature heavily in school syllabuses of the time.

Chemical conundrums

Meanwhile, on the other side of the Atlantic, an American chemist was making fundamental contributions to understanding the chemical bond, the mechanisms by which different elements link together to form larger structures and compounds. Born in Portland in 1901, Linus Pauling was one of only four people to win two Nobel Prizes – in his case, in Chemistry and the Peace Prize. His work was foundational in the development of modern chemistry and molecular biology, the discipline that now dominates biology with its emphasis on DNA, the chemical processes in cells and the complex molecular machinery that makes it possible for them to function.

Although in later life Pauling promoted concepts with little scientific basis, such as his support for vast doses of vitamin C as a counter to colds and flu, there is no doubt of the huge contribution he made to chemistry. Pauling was a prolific writer, but most of his books for the general public were on his more outlandish ideas. The title that ties to his career with most weight was *The Nature of the Chemical Bond and the Structure of*

Linus Pauling
THE NATURE OF THE CHEMICAL BOND, CORNELL UNIVERSITY PRESS, 1939

An advertisement from 1939 for Pauling's formative title beneath a photograph of him from 1947, examining a crystal.

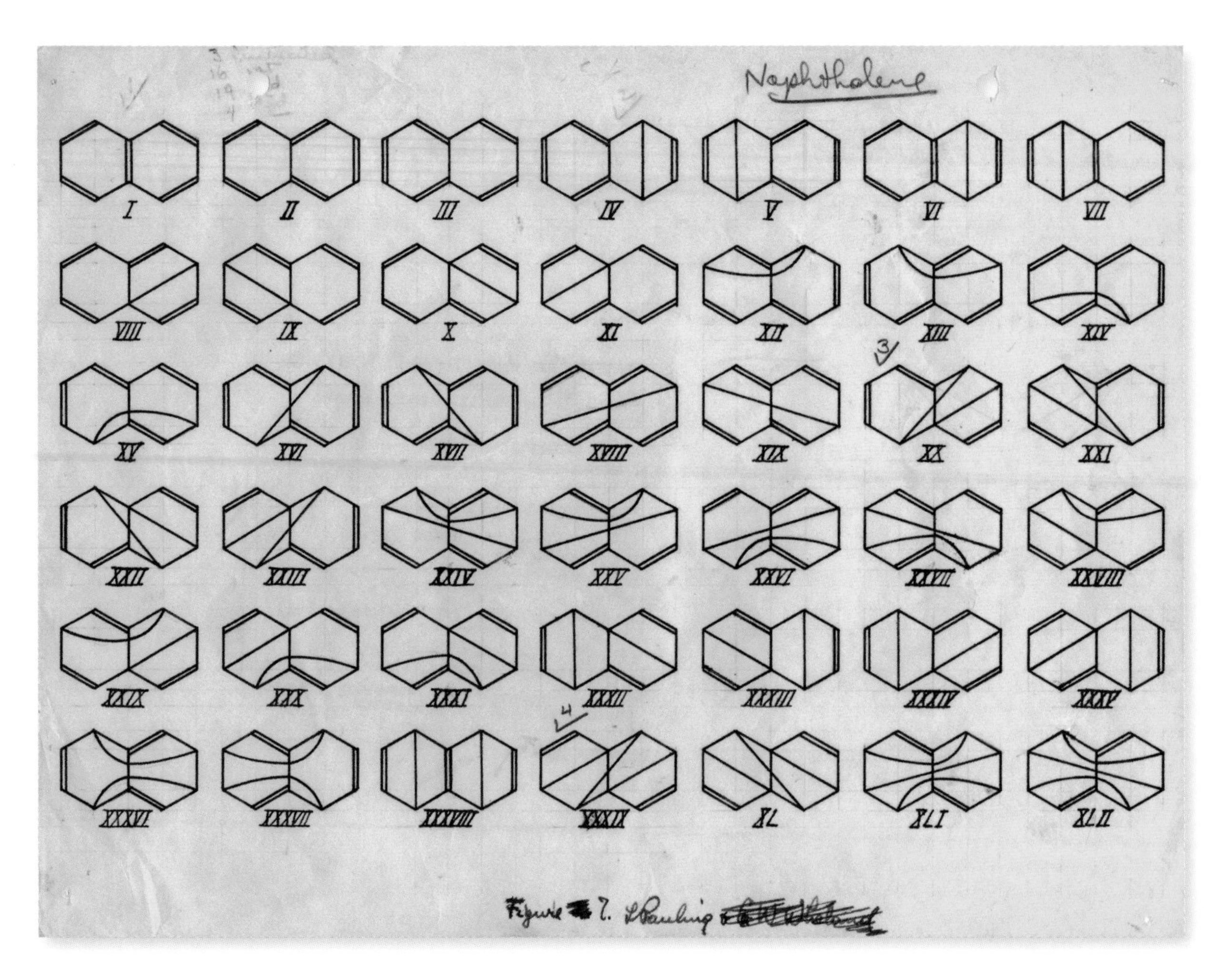

Linus Pauling
THE NATURE OF THE CHEMICAL BOND,
MANUSCRIPT, 1939

Reference illustration for naphthalene figure supplied by Pauling for his book.

166 *Complex Bond Orbitals*

TABLE 5-3.—OBSERVED MAGNETIC MOMENTS OF OCTAHEDRAL COMPLEXES OF TRANSITION ELEMENTS[a]

Hyperligated complexes	μ^b calculated	μ^b observed	Hypoligated complexes	μ^b calculated	μ^b observed
$K_3[Cr^{III}(CN)_6]$	2.83	3.3			
$K_3[Mn^{III}(CN)_6]$		3.0			
$K_4[Mn^{II}(CN)_6]$	1.73	2.0	$Mn^{II}(NH_3)_6Br_2$	5.92	5.9
$K_3[Fe^{III}(CN)_6]$		2.33	$(NH_4)_3[Fe^{III}F_6]$	5.92	5.9
$K_4[Fe^{II}(CN)_6]$	0.00	0.00	$(NH_4)_2[Fe^{III}F_5 \cdot H_2O]$		5.9
$Na_3[Fe^{II}(CN)_5 \cdot NH_3]$		.00	$[Fe^{II}(H_2O)_6](NH_4SO_4)_2$	4.90	5.2
$[Fe^{II}(dipyridyl)_3]SO_4$		.00			
$K_3[Co^{III}(CN)_6]$		.00			
$[Co^{III}(NH_3)_3F_3]$		.00[c]	$K_3[Co^{III}F_6]$	4.90	5.3[c]
$[Co^{III}(NH_3)_6]Cl_3$		.00	$[CoF_3(OH_2)_3] \cdot 1/2H_2O$		4.47[d]
$[Co^{III}(NH_3)_5Cl]Cl_2$		.00			
$[Co^{III}(NH_3)_4Cl_2]Cl$		.00			
$[Co^{III}(NH_3)_3(NO_2)_3]$		.00			
$[Co^{III}(NH_3)_5 \cdot H_2O]_2(C_2O_4)_3$		.00			
$[Co^{III}(NH_3)_4CO_3]NO_3 \cdot 3/2H_2O$		.00			
$K_2Ca[Co^{II}(NO_2)_6]$	1.73	1.9	$[Co^{II}(NH_3)_6]Cl_2$	3.88	4.96
$K_2[Pd^{IV}Cl_6]$	0.00	0.00			
$[Pd^{IV}Cl_4(NH_3)_2]$		.00			
$Na_3[Ir^{III}Cl_2(NO_2)_4]$		.00			
$[Ir^{III}(NH_3)_5NO_2]Cl_2$		.00			
$[Ir^{III}(NH_3)_4(NO_2)_2]Cl$		.00			
$[Ir^{III}(NH_3)_3(NO_2)_3]$		.00			
$K_2[Pt^{IV}Cl_6]$		.00			
$[Pt^{IV}(NH_3)_6]Cl_4$		.00			
$[Pt^{IV}(NH_3)_5Cl]Cl_3$		.00			
$[Pt^{IV}(NH_3)_4Cl_2]Cl_2$		.00			
$[Pt^{IV}(NH_3)_3Cl_3]Cl$		.00			
$[Pt^{IV}(NH_3)_2Cl_4]$		.00			

[a] The values quoted are taken in the main from W. Biltz, *Z. anorg. Chem.* 170, 161 (1928); D. M. Bose, *Z. Physik* 65, 677 (1930); and the *International Critical Tables*. Values for many other complexes are tabulated by P. W. Selwood, *Magnetochemistry*, Interscience Publishers, New York, 1956.

[b] In Bohr magnetons.

[c] Private communication from Prof. G. H. Cartledge of the University of Buffalo.

[d] H. C. Clark, B. Cox, and A. G. Sharpe, *J. Chem. Soc.* 1957, 4132.

have been investigated are hyperligated except that with fluorine, $[CoF_6]^{---}$, which is hypoligated. It is interesting that in the sequence $[Co(NH_3)_6]^{+++}$, $[Co(NH_3)_3F_3]$, $[CoF_6]^{---}$ the transition from hyperligation to hypoligation occurs between the second and third complex.

Bipositive cobalt forms hypoligating bonds with water and hyperligating bonds with nitrite groups.[47]

globin (hemin, ferroheme, hemochromogens) are in part ionic and in part covalent in structure (L. Pauling and C. D. Coryell, *Proc. Nat. Acad. Sci. U. S.* 22, 159 [1936]; 22, 210 [1936]; Coryell, Stitt, and Pauling, *loc. cit.* [43]).

[47] A detailed discussion of the magnetic moments and structure of complexes of cobalt(II) has been made by B. N. Figgis and R. S. Nyholm, *J. Chem. Soc.* 1959, 338.

5-5 *Magnetic Moments of Octahedral Complexes* 167

The magnetic method cannot be applied to tripositive chromium, the structures of the two extreme types having the same number of unpaired electrons and entering into resonance with each other. The chemical properties of the chromium complexes indicate that chromium, like the other iron-group elements, forms hyperligating bonds

FIG. 5-10.—A drawing representing the front half of the cubic unit of structure of the crystal prussian blue, $KFeFe(CN)_6 \cdot H_2O$. The structure can be described by reference to the eight small cubes that constitute the cubic unit of structure. Alternate corners of the small cubes are occupied by iron(II) and iron(III) atoms. The cyanide groups lie along the edges of the small cubes; each cyanide group forms a single bond with two iron atoms, defining the edge of the cube. Water molecules and potassium ions alternate at the centers of the small cubes. The structure can be described as a three-dimensional latticework of iron atoms and cyanide groups defining cubical cells that contain the water molecules and potassium ions.

with groups such as cyanide and hypoligating bonds with water and ammonia.[48] The complexes of the iron-group elements are discussed further in Section 9-7.

[48] That chromium forms d^2sp^3 hyperligating bonds with oxalate in the ion $[Cr(C_2O_4)_3]^{---}$ has been inferred by C. H. Johnson (*Trans. Faraday Soc.* 28, 845) [1932]) from the following facts: The chromium trioxalate complex and the cobaltic trioxalate complex can be resolved into optical isomers, whereas

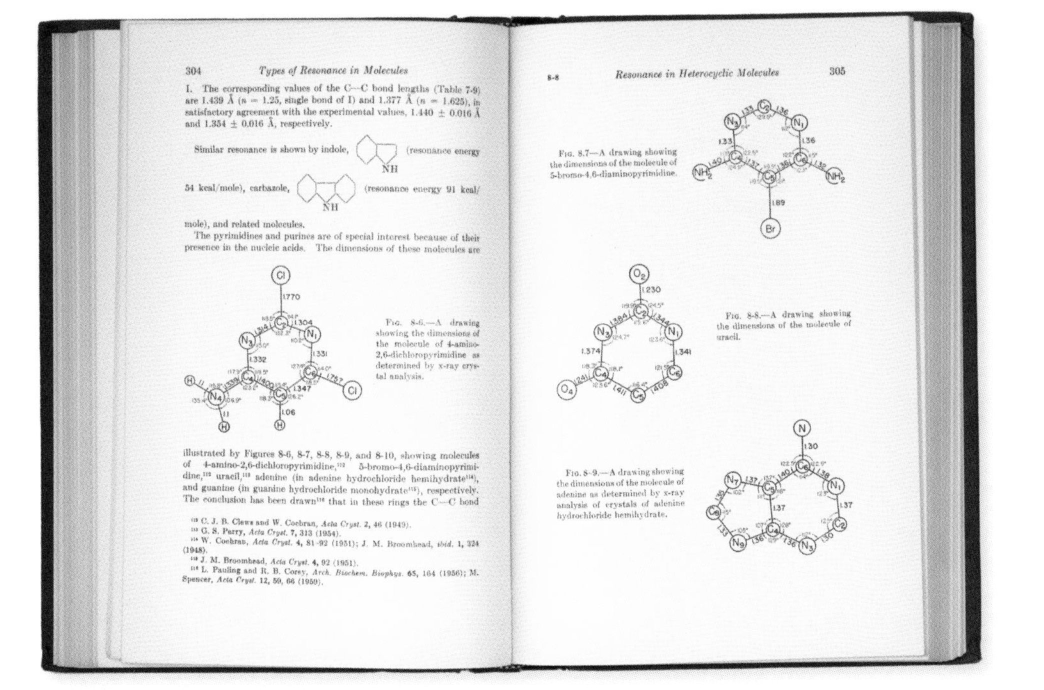

304 *Types of Resonance in Molecules*

I. The corresponding values of the C—C bond lengths (Table 7-9) are 1.439 Å (n = 1.25, single bond of I) and 1.377 Å (n = 1.625), in satisfactory agreement with the experimental values, 1.440 ± 0.016 Å and 1.354 ± 0.016 Å, respectively.

Similar resonance is shown by indole, [structure: NH] (resonance energy 54 kcal/mole), carbazole, [structure: NH] (resonance energy 91 kcal/mole), and related molecules.

The pyrimidines and purines are of special interest because of their presence in the nucleic acids. The dimensions of these molecules are

FIG. 8-6.—A drawing showing the dimensions of the molecule of 4-amino-2,6-dichloropyrimidine as determined by x-ray crystal analysis.

illustrated by Figures 8-6, 8-7, 8-8, 8-9, and 8-10, showing molecules of 4-amino-2,6-dichloropyrimidine,[112] 5-bromo-4,6-diaminopyrimidine,[112] uracil,[113] adenine (in adenine hydrochloride hemihydrate[114]), and guanine (in guanine hydrochloride monohydrate[115]), respectively. The conclusion has been drawn[116] that in these rings the C—C bond

[112] C. J. B. Clews and W. Cochran, *Acta Cryst.* 2, 46 (1949).

[113] G. S. Parry, *Acta Cryst.* 7, 313 (1954).

[114] W. Cochran, *Acta Cryst.* 4, 81-92 (1951); J. M. Broomhead, *ibid.* 1, 324 (1948).

[115] J. M. Broomhead, *Acta Cryst.* 4, 92 (1951).

[116] L. Pauling and R. B. Corey, *Arch. Biochem. Biophys.* 65, 164 (1956); M. Spencer, *Acta Cryst.* 12, 59, 66 (1959).

8-8 *Resonance in Heterocyclic Molecules* 305

FIG. 8.7—A drawing showing the dimensions of the molecule of 5-bromo-4,6-diaminopyrimidine.

FIG. 8-8.—A drawing showing the dimensions of the molecule of uracil.

FIG. 8-9.—A drawing showing the dimensions of the molecule of adenine as determined by x-ray analysis of crystals of adenine hydrochloride hemihydrate.

Linus Pauling
THE NATURE OF THE CHEMICAL BOND,
CORNELL UNIVERSITY PRESS, 1960

These pages from the third edition of Pualing's book (first published in 1939) show the structure of rhe crystal prussian blue (top) and the dimensions of various molecules (bottom).

Molecules and Crystals from 1939. Based on his 1931 paper that led to his Nobel Prize, the book also pulled in many of his thoughts on molecular structure.

The Nature of the Chemical Bond is a textbook – so an exception to the kind of books that had a wide influence by the mid-twentieth century – but it is the very rare example of a textbook that has continued to have worldwide recognition and in which a scientist covers his own discoveries. Along with a handful of other titles of the period, the book has more in common with the work of James Clerk Maxwell and scientists from previous centuries than the more common influential books of the twentieth century, increasingly written for the general public.

Strangely, though quantum physics has had far more impact on our lives than has relativity, there were no books on the subject from the great quantum physicists such as Niels Bohr, Werner Heisenberg and Erwin Schrödinger that had any great impact. Perhaps the most significant were Schrödinger's *My View of the World* from 1964 and Heisenberg's *Encounters with Einstein*, published in 1989, 13 years after his death. However, these were not approachable memoirs, but rather collections of essays on the scientific method and the impact of science on society, addressed to an academic audience. Books on quantum theory would not reach the public gaze until the 1980s. However, one of quantum physics' big names – Schrödinger – wrote a book that has been widely praised, in a field outside his own. This was *What is Life?*, written in English in 1944.

Austrian physicist Erwin Schrödinger was born in Vienna in 1887. From the 1920s he became one of the leading lights in quantum physics, winning the Nobel Prize in Physics in 1933. Attacked in his own country because of his opposition to the Nazi regime, he ended up in Ireland in 1940, continuing to work in Dublin until his retirement in 1955. Non-physicists might best associate him with his thought experiment that became known as 'Schrödinger's cat', while physicists remember him for his equation describing the behaviour of quantum particles. However, *What is Life?*, as the name suggests, links physics to biology. Though his book was aimed at the public, Schrödinger himself remarked that the content could not be described as popular – it was technical and heavy-going for the ordinary reader.

Based on a series of lectures Schrödinger gave in Dublin, the book deals with a key puzzle facing biologists. They were aware by this time that genetic information was somehow passed on through chromosomes, a series of microscopic blobs found inside cells. However, it was Schrödinger, basing his argument on the underlying physics of atoms and molecules, who suggested that the mechanism for this would have to be a particular type of molecule, which he described as an 'aperiodic crystal'. Most of the crystals we are familiar with, such as diamonds, have a simple, repeating structure. But Schrödinger proposed that in order to contain sufficient information, the crystal behind life would have to have a structure that didn't repeat – hence the 'aperiodic' part. (Think of a book. If it were like a traditional crystal, it would read something like ABA ABA ABA ABA.) The molecule playing this part was later identified as DNA, with the discovery of its aperiodic structure forming the basis of another key book from the twentieth century, *The Double Helix* (see pages 218–19).

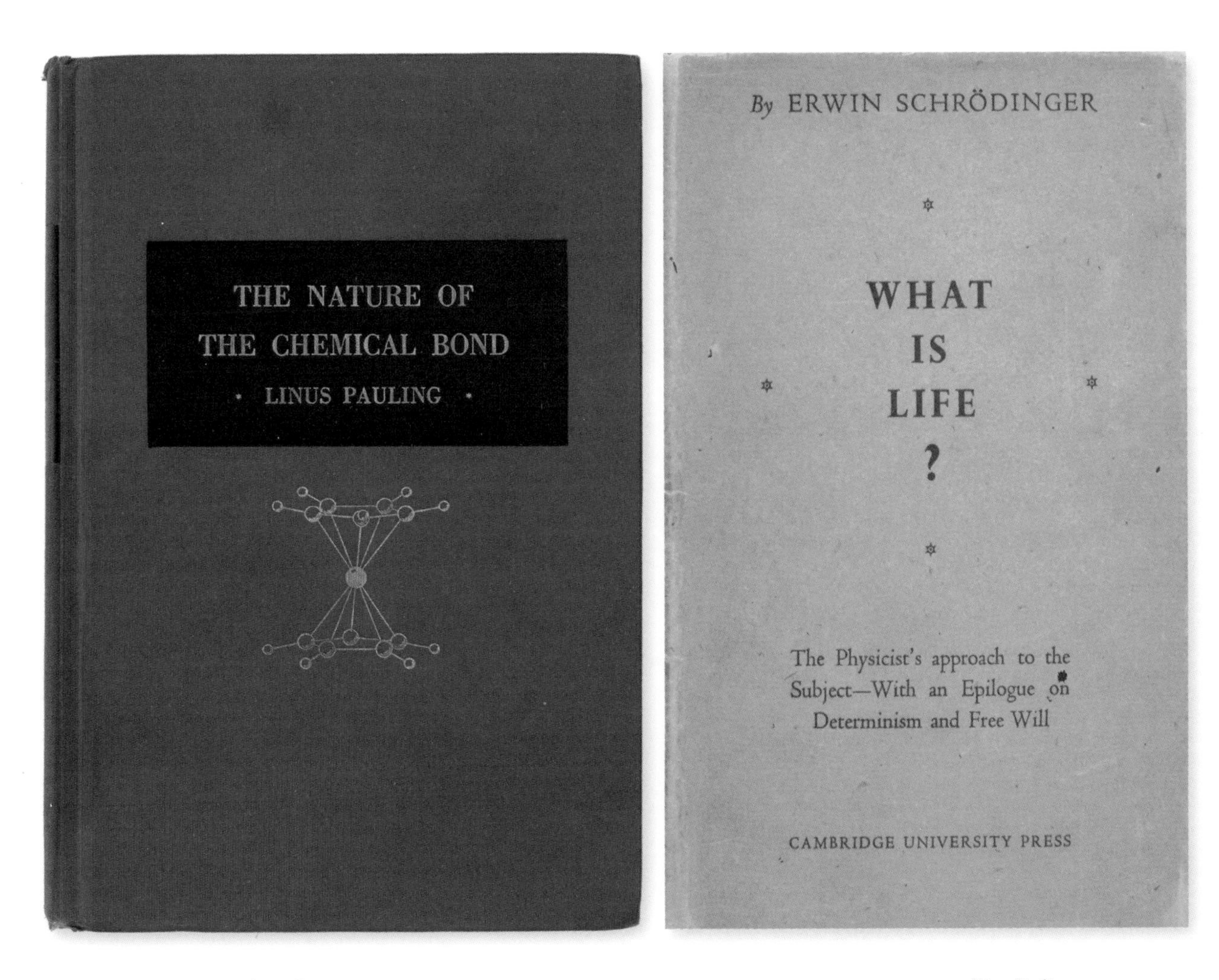

Linus Pauling
THE NATURE OF THE CHEMICAL BOND, CORNELL UNIVERSITY PRESS, 1960

The third edition of Pauling's book from 1939.

Erwin Schrödinger
WHAT IS LIFE?, CAMBRIDGE UNIVERSITY PRESS, 1944

The cover of Schrödinger's highly influential title.

Behaviour to the fore

As was the case with *What is Life?*, another important title from this period took a simplified look at a much more complex phenomenon, but in the case of Canadian scientist Donald Hebb's 1949 book, *The Organization of Behavior*, the subject was not life in general but the brain. A psychologist, Hebb was one of the earliest to explore the way that interactions between neurons in the brain resulted in the behaviours we associate with living organisms.

Hebb's biggest contribution to his field was 'Hebbian learning', which showed that when pathways in the brain are used repeatedly they are strengthened, forming a pattern that results in learning. The big breakthrough here was the idea that learning resulted in physical changes to the brain's structure, something that had not been previously realised. This was a book aimed at those in the profession, but it was not a textbook in the conventional sense, more a throwback to the earlier approach of a book used to share a new scientific idea with peers.

A useful contrast with *The Organization of Behavior*, in being far more approachable to the reader, was a book from an author whose achievements are now sometimes lost in revulsion for his politics (though it's fair to say that his titles are significantly less readable now than Arthur Eddington's, or even Galileo's). Austrian zoologist Konrad Lorenz was born in Vienna in 1903. His speciality, which won him the Nobel Prize in Physiology or Medicine at the age of 70, was ethology – like Hebb, Lorenz studied behaviour, but his subjects were animals.

Lorenz was not alone among scientists in supporting the Nazi regime in the Second World War. Heisenberg, for example, stayed on to run the German programme to develop nuclear weapons when many of his physicist colleagues fled the country. But Lorenz's war work involved more than just supporting his country's war effort. He published material supporting the racist narrative of the Nazis and worked for the 'Office for Racial Policy'.

Despite his politics, Lorenz's books on animal behaviour were immensely popular, in particular his 1949 *Er Redete mit dem Viehden Vogeln und den Fischen* (He Talked to the Cattle, the Birds and the Fish), which achieved international fame. The title is better known under its 1951 English title *King Solomon's Ring*, a reference to the legend that King Solomon had a ring enabling him to talk to the animals. In the book, Lorenz describes how he raised various species at home and what he learned about their behaviour and psychology as a result. Perhaps the most original concept he presented was the idea

Konrad Lorenz
ER REDETE MIT DEM VIEHDEN VOGELN UND DEN FISCHEN (KING SOLOMON'S RING)

Covers of a 1953 English translation published by the Reprint Society London and a 1963 German edition of Lorenz's 1949 bestseller on animal behaviour published by DTV Verlag.

of imprinting – the rapid process in which newborn animals, soon after birth, learn behaviour directly from their parents – but the book was also novel in helping the public understand more about what goes on in an animal's brain.

Getting philosophical

Books like Schrödinger's *What is Life?* and Hebb's *The Organization of Behavior* could be appreciated to some degree by the public, even if they were probably more valued in hindsight by other scientists. However, some would argue that they make straightforward reading indeed compared to a title published in 1962, *The Structure of Scientific Revolutions* by the American Thomas Kuhn, born in Cincinnati in 1922. This was one of the last significant original scientific books addressed to professionals in the field, rather than the public.

Kuhn's book covered the philosophy of science, rather than directly tackling science itself. To put it into context, we need first to consider another classic book in the philosophy of science, *Logik der Forschung* (The Logic of Scientific Discovery) by Karl Popper. Popper, an Austrian of Jewish descent born in Vienna in 1902, spent the majority of his academic life in England, where he moved via New Zealand. His book is now better-known in the re-written English version he produced in 1959 than the 1934 German original.

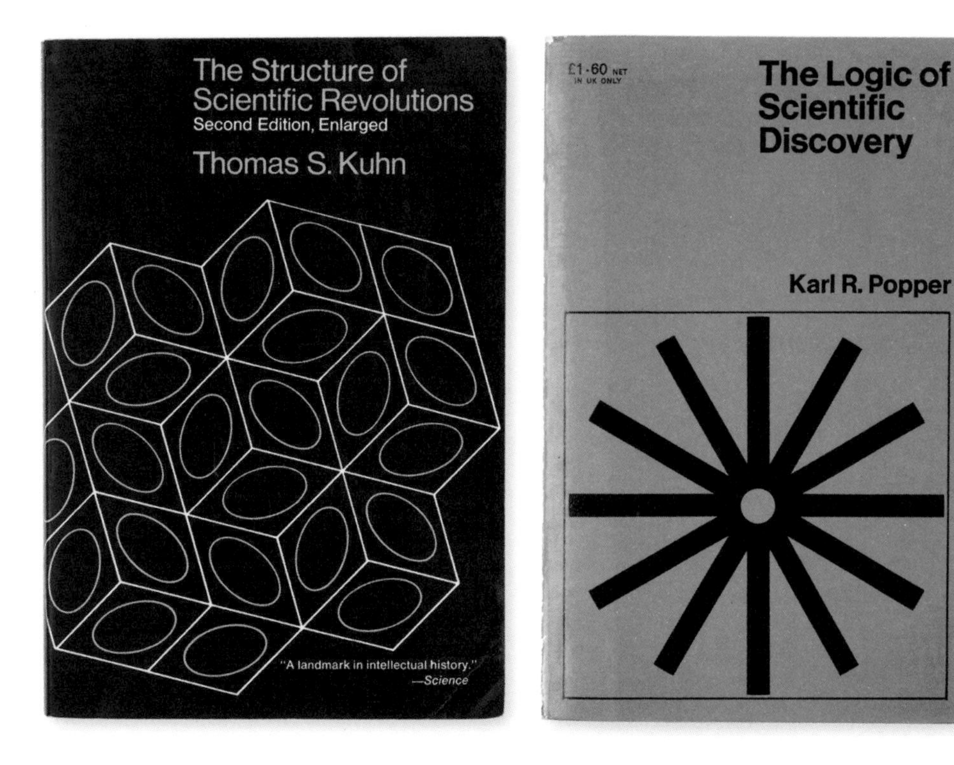

Thomas S. Kuhn
THE STRUCTURE OF SCIENTIFIC REVOLUTIONS, THE UNIVERSITY OF CHICAGO PRESS, 1970

Cover of Thomas Kuhn's 1962 bestseller in its second edition from 1970.

Karl Popper
LOGIK DER FORSCHUNG (THE LOGIC OF SCIENTIFIC DISCOVERY), HUTCHINSON & CO LTD, 1972

Cover of the sixth printing of the English version of Popper's 1934 German original.

Central to Popper's understanding of the scientific process was the idea of being able to falsify a proposition. For him, science could not involve statements that could not be falsified. The point underlying this was that science worked not by deduction, which produces absolute truths, but by induction, which can only at best give us the most likely outcome given current evidence. The classic example is the black swan. Europeans could reasonably make the scientific observation that all swans are white birds. However, this was open to falsification if black swans were discovered, which they duly were in Australia.

Popper's philosophy portrayed science as incremental – a little like the uniformitarian view in geology. Major changes, such as the Copernican model of the universe, would certainly occur, but Popper believed they weren't central to a process of devising new theories, testing them and looking for opportunities to falsify them. Kuhn, however, struck back with a clearer catastrophist equivalent – the concept of scientific revolutions, for which he introduced the term 'paradigm shift'.

Kuhn's book *The Structure of Scientific Revolutions* proposed far more than the not particularly surprising idea that there are sudden shifts in understanding when major new theories are introduced. In a postmodern view of science that now seems somewhat dated, Kuhn suggests that paradigm shifts involve such a major change in the way scientists look at the world that their words no longer have the same meanings. In his book, Kuhn emphasised that scientific viewpoints have their sociological element, being inevitably subjective, and speaks of new paradigms being 'incommensurable' with the old. This means that when, say, Newton and Einstein refer to 'gravity', the result of the paradigm shift means they are not referring to the same thing. According to Kuhn, this is not only in the sense that the different generations had a changed understanding of the same reality – Kuhn believed that reality itself is altered in a paradigm shift.

Few real scientists or members of the public accept (or are even aware of) more than a kind of Kuhn-lite, where revolutions in theory take place and we see the same universe in a different way – but it would be impossible to underestimate the impact that Kuhn's work has had on the scientific zeitgeist. It transformed the way that we look at the history of science. The very fact that the book sold over a million copies demonstrates that, despite being technical, it had a reach that went far beyond philosophers of science and even scientists to a wider academic market. (If anything, it received greater appreciation from the humanities than the sciences.)

The environmental balance

If Thomas Kuhn's book was to have a huge influence on academia, a title published in the same year, 1962, was to have an even larger impact on the world, bringing a relatively new scientific topic to the public eye: environmentalism. The book would change public policy, raise awareness, and according to some observers would be indirectly responsible for the deaths of millions of people. And it was written by a woman.

The author was Rachel Carson. Born in Springdale, Pennsylvanian in 1907, the American marine biologist had already had considerable success with two books on sea life before publishing *Silent Spring* in 1962. The title referred to the impact that the insecticide

Rachel Carson

'. . . what we have to face is not an occasional dose of poison which has accidentally got into some article of food, but a persistent and continuous poisoning of the whole human environment . . .'

Silent Spring

Rachel Carson
SILENT SPRING,
PENGUIN PRESS, 1971

Carson's title has been constantly in print since publication in 1962. But by the time this Penguin edition was published DDT use was already in decline.

Rachel Carson
SILENT SPRING
(PRIMAVERA SILENZIOSA),
GIANGIACOMO FELTRINELLI
EDITORE, 1966

An Italian translation of *Silent Spring* published by Feltrinelli's Universale Economica imprint.

DDT was having on birds. Carson suggested that, should the use of DDT continue, there could come a point where spring was not greeted by birdsong, but by silence.

Part of the appeal of Carson's writing, compared with the stodgy, textbook-like work of many science books written for the public before it, was her near-poetic writing style, typified when she envisaged a bird-free future: 'The birds – where had they gone? Many people spoke of them, puzzled and disturbed. The feeding stations in the backyards were deserted. The few birds seen anywhere were moribund; they trembled violently and could not fly. It was a spring without voices.'

DDT (dichlorodiphenyltrichloroethane) had proved an extremely impressive insecticide, which vastly reduced the incidence of the deadly typhus disease in Europe and had been adopted by the World Health Organisation as the primary weapon for fighting malaria-carrying mosquitoes. Yet in the political response to Carson's book, DDT usage was widely restricted. In 1963, the year after the book's publication, the US president set up an investigation specifically to check out the validity of *Silent Spring's* message. DDT was effectively banned in the US in 1972 and since then controls have been instigated around the world. DDT is thought to have saved around 25 million lives since its invention, and some believe it likely that many more could have been saved had it not been for the impact of *Silent Spring*. Malaria still kills over a million people a year, many of them children.

Ironically, Carson does not suggest that DDT should be taken off the shelves, but rather that its usage should be carefully targeted. When *Silent Spring* was published, DDT was being used profligately as an agricultural pesticide, and it was this that was causing real problems. The environmental impact of DDT should not be played down – its ban in the US is thought to have been a major factor in the recovery of the bald eagle from near extinction, for example – yet DDT did prove extremely effective when used in small areas to wipe out disease-carrier concentrations. While this was continued to some degree, the backlash against the insecticide was so strong that opportunities to use it before DDT-resistant mosquitoes became more common were missed.

Despite Carson's balanced argument, the tenor of the book is emotive, and comments in it such as, 'Can anyone believe it is possible to lay down such a barrage of poisons on the surface of the earth without making it unfit for all life? They should not be called "insecticides", but "biocides",' made *Silent Spring* a powerful voice against the insecticide.

Without doubt, *Silent Spring* marks the arrival of a new kind of science book. A polemic with a message, it was written by a scientist – but on a topic outside her area of expertise. What Carson contributed was a skill for storytelling: the ability to produce a narrative that would carry the reader along – a capability that would soon come to be expected if a science book were to be considered good quality.

It's interesting that one of DDT's great successes was in practically eradicating the scourge of typhus from Europe, as another much earlier science title focused on this horrible disease. This was *Rats, Lice and History* by the American bacteriologist Hans Zinsser. Born in New York in 1878, Zinsser had isolated the typhus bacterium (not to be confused with typhoid, which is so named because it is 'typhus-like'). His 1935 book is quirky in the extreme. Its subtitle is 'after twelve preliminary chapters', chapters which Zinsser spends darting here and there, always seemingly about to describe typhus before heading off to consider the nature of biographies or the lifecycles of the rat and louse.

RACHEL CARSON, 1962

A photograph of Carson in the woods near her home, taken as part of a photoshoot for *Time* magazine.

Hans Zinsser
RATS, LICE AND HISTORY, GEORGE ROUTLEDGE & SONS, 1937

The third printing of Zinsser's book with its unique 'biography of a disease' approach, which, despite Zinsser's protestations, was definitely popular science.

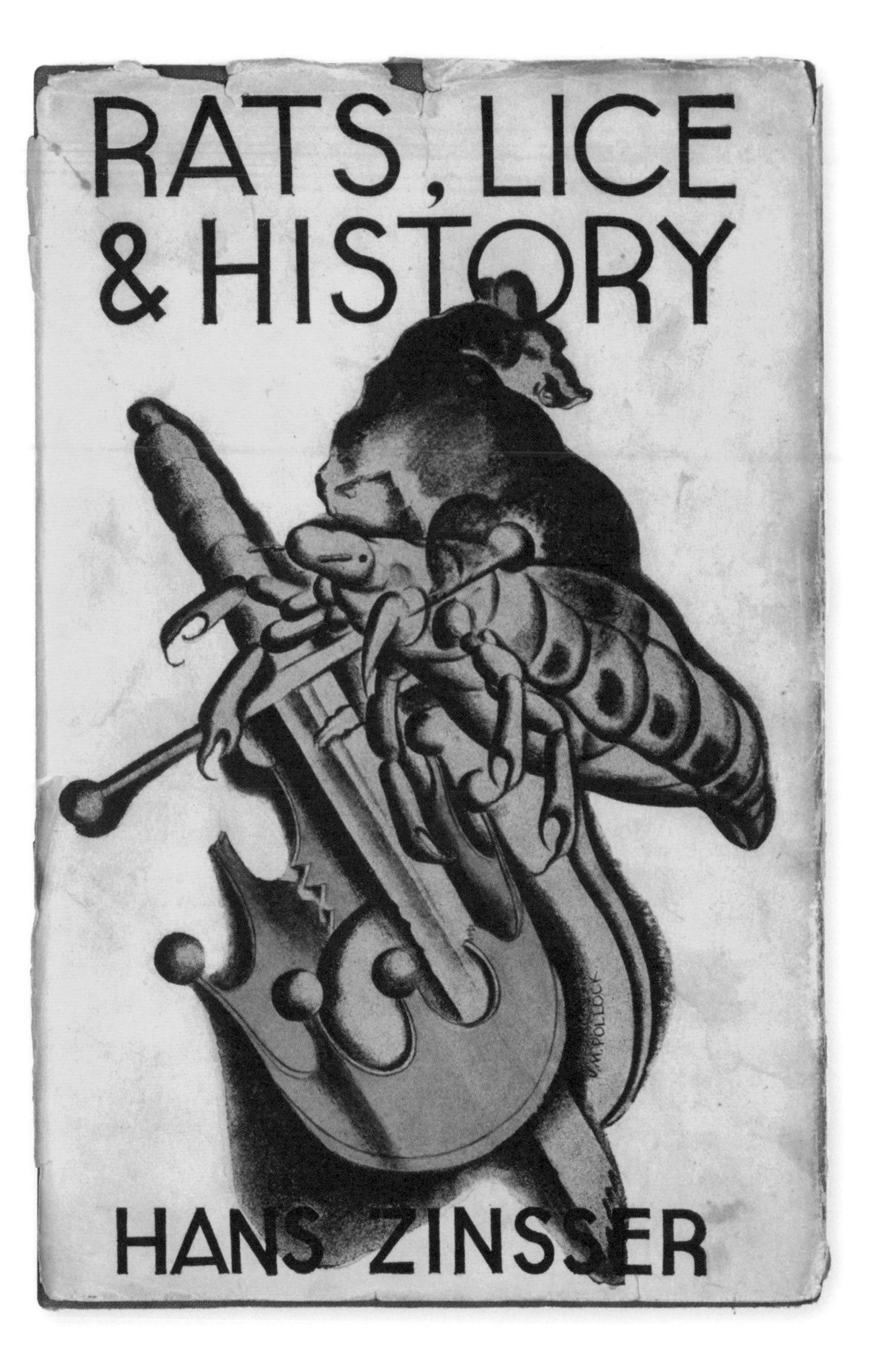

Zinsser spent time on discussing what a biography was, and asserting that he was in the business of writing a biography of a disease, because he was adamant that his book should not be considered popular science. (It was.) He grumpily remarks, 'To describe [the work of medical teams] belongs to technical literature. To attempt to do so in this book would lead us into "popular science", a form of production which we detest and have endeavoured to avoid.' It's possible that Zinsser had in mind as 'popular science' the kind of Olympian, condescending title that was common in the genre in the early twentieth century. Zinsser wanted his book to be different – to give a human viewpoint, as we have come to expect of a good popular science title.

As a result, Zinsser ends up straying far and wide, often deploying distinctive humour, even spending a while discussing the work of T. S. Eliot. His approach makes *Rats, Lice and History* an eccentric but novel read, giving a hint of what might come from taking bolder and more interesting approaches to communicating science to the masses.

The bongo-playing genius

Carson and Zinsser both put considerable effort into writing for the public. Strictly speaking, American physicist Richard Feynman never did so, even though one of his books was a huge popular hit. This was a man that most physicists hold in awe. Born in New York in 1918, Feynman, like Faraday and Tyndall before him, achieved his literary success from the publication of his lectures and talks, rather than from writing books in the conventional sense.

A Nobel Prize winner for his work on quantum electrodynamics (QED) – which describes the interaction of light and matter at the quantum level – Feynman's biggest popular success was the 1985 collection of autobiographical anecdotes *Surely You're Joking, Mr. Feynman!*. These stories were not written down by Feynman, but edited from conservations with biographer Ralph Leighton. The accuracy of Feynman's tales, from his adventures playing the bongos to his safe-breaking activities while working on the Manhattan Project to design the first atomic bomb, has sometimes been questioned – but there's no doubt he told a great story, and gave a fascinatingly human side to the scientific developments of that period.

Published just three years before Feynman's death, *Surely You're Joking* was a late entry amongst his books. The same year, he also published his most effective quasi-popular science book, *QED: The Strange Theory of Light and Matter*. Based on a series of public lectures, it starts with a superb introduction which includes some classic Feynman comments. For example:

> **[Y]ou think I'm going to explain it to you so you can understand it? No, you're not going to be able to understand it. Why, then, am I going to bother you with all this? Why are you going to sit here all this time, when you won't be able to understand what I am going to say? It is my task to persuade you not to turn away because you don't understand it. You see, my physics students don't understand it either. This is because I don't understand it. Nobody does.**

Richard Feynman
THE FEYNMAN LECTURES ON PHYSICS, ADDISON-WESLEY, 1966

A paperback set of Feynman's 'red books' from three years after initial hardback publication. Feynman (above) lecturing at CERN, Geneva in 1965.

Although the main part of the book gets relatively technical for a popular title, it does so without resorting to mathematics and with Feynman's typical man-of-the-people style. This was also much in evidence when Feynman was part of the Rogers Commission looking into the causes of the Challenger Space Shuttle disaster of 1986. Unhappy with the bureaucratic and heavily controlled approach of the commission, Feynman did his own evidence gathering, and then, at a televised session of the commission, used his glass of iced water to cool a section of the rubber O-ring used to seal joints on the shuttle, demonstrating that O-rings lost flexibility in the cold, reducing their ability to seal, and so causing the shuttle's catastrophic failure.

It is because so much of Feynman's personality comes through even in a textbook, that his most influential book is arguably *The Feynman Lectures on Physics*, published in 1963. Known amongst physicists as 'the red books' (they were first published with plain red covers), these three volumes cover his lectures for an undergraduate course given at the California Institute of Technology. Remarkably, over 1.5 million copies of this decidedly technical title have been sold in English alone.

When I first came across the red books as a physics undergraduate at Cambridge in the 1970s, I was swept away, as so many others have been, by Feynman's conversational style and his very different way of presenting much of the material. What he is writing about is not always easy to grasp – he pulls no punches mathematically – yet the accompanying text is as far as it is possible to get from a typical dull textbook that simply presents a collection of facts. Physicists around the world revere the red books like no other textbook.

Man the ascending animal

Feynman's highly technical red books are very different from the popularist work of English author Desmond Morris, but Morris would also prove highly influential in his own way. Born near Swindon in 1928, Morris was a zoologist who achieved fame in the UK as the presenter of early television programmes on nature, notably the 1950s and 1960s series *Zoo Time*, a weekly family show that used specimens from Regent's Park Zoo in London to explore natural history.

However, Morris's breakthrough title was not the inevitable 'book of the television series' with which we now get deluged. Instead, it was far more subversive. *The Naked Ape* examined the human species from a zoological viewpoint. Despite the widespread intellectual acceptance since Darwin's time that humans were just another ape, Morris's book went for the gut, bringing the message home. Coming out in 1967, when the sexual revolution was still shocking to many, it was also widely considered subversive because of its open discussion of human sexuality. Bear in mind that this was only seven years after Penguin Books had been subject to a public prosecution for publishing D. H. Lawrence's novel *Lady Chatterley's Lover*. The times might have been a-changing, but not particularly quickly in conservative England (or, for that matter in the US, where *The Naked Ape* was the subject of a court case after being one of a handful of titles removed from a school library because they were considered 'anti-American, anti-Christian, anti-Semitic, and

Desmond Morris
THE NAKED APE,
CORGI, 1969

First published by Jonathan Cape in 1967 with a neutral black cover, this paperback edition features an image that was unlikely to do anything to decrease the book's controversial nature. Above, Desmond Morris (and friend) on the Zoo Time television show in 1956.

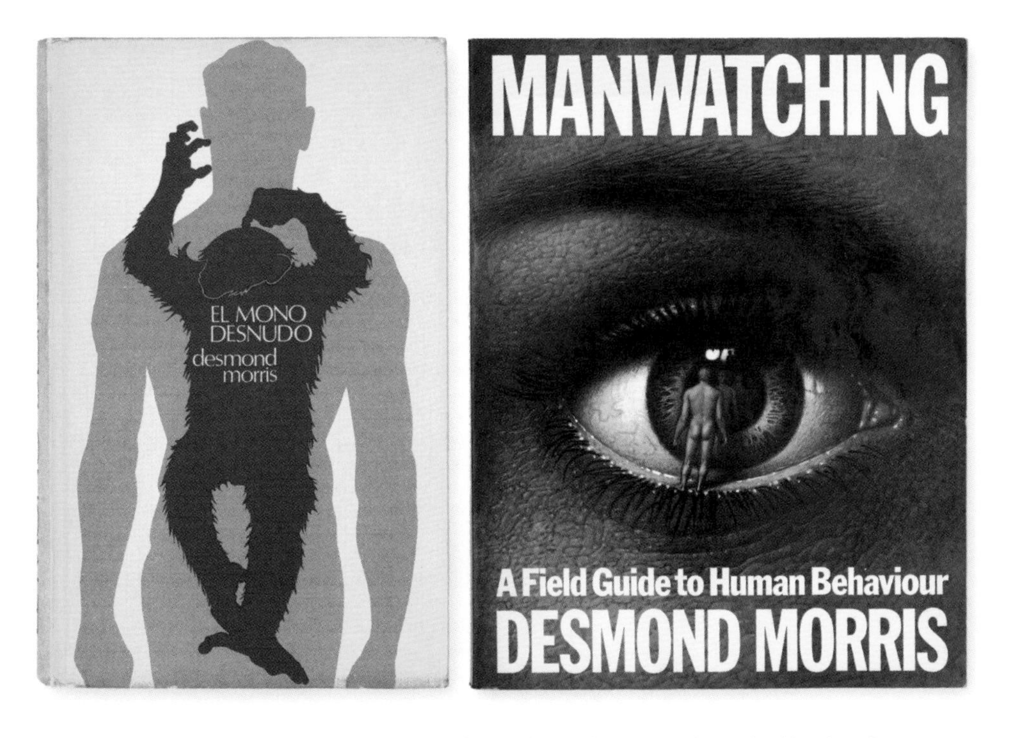

Desmond Morris
THE NAKED APE
(EL MONO DESNUDO),
CIRCULO DE LECTORES,
1969

A rather less controversial Spanish cover for *The Naked Ape*.

Desmond Morris
MANWATCHING, TRIAD/
PANTHER BOOKS, 1978

This paperback edition was published a year after the hardback Jonathan Cape edition.

just plain filthy'). The cover of *The Naked Ape* alone, featuring the naked backs of a man, woman and child, would have marked it out as a book that could not comfortably be consumed in public.

A particularly important part of *The Naked Ape's* contribution to the public understanding of evolution is the way that it brings evolutionary factors into explanations of human behaviour. Evolution had largely been described to the public as a mechanism for explaining physical change, but Morris described how human social interaction, sexual behaviour, attitude to child rearing, urge to explore and tendency to fight could all have been shaped by evolutionary pressures.

Morris went on to write a whole range of titles, the majority of which could be seen as sequels to *The Naked Ape*. Perhaps the most effective was *Manwatching: A Field Guide to Human Behaviour* from 1978. This large-format, heavily illustrated book carried forward the behavioural theme, always linking it to evolutionary and biological information, and made the most of its full-colour illustrations. For example, in one experiment the reader is shown a pair of apparently identical faces and asked to decide which is more attractive; it is then revealed that the photograph that would most frequently be selected had been modified to enlarge the pupils of the model's eye, demonstrating a natural reaction to such physical cues. The success of *Manwatching* led Morris into a whole series of spin-offs including *Bodywatching*, *Peoplewatching*, *Babywatching* and the increasingly unlikely *Dogwatching*, *Catwatching*, *Horsewatching* and *Animalwatching*.

In many ways, *Manwatching*, with its large format and illustrations, feels like a 'book of the television show' – but it wasn't. Some such books have been quite successful in

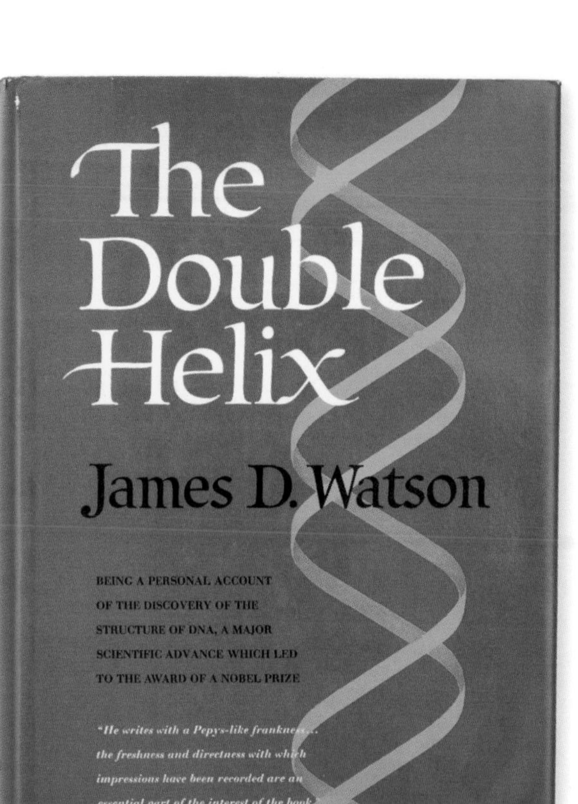

Jacob Bronowski
THE ASCENT OF MAN,
BRITISH BROADCASTING CORPORATION, 1975

The rather uninspiring cover of Bronowski's impressive book-of-the-series in an edition issued two years after it was first published in 1973.

James D. Watson
TEH DOUBLE HELIX,
ATHENEUM, 1968

An early copy of Watson's popular and very personal history of science bestseller, never out of print since.

their own right, though many feel like the take-home glossy programme from a night at the theatre. However, one title does deserve consideration. This was *The Ascent of Man* from 1973.

Written by Polish-English mathematician Jacob Bronowski, born in Łódź in 1908, the book is pretty much a transcript of his television series, heavily illustrated in colour. What Bronowski delivers is a celebration of human achievement (the title is an intentional play on Darwin's *The Descent of Man*), making it clear that the sciences cannot be treated in isolation, but have to be seen as part of our wider cultural development. Bronowski shows how science does not only emerge from our culture and creativity, but how it has also shaped our culture: in many ways, he seems to suggest, it is science that makes us truly unique as a species.

Bronowski shows that while there is no doubt that Desmond Morris was right in categorising us as just another ape, our scientific achievements make the human species very special. He does not give a rose-tinted viewpoint of humanity's capabilities – Bronowski lost many of his family in Auschwitz – yet he is still able to celebrate the remarkable success of the sciences and their development, intertwined with that of the arts.

The heart of biology

By the time Bronowski was writing, a new book was available that looked back to a dramatic scientific discovery in 1953. This was *The Double Helix*, published in 1968, written by one of the discoverers of the structure of DNA, James Watson. Like its author, *The Double Helix* would result in considerable controversy, yet few would dispute its significance in the history of science writing.

American molecular biologist James Watson, born in Chicago in 1928, was one of the four scientists who cracked the structure of DNA in 1953. Along with Francis Crick, Rosalind Franklin and Maurice Wilkins, Watson won the race to uncover how the complex molecule DNA acts as a store for the genetic code. The aperiodic structure predicted by Schrödinger's *What is Life?* (see page 204) was discovered to exist in the pattern of four chemical compounds known as bases, which link the double helix structure of DNA like the rungs of a spiral staircase.

The Double Helix tells the story of the discovery, and, perhaps even more importantly, it takes the style of science titles further away from that of a textbook. While *Silent Spring* was a well-written polemic for the public, *The Double Helix* turns the discovery of science into drama. It brings in personalities, it shows the dark side of science in the desperate race to get there first, and it puts narrative at the heart of a science book. It is excellent storytelling. It would take a while, but within two decades of *The Double Helix* being published many great popular science writers had picked up the idea that narrative is as important as content.

James D. Watson
THE DOUBLE HELIX
(BIOLOGIE MOLÉCULAIRE DU GÈNE), EDISCIENCE À PARIS, 1968

A French first edition of Watson's book. James Watson (left) and Frances Crick pose in 1953 with an early model of the structure of DNA at the Cavendish Laboratory, Cambridge.

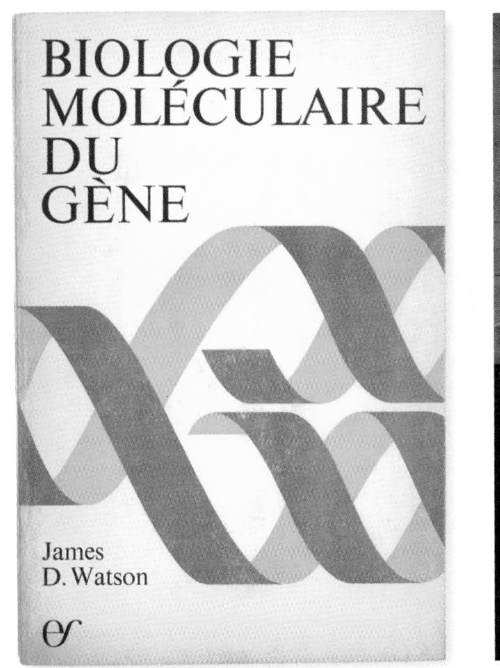

Of course, it helped the huge success of the book that Watson was involved in one of the greatest scientific breakthroughs of the century. Yet *The Double Helix* also highlights the dangers when a scientist writes about his or her own work. The discovery was shrouded in controversy anyway. Rosalind Franklin was excluded from the Nobel Prize in Physiology or Medicine, which was awarded in 1962 to Watson, Crick and Wilkins. As it happens, Franklin died before the prize could be awarded, and it is never awarded posthumously. But the prize is limited to three recipients, and some observers suspect that the infamously conservative Nobel committee would still have chosen to exclude Franklin had she survived.

There was also a dispute between Watson and Crick, working in Cambridge, and Franklin and Wilkins, working in London. Watson and Crick produced the theory, but it was Franklin and Wilkins' work on X-ray diffraction images of DNA that revealed the structure. In telling the story, Watson makes no effort to be objective – it is a tale of triumph from Watson's perspective. It makes the book interesting reading, but for a more balanced account it helps to have read contributions from Wilkins (in his book *The Third Man of the Double Helix* from 2003) and from Franklin's biographers (she died too soon to write her own book) such as Anne Sayre, whose *Rosalind Franklin and DNA* was published in 1975 – though the latter has been criticised by Franklin's sister for overstating the level of sexism she faced.

THE BODY HAS A HEAD
GUSTAV ECKSTEIN

Gustav Eckstein
THE BODY HAS A HEAD,
HARPER & ROW, 1969

A first edition cover of Eckstein's book alongside a photograph of Eckstein in his lab at the University of Cincinnati, feeding one of his favourite pigeons, Red.

By the time we reach the end of the 1960s it might seem that the cultural revolution would have led to a total transformation of writing style from the stuffy formality of previous ages. This did come, but it took time. A good example of a transitional work was *The Body Has a Head* by American medical doctor and psychologist Gustav Eckstein. Born in Cincinnati in 1890, Eckstein was old enough to have one foot in the past, yet by 1969 when his book came out, he was aware that there was a need for a new approach.

In his exploration of the physiology of the human body, no one can accuse Eckstein of being cold and clinical. His writing is full of literary flourishes, never using one word where a whole phrase could be squeezed in. Yet despite this, he sometimes manages a dazzling turn of speed. In the first 20 pages, the reader is transported from the earliest ancient Greek writers to Descartes describing the body as a machine.

The strange, staccato style Eckstein adopts is sometimes closer to poetry than prose. This is illustrated well by his first words on the male sexual role: 'Into the town comes the swashbuckler, has something to sell. That is his role, or his illusion. Assault is his physiology.' There is no doubt that Eckstein was eccentric. He managed to get a play he wrote performed on Broadway, which one review described as 'anti-entertainment'. And visitors to his lab would be greeted by Eckstein wearing a large, battered straw hat, worn to protect him from the droppings of the many canaries that flew free inside.

The Body Has a Head now feels like a period piece. For example, this book was published just seven years before Richard Dawkins' *The Selfish Gene*, yet genetics receives just three lines of text. Nonetheless, the book was a big success, because this was a subject that combined great public interest with little previous coverage. The title was both influential in exposing the public to the details of human biology and in helping other scientists and writers to realise how far a science book could deviate from conventional, clinical and, frankly, dull writing styles.

The best immediate contrast to *The Body Has a Head* is *Le Hasard et la Nécessité* (Chance and Necessity) by the French biologist Jacques Monod, published in 1970. Born in Paris in 1910, Monod won the Nobel Prize in Physiology or Medicine for his work on the interaction of genes with enzymes and viruses. Monod was probably best known outside the scientific field for his humanist views and a philosophy that put science at the forefront of the interaction between humanity and the world. However, his book prefigured more familiar titles in English-speaking countries such as those of Richard Dawkins and Daniel Dennett in exploring the implications of evolution and genetics, emphasising that evolution does not have goals, but rather is based on randomness – the 'chance' in the title of the book.

Monod's book shows how random mutations were responsible for creating humanity, rather than any deistic guidance. He also strongly refutes the Marxist concept of dialectical materialism, which he explains misunderstands evolution, seeing it as a mechanism for higher levels of existence to emerge from lower. Monod is clear that the randomness of the evolutionary process meant that it's perfectly possible to evolve a 'lower' level of existence. While the science in this book is good, Monod's vision is of a cold, purely scientific viewpoint that encourages a move from democracy to technocracy. Monod was way ahead of Eckstein in his science, but lacked Eckstein's ability to address the humanity of his readers.

Jacques Monod
Le hasard et la nécessité
essai sur la philosophie naturelle
de la biologie moderne

Jacques Monod
Prix Nobel
SEUIL

aux Éditions du Seuil, Paris

Jacques Monod
LE HASARD ET LA NÉCESSITÉ, EDITIONS DU SEUIL, 1970

The first edition cover of Monod's book (emphasising his 1965 Nobel Prize) alongside a rather stiffly posed photograph of Monod from the same year.

Crystal balls

The middle of the twentieth century was a period of upheaval in science and also in science writing. The year 1970 saw the publication of what would be by far the most influential title in a genre that straddles science, history, politics and economics: futurology. Since ancient times humanity has sought to predict the future. While original attempts depended on the occult, with no scientific basis, the development of probability and statistics have enabled us to make our best guesses about the way systems might develop. We will discover in the next chapter that such attempts will always be limited by aspects of chaos (see page 237). But this has not stopped a large number of popular books being published attempting to describe humanity's destiny.

Such titles were not new in 1970. There was a significant fad, for example, around the end of the nineteenth century for books describing a future world of technology, and H. G. Wells would put forward his own 'future history' in the 1933 novel *The Shape of Things to Come*. But such titles were fiction. Although modern attempts at futurology are equally fictional from a scientific viewpoint, they have gained traction as serious works, thanks principally to *Future Shock*.

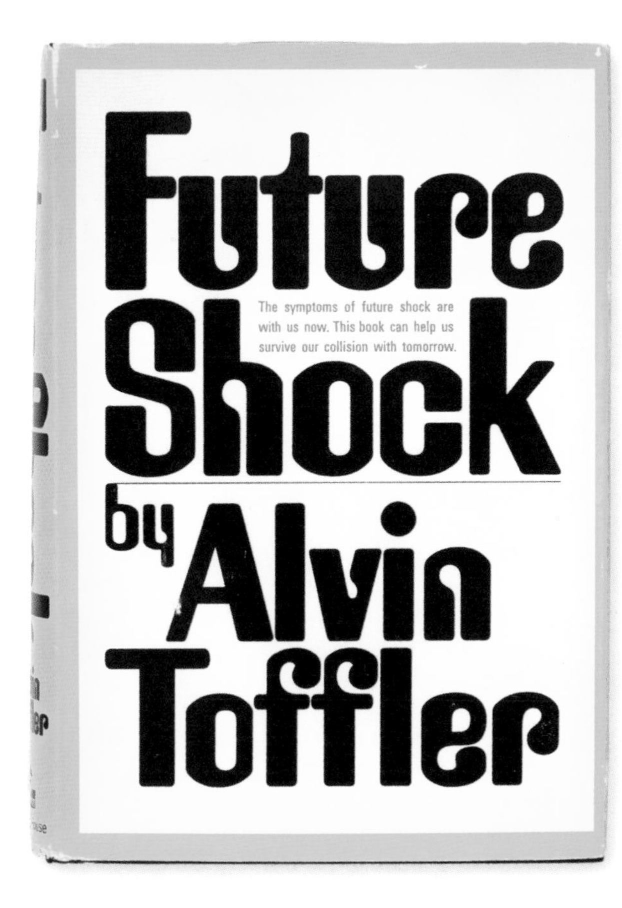

Alvin Toffler
FUTURE SHOCK,
RANDOM HOUSE, 1970

The first edition cover of Toffler's bestseller, which made him a media star, alongside a photograph of Toffler in New York from the following decade.

This was the work of Alvin Toffler, a New York-born writer and 'futurist', whose biggest impact was probably popularising the term 'information overload'. *Future Shock* sold over 6 million copies and was lapped up widely. Central to the book was the idea that the rapidity of change since the Industrial Revolution was overwhelming to humans (a message that worked better in a science-fiction novel inspired by Toffler's book, John Brunner's *Shockwave Rider*). Like the output of many soothsayers of the past, however, the contents of Toffler's book don't stand up to modern scrutiny. He imagined, for example, a throwaway society where women would wear one-use paper dresses. There was no prediction of the green ideas that have transformed attitudes to disposability. At the time of writing, single-use plastic is being demonised: Toffler expected disposable products to continue to increase in popularity.

It's not that Toffler got everything wrong – he was probably strongest on information technology, which was something of a specialty – but the impact of the book demonstrates powerfully the general direction of science writing, which has moved from emphasising the details of science to putting them into context. For Toffler and those who followed with other futurological titles, the context was not the past, or even the present, but the future.

The genetic revolution

As a result of the work on DNA and discoveries of the action of sections of DNA strands called genes – corresponding to the hereditary information predicted by Mendel's work over 100 years previously (see page 168) – the public's interest in genetics soared during the 1970s (something that Toffler also didn't foresee). And never more so than with the publication of *The Selfish Gene* by English zoologist and evolutionary biologist Richard Dawkins.

Born in Nairobi, Kenya, in 1941, Dawkins is now probably better known as a strident atheist, whose 2006 book *The God Delusion* has become a worldwide phenomenon. However, it's hard to remember now how much impact *The Selfish Gene* had when it was published in 1976. The concept in the title is overdramatised – Dawkins suggests that genes are selfish in the sense that from their 'viewpoint', living organisms are nothing more than a vehicle for ensuring the survival of the gene. However, what proved important about the book was that it revived and changed the discussion on evolution.

As we have seen, there was surprisingly little negative reaction following the publication of Darwin's books. But by bringing evolution to the fore, Dawkins helped bolster the anti-evolutionary feelings in religious groups by combining scientific theories with his abrasive brand of atheism. Ironically, for some years Dawkins held the post of Professor for Public Understanding of Science at Oxford University. Without doubt, his books, such as *The Selfish Gene*, did improve public understanding of science, yet Dawkins probably did significantly more damage to understanding with his confrontational style.

Even so, *The Selfish Gene* gave the public one of the earliest descriptions of a genetic view of evolution (now the standard way to look at it), rather than from the viewpoint of species. Species are arbitrary labels, devised for convenience of fitting with the kind of structure that was developed by Linnaeus (see pages 128–9). Genes are actual physical distinctions. And by taking the genetic viewpoint, Dawkins was also able to examine the implications for our tendency to provide more social support for close biological relations in whom we have the closest genetic investment.

Of significantly less importance (though it was praised at the time), the book also introduced the word 'meme' – a term Dawkins invented to suggest that ideas and behaviours can be described in an equivalent way to genes. As they are passed from person to person, he suggested, ideas evolve in a similar way to biological organisms (though much quicker). Now the concept is not taken so seriously, and the word has been repurposed to describe images with witty captions, which are spread on social media.

The Selfish Gene was highly influential on public attitudes in some countries, though it isn't quite clear why it was described as the most influential science book of all time in a 2017 poll by the UK's Royal Society. As we have seen, exactly what constitutes a science book has changed through the years, but it's hard to believe that some earlier science books with a much longer-lasting impact have not been more influential than *The Selfish Gene*.

Dawkins has sometimes been accused of taking the joy out of nature. He countered this effectively in his 1998 title *Unweaving the Rainbow*, one of a range of books that followed *The Selfish Gene*. The title refers to the poet John Keats' argument in his poem *Lamia*:

Richard Dawkins
THE SELFISH GENE,
OXFORD UNIVERSITY PRESS,
1976

Reminiscent of a science-fiction novel cover of the period, the first edition of Dawkins' bestseller seems designed to cover up its origins from Oxford University Press – an academic publisher.

James Lovelock
LE NUOVE ETA' DI GAIA (GAIA), BOLLATI BORINGHIERI EDITORE, 1991

A rather elegant Italian edition from 12 years after *Gaia*'s first publication in 1979, alongside a 1980 photograph of Lovelock in his Cornish garden with the fluorocarbon detector he invented.

Do not all charms fly
At the mere touch of cold philosophy?
There was an awful rainbow once in heaven:
We know her woof, her texture; she is given
In the dull catalogue of common things.
Philosophy will clip an Angel's wings,
Conquer all mysteries by rule and line,
Empty the haunted air, and gnomed mine –
Unweave a rainbow, as it erewhile made
The tender-person'd Lamia melt into a shade.

Keats suggests that by explaining and hence 'unweaving' the rainbow, Newton and other philosophers make the wonderful dull. As others, notably Richard Feynman, have made clear, scientists argue the reverse. Science does not stop us appreciating the beauty of nature, but we can add to this the insight that understanding brings. This was Dawkins' message in *Unweaving the Rainbow*.

From Gaia to Gödel

While Dawkins was enthusiastic not to overlook the beautiful, it's likely he would not have been comfortable with some of the spiritual implications that have been loaded onto a book published in 1979, a few years after *The Selfish Gene*. This was *Gaia* by James Lovelock, an independent English scientist born in Letchworth in 1919. *Gaia* covers an environmental theme – but takes in a far bigger picture. Admittedly, Lovelock does not avoid the potential for his book being considered spiritual by giving it the name of the Greek goddess of the Earth. But his 'Gaia hypothesis' is purely scientific.

Lovelock came up with the Gaia principle while working on instrument design for NASA. His career had been anything but conventional. He couldn't afford to go to university after school and studied in the evening until he had raised enough money to begin a chemistry degree. He went on to undertake medical research during the Second World War, which led to a PhD in medicine. After this, he experimented successfully on lowering the body temperature of rodents to freezing point and then reanimating them, before working for NASA on instruments used in planetary exploration. Lovelock has never had a university affiliation, preferring to work on his own, in the manner of a natural philosopher of an earlier generation.

When Lovelock was developing ideas for detecting life on Mars for NASA during the 1960s, he began to construct the Gaia hypothesis. This theory suggests that the whole of the Earth – both its living and non-living parts – acts as a self-regulating system, that could be likened to a single living being, just as the independent cells of the body act together to form a single organism.

When *Gaia* was published, it won a lot of support from environmentalists, who liked the message that the we are part of a larger whole, though they did not necessarily think through the implication of a self-regulating system where, for example, living species were not important per se. The hypothesis drew significant criticism from scientists (including Dawkins), who pointed out that the Earth didn't have the feedback mechanisms that an organism has between its parts.

While it's possible to argue the pros and cons of Gaia as a scientific theory, there is no doubt that the book *Gaia* had a large impact on both environmentalism and enthusiasm for environmental sciences. It seems likely that both sides of the argument were too extreme – it is probably an exaggeration to describe the Earth as an organism, but the opponents of Gaia overlooked genuinely useful observations on self-regulating systems, which they struggled to understand.

Published the same year as *Gaia* was a book that probably rivals Stephen Hawking's *A Brief History of Time* for the proportion of readers who have started but never managed to complete it. Written by Douglas Hofstadter, an American professor of cognitive science born in New York in 1945, *Gödel, Escher, Bach* is a sweeping fusion of ideas on what its subtitle describes as 'An Eternal Golden Braid'. It has been equally applauded for its depth of thinking and criticised for being near-incomprehensible.

The very title tells the reader that this is no ordinary science book. We get the names of a mathematician who proved that no system of mathematics can be complete (Gödel), an artist whose work is regularly used to illustrate ideas of symmetry in maths and physics lectures (Escher), and a musician who was known for his mathematical approach to composing (Bach). If Bronowski brought us the fusion of science and culture in *The Ascent of Man*, then Hofstadter takes us deeper into the melding of the disciplines in exploring the nature of knowledge and systems.

Although the work of the three title characters is featured, the book is not an attempt to find crossovers but rather pulls all areas of human thought together to try to understand how our mental processes work. It explores where the ability to think comes from in the collection of cells that makes up a brain. As a piece of writing, it's arguable that *Gödel, Escher, Bach* is unnecessarily complex in its structure and more than a little self-aware of its own cleverness. Hofstadter plays around with narrative flow, presents different discussions and puzzles and does very little to help the reader get the message.

Perhaps the most interesting aspect of *Gödel, Escher, Bach*, looking back on it, is how its communication style fits into the wider development of science book writing. We have seen how, since the time of Newton, science books have moved from being technical communications between practitioners, to becoming a voice from on high where experts deign to communicate to lesser beings, typical of the early twentieth century, through to the kind of engaging narrative that was just starting to emerge in the 1970s. These newer authors understood how to communicate well – something that had been lacking from the training and experience of the earlier academic writers.

By contrast, Hofstadter deliberately obscures his message, turning a science book into something that is best considered a work of art that happens to have scientific content. In this respect the book is highly unusual. It's not a form that works well to communicate science, but remains of interest as a piece of writing. Perhaps the only other title that came close to succeeding with this approach was *Infinite Jest*, the 1996 novel by American author David Foster Wallace. Although this is more clearly fiction, it also plays with the form of the novel and brings in aspects of science and the mathematics of infinity.

Douglas R. Hofstadter
GÖDEL, ESCHER, BACH,
PENGUIN, 1984

A Penguin edition from five years after the original publication of the book that fascinated and bemused in equal measures.

By the end of the 1970s, science books by female authors were still relatively uncommon. In both science and science writing, a major structural change had yet to take place. On my wall I have my *Cavendish Laboratory Part II Students* photograph from 1975, showing all the students starting the final year of an undergraduate natural sciences degree specialising in physics at Cambridge University. Of around 200 students, perhaps half a dozen are female. Similarly, in the US in the 1970s, fewer than 15 per cent of students in physics, maths, computer science and engineering were female. The causes of this disparity have been much discussed, but it seems likely there was still a strong cultural assumption that some disciplines were more 'suited' to female talents and some to male – an assumption that has lagged well behind data disproving it.

Things were significantly better in the biological sciences (in the 1970s, for example, around 40 per cent of US biology undergraduates were female and the figure is now closer to 60 per cent), but given that science writers tend to be either working scientists or have science degrees, it is not entirely surprising that the vast majority of science books before 1980 were written by men. Thankfully, though, this was about to change, as we shall see in the next chapter.

5

THE NEXT GENERATION

TRANSFORMING UNDERSTANDING

As science writing moved into the 1980s, it continued to evolve. The public's attitude to science was changing, from reverence to a mix of fascination and scepticism. The importance of science was greater than ever, and the public wanted better to explore the context and subtext of scientific discoveries. Many of the best books written in the period covered by this chapter – from 1980 to the modern day – have been the work of science writers, rather than working scientists. Where the previous chapter closed with a book that seems designed to make its message more obscure, this period has seen accessible science gradually come to the fore. Science books have grown up.

Interpreting quantum physics

The move to making science approachable did not mean, however, that science writing would entirely avoid obscure subjects (or even obscure presentations). Some of the old school were still active, and none more so than rebel physicist, David Bohm. Born in Wilkes-Barre, Pennsylvania in 1917, Bohm specialised in quantum physics. He could have been one of America's leading scientists, but despite his contribution to the Manhattan Project, his left-wing political views made him a target of Senator McCarthy's House Un-American Activities Committee. Bohm moved to Brazil in 1951, and later to Israel, before settling in the UK.

For physicists, Bohm's most important book was his textbook *The Special Theory of Relativity* from 1965, yet his most original book, and one that reached a more eclectic audience, came later in life. This was *Wholeness and the Implicate Order* from 1980. The book builds on Bohm's unusual interpretation of quantum theory to give a mixed physical and philosophical view of the workings of the universe as a whole.

Quantum physics is the only area of science where there is a significant focus on interpretations. The basics mathematics of quantum theory is stunningly precise at predicting the behaviour of electrons and atoms, molecules and photons. Richard Feynman observed that it was so accurate it was like predicting the distance from New York to Los Angeles to the width of a human hair. But the theory says nothing about what actually happens inside a quantum system. The physics is a black box where we put the numbers in, turn the handle and get a correct result out, but with no idea why these results emerge.

Ever since the early days of quantum theory, physicists have come up with interpretations that try to explain what is going on. The original, and still probably most used, is the Copenhagen interpretation, named after the home city of founding quantum physicist Niels Bohr. It says there is nothing but probability before a particle is observed – there is no underlying reality. Often described as 'shut up and calculate', this approach did not satisfy everyone. David Bohm picked up on the work of French quantum physicist Louis de Broglie, suggesting that each particle had a real (if unknowable) location, and had an associated 'pilot wave' which guided its movement.

The problem with Bohm's theory, and the reason that it has never been widely accepted, is that it requires everything everywhere to have immediate influence on the rest of the universe. For his interpretation to work, the universe has to be an interconnected whole, transcending the lightspeed limit of Einstein's special relativity. It was this that led Bohm to write *Wholeness and the Implicate Order*, which put forward the idea that there are two levels of reality: explicate reality, the level we usually xperience; and implicate reality, which underlies everything and links everything together, and is responsible for still unexplained phenomena such as consciousness. Bohm's ideas are even harder to penetrate than those of Douglas Hofstadter in *Gödel, Escher, Bach* (see page 228) but there is no doubt that this was an influential book, reaching well beyond the physics community.

Bohm's title does not explore the nature of quantum physics itself, a topic that was relatively infrequently touched on until the following century. Feynman's book *QED* (see page 213), for example, only covered a relatively small subset of the topic. However, English science writer John Gribbin gave quantum physics a new appeal in 1984 with his best-known book, *In Search of Schrödinger's Cat*. Like Desmond Morris and Richard

David Bohm
WHOLENESS AND THE IMPLICATE ORDER,
ROUTLEDGE & KEAGAN PAUL, 1980

A first edition of Bohm's title with a cover that makes no concessions to the reader.

John Gribbin
IN SEARCH OF SCHRÖDINGER'S CAT,
BANTAM BOOKS, 1984

Quantum physics had very limited popular exposure before Gribbin's book.

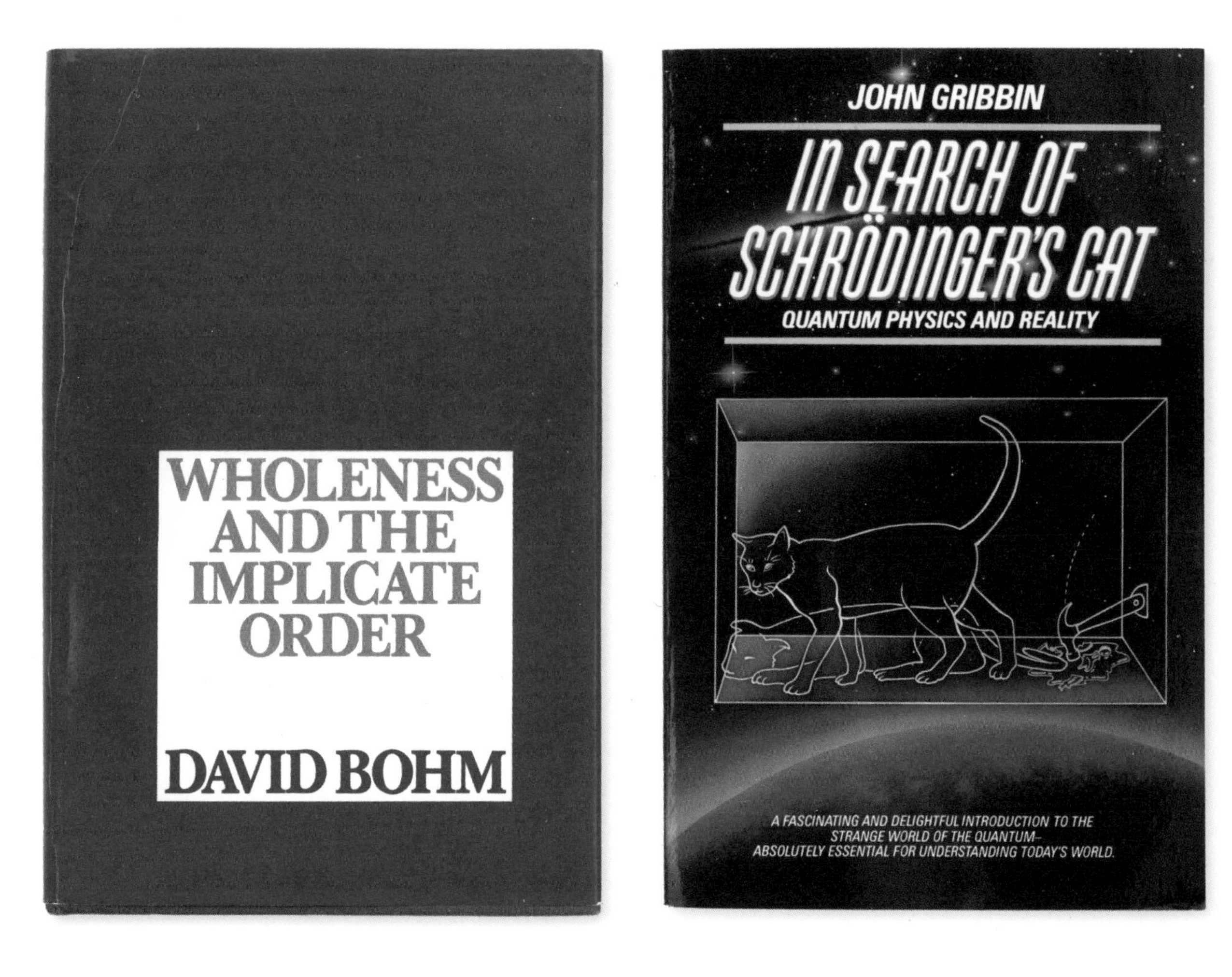

Dawkins, Gribbin started his life as a working scientist, but became an extremely prolific science writer, continuing to the present day. The cat in question is the poor hypothetical animal that appeared in the Austrian physicist's thought experiment back in 1935.

The idea of Schrödinger's cat was to highlight the oddity of a concept called superposition. This says that a quantum particle that could be in one of two possible states will not be in either state until observed; instead, it exists purely as probabilities of each possibility. In the case of the cat, Schrödinger envisaged an experiment where a detector released poison gas into a box when a quantum particle decayed. When this happened, the cat in the box would die. But before the particle was observed, it would be in a superposition of both decayed and not-decayed states – meaning that until the box was opened, the cat would be neither dead nor alive.

In practice, this was a throw-away concept that doesn't deserve the attention that has been given to it – but the image of the dead-or-alive cat proved popular and Gribbin cleverly hangs his book on it, even though he covers far more of quantum physics. Gribbin's is one of the earlier examples of what might now be considered the standard framework for a popular science book. It explains the science – in this case quantum physics – largely without using mathematics, but with a considerable amount of analogy, and it puts the science into context, giving it a significant amount of human interest by including some history of science too, telling us about the people who developed the theory and how they came up with their ideas. *In Search of Schrödinger's Cat* is the archetypal approachable popular science book.

Confusion in the brain

Oliver Sacks
THE MAN WHO MISTOOK HIS WIFE FOR A HAT,
SUMMIT BOOKS, 1985

Despite being a collection of unconnected essays, Sacks' book struck a chord with readers thanks to his storytelling ability.

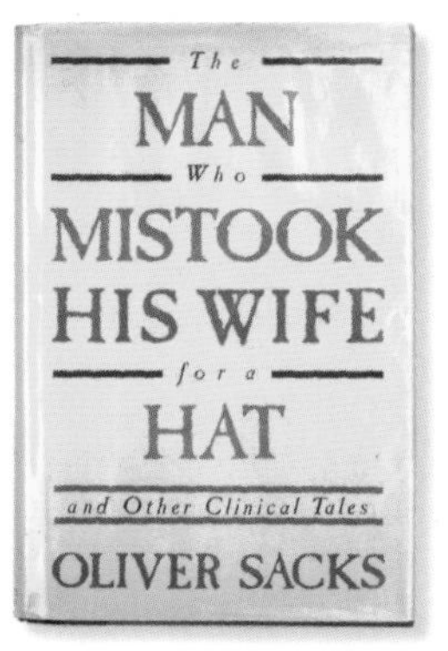

Something that comes through strongly in the influential texts since 1980 is the dominance of two fields: physics and human science. *Schrödinger's Cat* was followed up in the next year by a classic human science title, *The Man Who Mistook His Wife for a Hat*. Written by English neurologist Oliver Sacks, the book plays on our human interest in the suffering of others, exploring the cases of a number of Sacks' patients.

Unlike most successful science books, this 1985 title is a collection of unconnected essays, each exploring a different case study. The studies cover individuals suffering from a range of conditions, including the visual agnosia of the title story (an inability to visually recognise objects), the inability to remember anything that has happened since the end of the Second World War, and the experience of twin autistic savants. Sacks uses storytelling surprises to keep the reader engaged, such as in his account of the reactions of the inhabitants of a ward for the mentally ill to a televised speech, where the elements that the individuals took from the speech were totally different, informed by their conditions.

There is a danger that a book like this becomes the literary equivalent of a freak show. This was even more of a risk with the 2004 title *Mutants* by evolutionary biologist Armand Leroi, which showed how we are all genetic mutants, using as examples historical extremes of human mutation who were exhibited and vilified. Leroi manages to avoid the freak-show effect by keeping human development as the focus of the book, rather than the mutants. It's harder to get away from the individuals in Sacks' book, because

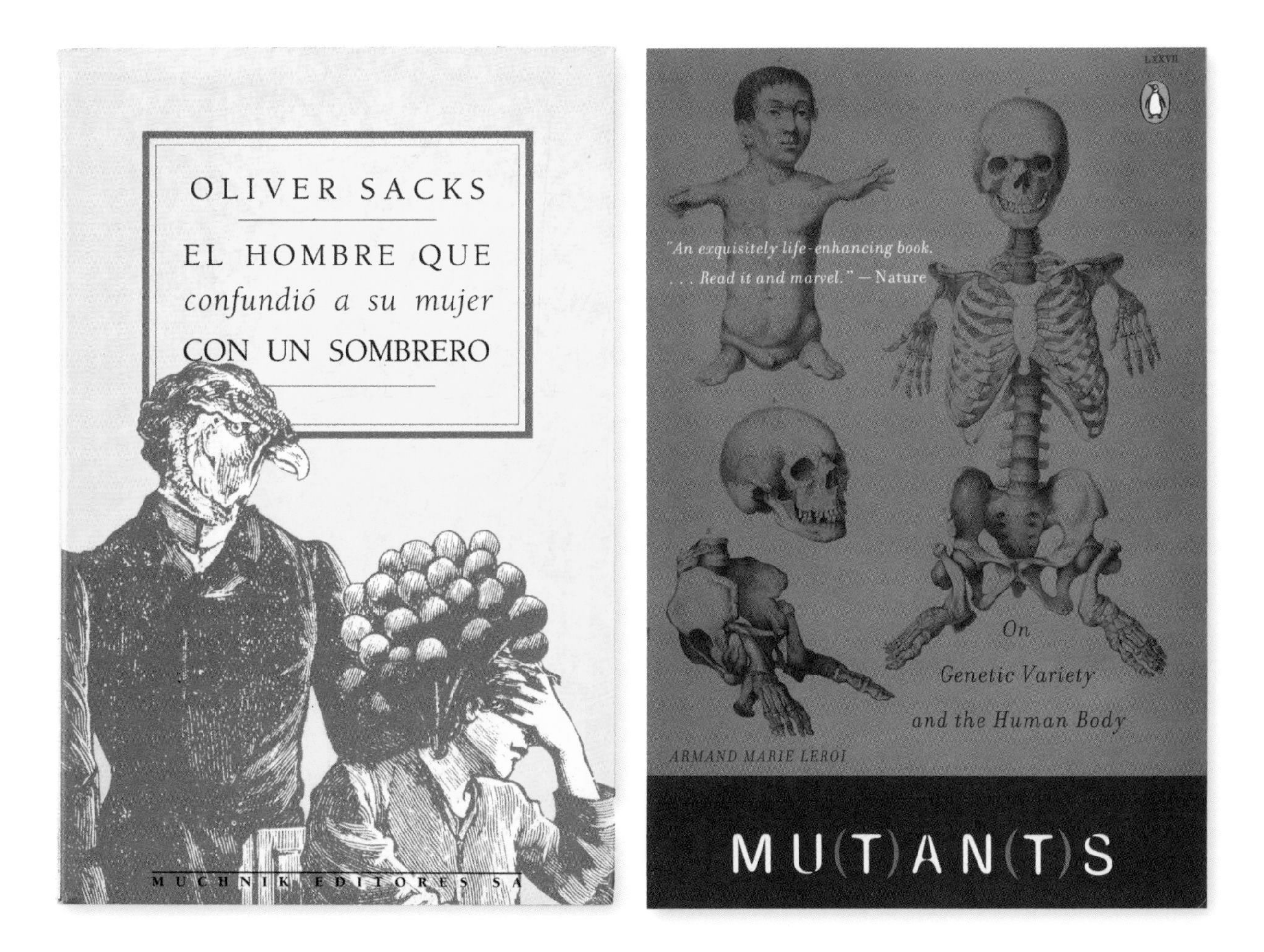

here the disorders themselves are his topic. Even so, Sacks manages to use his case studies to explain aspects of the functioning of the brain without blatant exploitation. The result was a bestselling title that appealed to a much wider public than a normal medical text.

Armand Leroi
MUTANTS, PENGUIN, 2005

A US Penguin edition of the UK original published in 2004 by HarperCollins.

Oliver Sacks
THE MAN WHO MISTOOK HIS WIFE FOR A HAT (EL HOMBRE QUE CONFUDIÓ A SU MUJER CON UN SOMBRERO), 1991

A Spanish edition published by Muchnik Editores.

Science stories and scientific legends

As we have seen, the co-discoverer of the structure of DNA, James Watson, wrote a popular book in 1968. *The Double Helix* (see pages 218–19) took a story- and biography-driven approach that would come to typify the best American popular science titles, in contrast to European equivalents that tended to be driven more by the science. Typifying this 'human-interest story' approach was a book published in 1987 called *Chaos: Making a New Science* – the first book by a mainstay of US popular science, James Gleick.

James Gleick
CHAOS, VIKING PENGUIN INC., 1987

A dramatic cover suggestive of fractal patterns from the first edition of Gleick's successful story-driven approach to a mathematical topic.

Born in New York, Gleick remains unusual in popular science circles in having an arts background rather than a science degree. This no doubt contributed to his use of the distinctive story-driven approach seen in the opening of *Chaos*: 'The police in the small town of Los Alamos, New Mexico, worried briefly in 1974 about a man seen prowling in the dark, night after night, the red glow of his cigarette floating along the back streets. He would pace for hours, heading nowhere in the starlight that hammers down through the thin air of the mesas.' It might have been the opening of a noir detective novel. If the staid old 'authority preaching to the masses' approach of popular science had been pushed aside by the likes of Watson and Gribbin, it was now totally shattered.

Another impressive aspect of Gleick's book was that it made popular maths acceptable. Up to this point, mathematics had been a practical subject, either taking the reader through a method and examples, or being an unfortunate adjunct to science that made popular titles harder to read. In *Chaos*, Gleick showed that, done right, maths could be as appealing as the other sciences. It remains the hardest topic to make accessible, yet *Chaos* would make it possible for a range of other popular maths books to do well.

Gleick's topic, the development of chaos theory, was timely. Chaos in mathematical terms is not the inchoate mess implied by the general usage of the word. Chaotic systems obey clear rules. However, interactions between parts of the system are so disruptive that the tiniest change in the way a process starts can have a big influence on its progress. This is why weather forecasting is so difficult. In fact, chaos theory began when American meteorologist Edward Lorenz decided to re-run an early computerised forecast. The process took hours, so rather than start it at the beginning, he used the values from a printout made after the program had run for some while. However, the computer missed a couple of decimal places off the printout compared to the values it used internally, and the tiny fractional change totally altered the forecast.

At the time Gleick wrote the book, chaos theory seemed about to transform the way that mathematics was used in science, and Gleick was able to make it sound impressively important (it's no surprise that in the movie *Jurassic Park*, made just a few years after *Chaos* came out, Jeff Goldblum's sceptical character Dr Ian Malcolm was an expert in chaos theory). As it happens, that early promise fizzled out. Awareness of the impact of chaotic systems has been hugely important, but the direct applications that Gleick's book foresaw have not come to pass. Even so, *Chaos* remains a great read and, crucially, took the development of science writing to the next level.

The following year, a book was published that had a huge impact on the popular science market and would make its author a media star. The book was *A Brief History of Time* by Stephen Hawking. Interestingly, in contrast to *Chaos*, the book's approach is that of an authority speaking from on high – though Hawking softens this by adding a few personal comments and touches (added during repeated edits as the publisher tried to make the book more accessible).

The famous English physicist worked on highly esoteric aspects of black holes and cosmology, yet his personality and his management of his debilitating medical condition won over the public. *A Brief History of Time* became the bestselling popular science title in modern history, until it was eclipsed 15 years later by Bill Bryson's *A Short History of Nearly Everything* (see pages 245–6). Despite the title, Hawking's book tells the reader

Stephen Hawking
A BRIEF HISTORY OF TIME (DAL BIG BANG AI BUCHI NERI), 1988

The first edition cover of Hawking's bestseller published by Bantam Press, alongside an Italian translation published by Rizzoli in the same year that reverses title and subtitle to make this *From the Big Bang to Black Holes.*

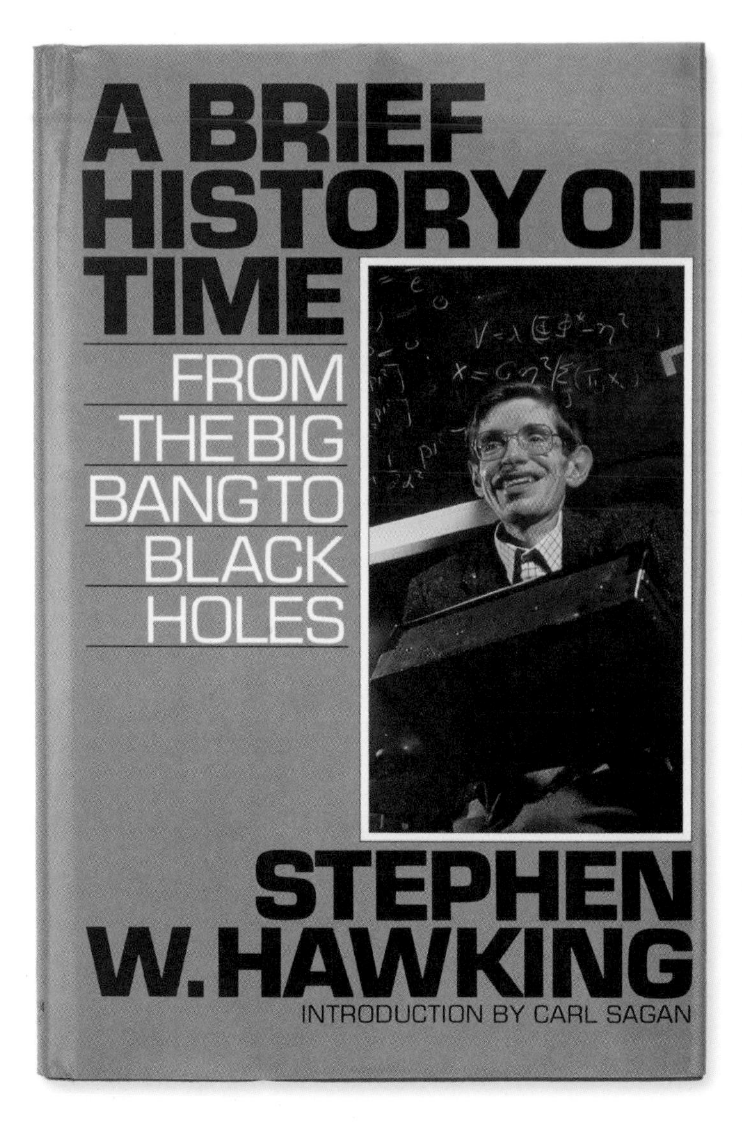

Stephen Hawking
A BRIEF HISTORY OF TIME
(EINE KURZE GESCHICHTE DER ZEIT DIE SUCHE NACH DER URKRAFT DES UNIVERSUMS), ROWOHLT, 1988

The first German edition of this popular science classic where the subtitle has become 'the search for the primal force of the universe', alongside a photograph of Stephen Hawking from the 1980s.

very little about time itself. Instead, it starts with the way that relativity intertwines space and time, then goes on to cover cosmology.

A Brief History of Time is not a long book, which may make it seem odd that it has such a reputation for being far more frequently bought than read. In reality, many purchasers may well at least have started it, as Hawking eases the reader in gently with a warm story of a scientist (possibly the philosopher Bertrand Russell) giving a lecture on astronomy, only to be told by an old lady in the audience, with shades of Terry Pratchett, that the world was a flat plate on the back of a giant tortoise.

Hawking took note of his publisher's repeated warning that every equation included halves the number of readers. But the book lacks context and narrative, and I suspect many readers gave up when Hawking got onto the relativistic concept of light cones, which can confuse more than they explain. It feels odd to describe *A Brief History* as a great science book, given so many readers have given up on it – the essence of a great science book is its ability to communicate science with the desired audience, in which requirement this title clearly fails. However, what Hawking's book did was to bring popular science to the attention of publishers, impressing them by its sales, and its transformation of the popular science market has enabled far more effective books to published.

The TV effect

It's arguable that the success of *A Brief History* also encouraged a short flowering of high-quality science programming on television, which would themselves have spin-off books. There has always been a difficult balance in presenting television science. Early attempts were, like early popular science books, very much about an authority lecturing the audience. But television also produced lowest-common-denominator shows, which spent most of their time on pretty images and hand-waving statements with little coverage of the science itself.

The exception to this has been natural history, which lends itself to effective visual presentation more than any other scientific topic. As a result, for example, of the success of British naturalist and broadcaster David Attenborough's many series, such as *Life on Earth* and *Blue Planet*, we have seen a number of visually pleasing books. It's arguable, though, that these titles – large format and heavily illustrated – aren't true science books. Where a science book covers far more, in much greater depth, than a documentary, these 'book of the series' titles simply follow the structure of the show and illustrate the episodes in it. And because these programmes don't usually have the sophistication of scripting of Bronowski's *The Ascent of Man* (see page 218), the result is limited as science writing.

DAVID ATTENBOROUGH WITH MOUNTAIN GORILLAS, 1979

A scene in Rwanda, on location for the BBC television series *Life on Earth.*

Natural history apart, only a handful of television series have had much literary impact. Carl Sagan's 1980 *Cosmos* series in the US produced a very popular spin-off book, though its history of science was distinctly weak, claiming that nothing happened between the fifth and the fifteenth centuries, which is clearly untrue. Meanwhile, in the years following the publication of *A Brief History of Time*, the BBC's Horizon programme had one major written success. This was a spin-off from the show *The Proof*, presented by English science writer Simon Singh. Despite using the same research as the television

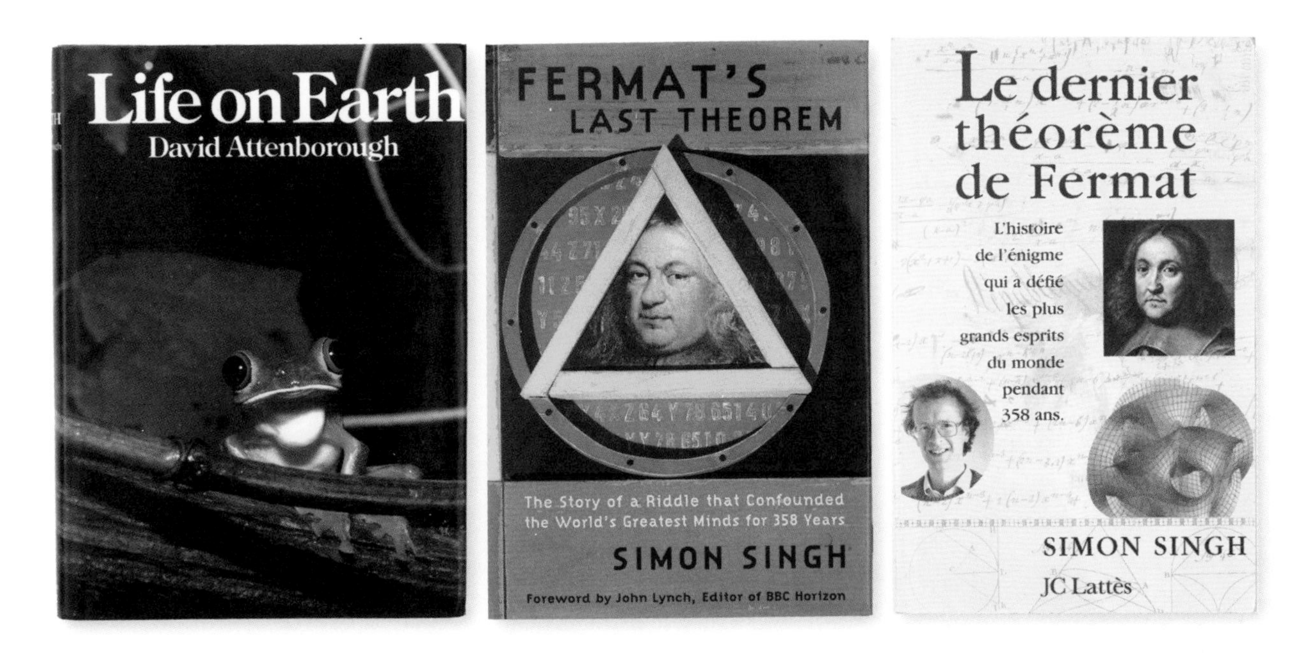

show, Singh's 1997 book on the same subject, *Fermat's Last Theorem*, was excellent, because it acted as a standalone title in its own right, rather than as a coffee-table 'book of the series', containing far more than the documentary ever could.

Singh achieves something of a miracle with this title. Not only is *Fermat's Last Theorem* about the frequently inaccessible topic of maths, it covers a particularly obscure aspect of it. Yet by providing an effective narrative, Singh makes the topic fascinating – and as a result he had the UK's first ever number-one non-fiction bestseller with a maths title. The theorem in question shows that it isn't possible to have three positive integers, each cubed or raised to (the same) greater power, where adding two together made the third. Not earth-shattering mathematics. But what has intrigued mathematicians since the seventeenth century is the way that the theorem was teasingly announced.

In 1637, French amateur mathematician Pierre de Fermat scribbled something in the margin of a copy of *Arithmetica* by the third-century Greek philosopher Diophantus (see page 61), which contained an early version of algebra. Diophantus was discussing a similar (solved) problem applied to the squares of numbers. Fermat noted that he had proved it was impossible to do this for cubes and higher powers. 'I have discovered a truly marvellous proof of this, which this margin is too narrow to contain', he noted.

Ever since, attempts have been made to find Fermat's 'proof', strictly a conjecture (as he gave no proof), but universally referred to as Fermat's last theorem. Remarkably, it was not until 1994 that the theorem was actually proved by English mathematician Andrew Wiles, using 100-plus pages of mathematics that were far beyond anything Fermat would have understood. (It is generally thought that Fermat was, at best, over-confident.) In his book, Singh takes this idea and builds it into a satisfying story, mixing the history of the concepts involved, from the earliest days of algebra to Wiles's epic solution.

David Attenborough
LIFE ON EARTH,
COLLINS/BBC BOOKS, 1991

Spin-off title from the popular television series.

Simon Singh
FERMAT'S LAST THEOREM
(LE DERNIER THÉORÈME DE FERMAT)

The first edition of Singh's excellent 1997 mathematical title published by Fourth Estate, alongside a 1998 French translation published by JC Lattès.

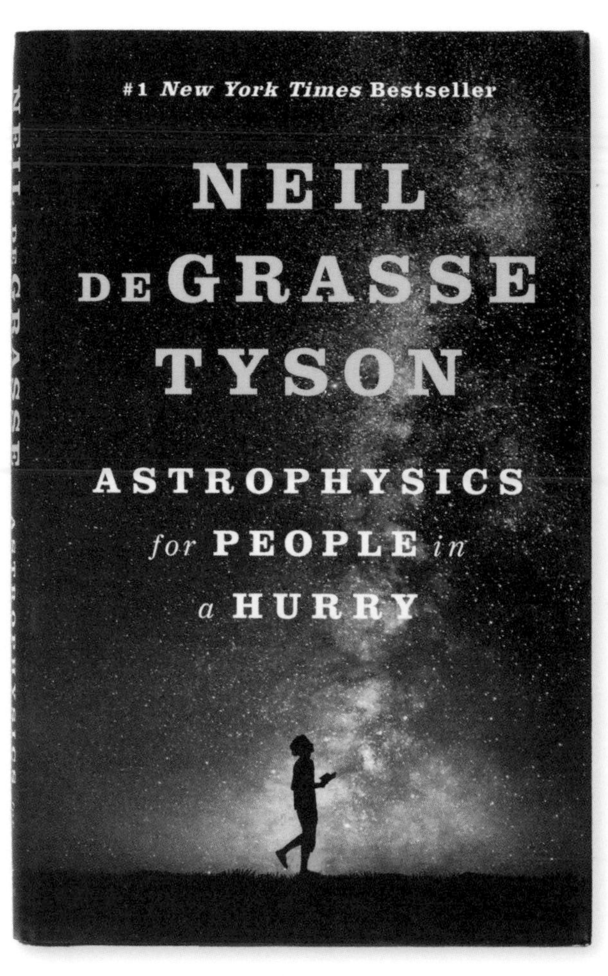

Neil deGrasse Tyson
ASTROPHYSICS FOR PEOPLE IN A HURRY,
W.W. NORTON & COMPANY. 2017

Popularity from presenting television shows helped Tyson, an astronomer, to gain a wider audience for his explanation of the basics of astrophysics.

Brian Cox and Jeff Forshaw
WHY DOES $E=MC^2$?,
DA CAPO PRESS, 2010

Although Cox had significant success with spin-off books from his television shows, his best titles have been standalone books going into far more depth, such as this one co-authored by Forshaw.

While we're in the area of books linked to television science shows, two notable, more recent ventures include the reboot of *Cosmos*, presented by American astronomer Neil de Grasse Tyson, and a number of UK science series presented by physicist Brian Cox. Although these inevitably spawned associated coffee-table books, far more interesting were the titles that both presenters put out in their own right on the back of their television exposure. For Tyson, the best example was the 2017 *Astrophysics for People in a Hurry*, while Cox has co-authored a number of titles with physicist Jeff Forshaw, notably *Why Does* $E=mc^2$*?* from 2010 and *The Quantum Universe* from 2012.

Where Tyson continues to write at the level of the television show, but adds more content, Cox takes the bolder step of writing books that require considerable effort to read. As a result, they really reward the reader by giving more depth to the physics than would normally be found in such a title. In both cases, though, what we have here are examples of modern science books by scientists – not covering their own tiny specialist area in technical detail as a scientist of a previous generation might have done, but introducing the general reader to the broader field.

Human stories and the innocent abroad

Science and technology go hand in hand, never more so than when determining longitude – the east-west position of a location on the Earth – a process that was inextricably tied to advances in astronomical science. This became increasingly important from the fifteenth century as long sea journeys were made. In essence, given that the position of the Sun at a particular time in the sky indicates the longitude of the observer, the essential requirement was to accurately fix the time, compared with a known location.

Astronomical attempts to solve the problem – an important enough puzzle that the British government in 1714 offered a significant reward for its solution – used observations such as the positions of the moons of Jupiter as celestial timepieces. However, the ultimate effective approach was to devise a clock that could survive long sea voyages and still keep accurate time. This was not easy, as most accurate clocks of the period used pendulums, useless on a rocking ship.

Describing the solution to this problem was *Longitude*, published in 1995 by American science writer Dava Sobel. Sobel has specialised in titles exploring the history of science through the viewpoint of an individual character. (Her other best-known title, *Galileo's Daughter* from 1999, uses Galileo's relationship with his illegitimate daughter Virginia to explore his work.) *Longitude* tells the story of English clockmaker John Harrison, his attempts to beat the longitude challenge, and his fight with the authorities reluctant to pay him his due.

Like the best popular science, Sobel's work gives plenty of information about the problem of measuring longitude and the methods of discovering it, but does so in a way that leads the reader along using the story of Harrison's life and work. Harrison was no overnight success – it took him many years to perfect his shipboard chronometer, and even when he did so, the British Board of Longitude was reluctant to award him the £20,000 prize, which the accuracy of his clocks should have won. This was a huge

John Harrison
THE MOVEMENT OF H.4,
INK ON PAPER,
CA. 1760–72

The movement of H.4, one of the series of longitude chronometers designed and drawn by John Harrison, described in Dava Sobel's *Longitude.*

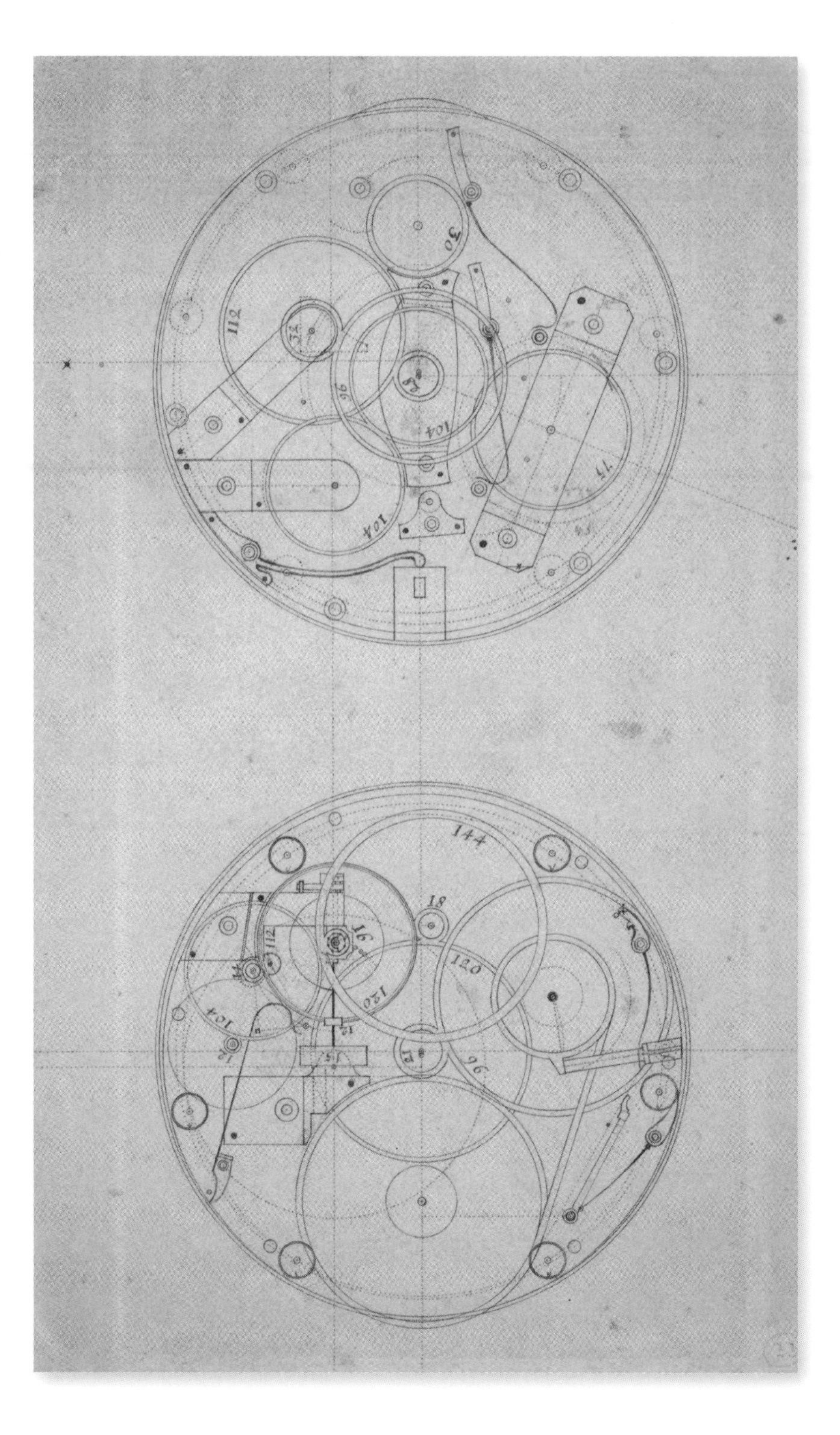

Dava Sobel
LONGITUDE,
FOURTH ESTATE, 1996

The first UK edition of Sobel's 1995 title on Harrison's struggles to win the Longitude Prize, described on the cover as 'the greatest scientific problem of his time'.

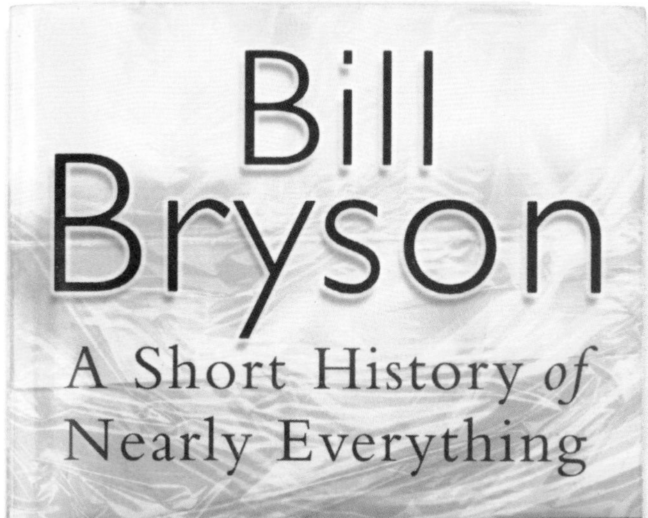

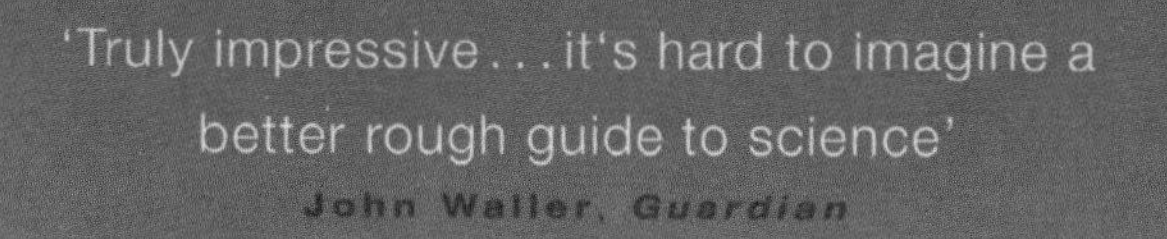

Bill Bryson
A SHORT HISTORY OF NEARLY EVERYTHING,
BLACK SWAN, 2004

First published in 2003, this accessible guide to science is now the highest-selling popular science book in the twenty-first century.

amount of money at the time – the equivalent in 2018 of around £2.8 million ($3.5 million) in purchasing power, or a remarkable £35.5 million ($45 million) in terms of labour value (based on what the average wage of the time would buy).

Harrison eventually got the majority of the prize money from the Board, which was topped up when he was 80 by the British parliament after Harrison petitioned them at the suggestion of King George III. What Sobel demonstrates is the importance of context for the general reader – she brings the science and technology alive through the circumstances of the lives of those involved in its development.

This approach is fine when looking back at how a scientific or technological breakthrough was made in the past, and such historical context is also very important when explaining the latest scientific breakthroughs. But with new developments, it can also help if the scientists involved in the discovery are interviewed in the process of writing the book. The most extreme example of this approach is Bill Bryson's *A Short History of Nearly Everything* from 2003, in which Bryson takes on the role of the ignorant but curious everyman, interviewing a string of scientists to discover more about different topics. The result was the bestselling modern science book yet to be published.

Three physics heavyweights

The success of Bryson's book did not mean that such lightweight overviews would come to dominate the field. Entirely different in almost every regard was Lee Smolin's important physics title from 2006, *The Trouble with Physics*. Smolin is an American theoretical physicist regarded by some as a maverick. Not only is the book written by an expert, it focuses on one specific area of science – string theory – and provides an analysis which, while approachable by the general reader, does not pull any punches.

Smolin gives the reader a good summary of string theory, the leading attempt to unify the two central but incompatible aspects of physics, quantum theory and the general theory of relativity. However, the power of his book is in communicating to the general reader the worrying aspects and flaws of string theory. All too often, science books for the public smooth over problems and present as fact what is no more than hypothesis. But Smolin opens up a can of worms that the physics community would have preferred to keep hidden. Smolin's was not the only book to do this – another physicist, Peter Woit, produced a more technical book on the same subject with the catchy title *Not Even Wrong* in the same year – but Smolin's has had the bigger impact because of its accessibility.

Smolin shows how the physics community has been beguiled by the elegance of string theory. Since the 1980s, many physicists have built their careers on this theory, and so have been highly reluctant to consider any other possibility. And yet the theory makes no useful predictions that can be tested. Smolin shows how much the focus on string theory has suppressed original thinkers and prevented the development of alternative theories.

Equally meaty in its contribution to an understanding of science, but focusing more on cosmology, was *From Eternity to Here* by American physicist Sean Carroll, published in 2010. This was, in many ways, what Hawking's *A Brief History of Time* should have been for its title to be accurate. As we have seen, one thing that Hawking's book did not

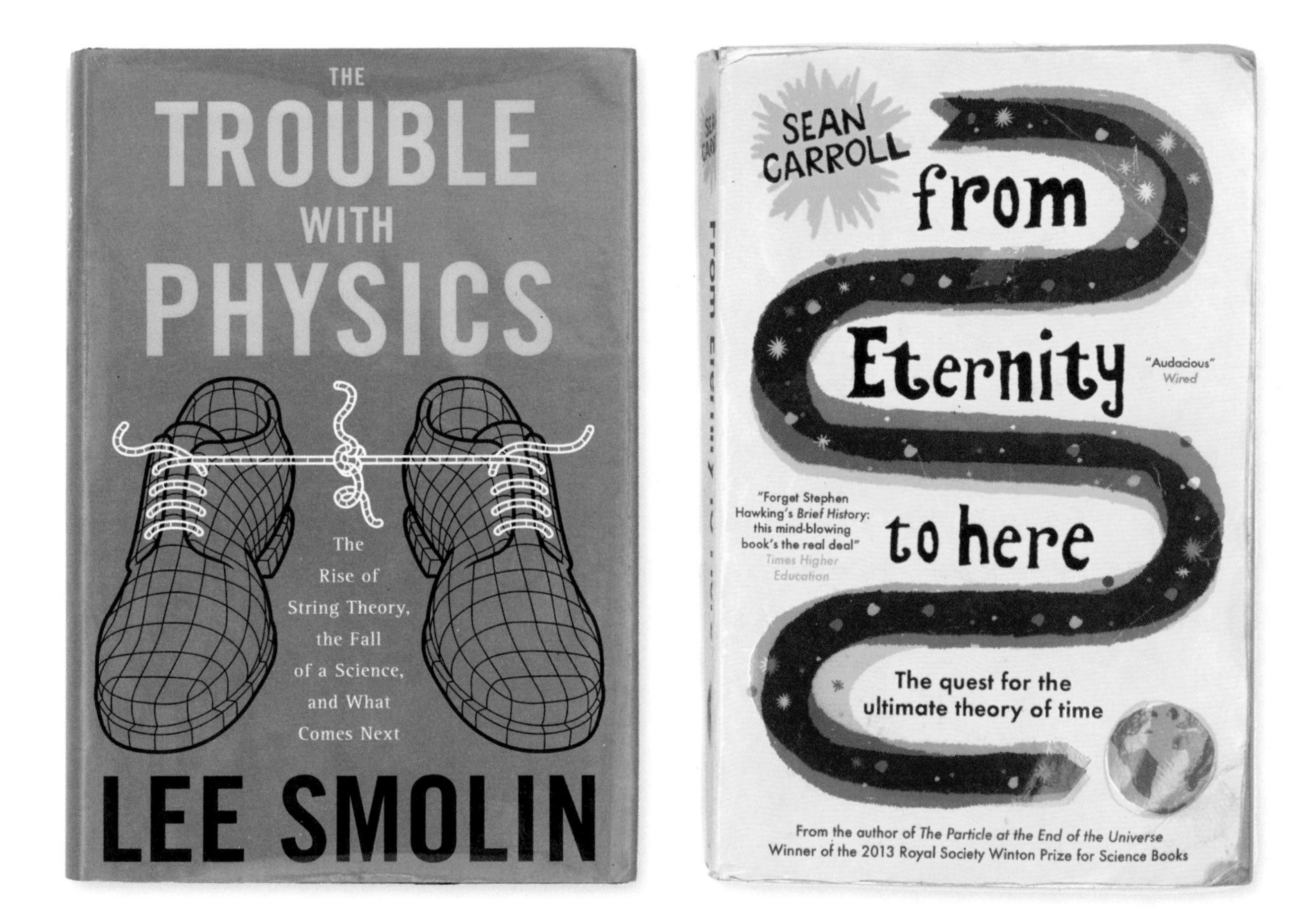

help with, frustratingly, was the nature of time itself. Carroll does this for us in a book that goes into considerably more depth than Hawking's, dealing with aspects of thermodynamics, relativity and quantum theory, yet managing to be significantly more comprehensible.

That the public has an appetite for science if it is presented right was demonstrated impressively in 2014 with the reception of another physics book, also written by a physicist, Carlo Rovelli's *Sette brevi lezioni di fisica* (Seven Brief Lessons in Physics). Hugely popular in Italy, Rovelli's book was one of the very few modern science books written in a language other than English that has gone on to be a worldwide bestseller. It is easy to see why Rovelli's book became such a breakout success. It is very short, made up of seven essays stitched together and has spread the word about physics to a wide audience.

There seem to be three reasons for the book's success. While Rovelli does not do much in terms of bringing in the stories of those involved in science, he gives the book a human touch that is reminiscent of works like Erasmus Darwin's *The Botanic Garden* (see page 138). Rovelli makes use of a highly poetic prose style and imbues the text with his own

Lee Smolin
THE TROUBLE WITH PHYSICS, HOUGHTON MIFFLIN COMPANY, 2006

Smolin's book takes on the underlying problems of the influential string theory in physics.

Sean Carroll
FROM ETERNITY TO HERE, ONEWORLD PUBLICATIONS, 2015

First published in 2010, Sean Carroll's first popular title gives a lucid exploration of the physicist's view of time.

Carlo Rovelli
SETTE BREVI LEZIONI DI FISICA (SEVEN BRIEF LESSONS IN PHYSICS)

The original Italian edition (above) published in 2014 by Adelphi Edizione, alongside the elegant English translation (right) published by Allen Lane in 2015.

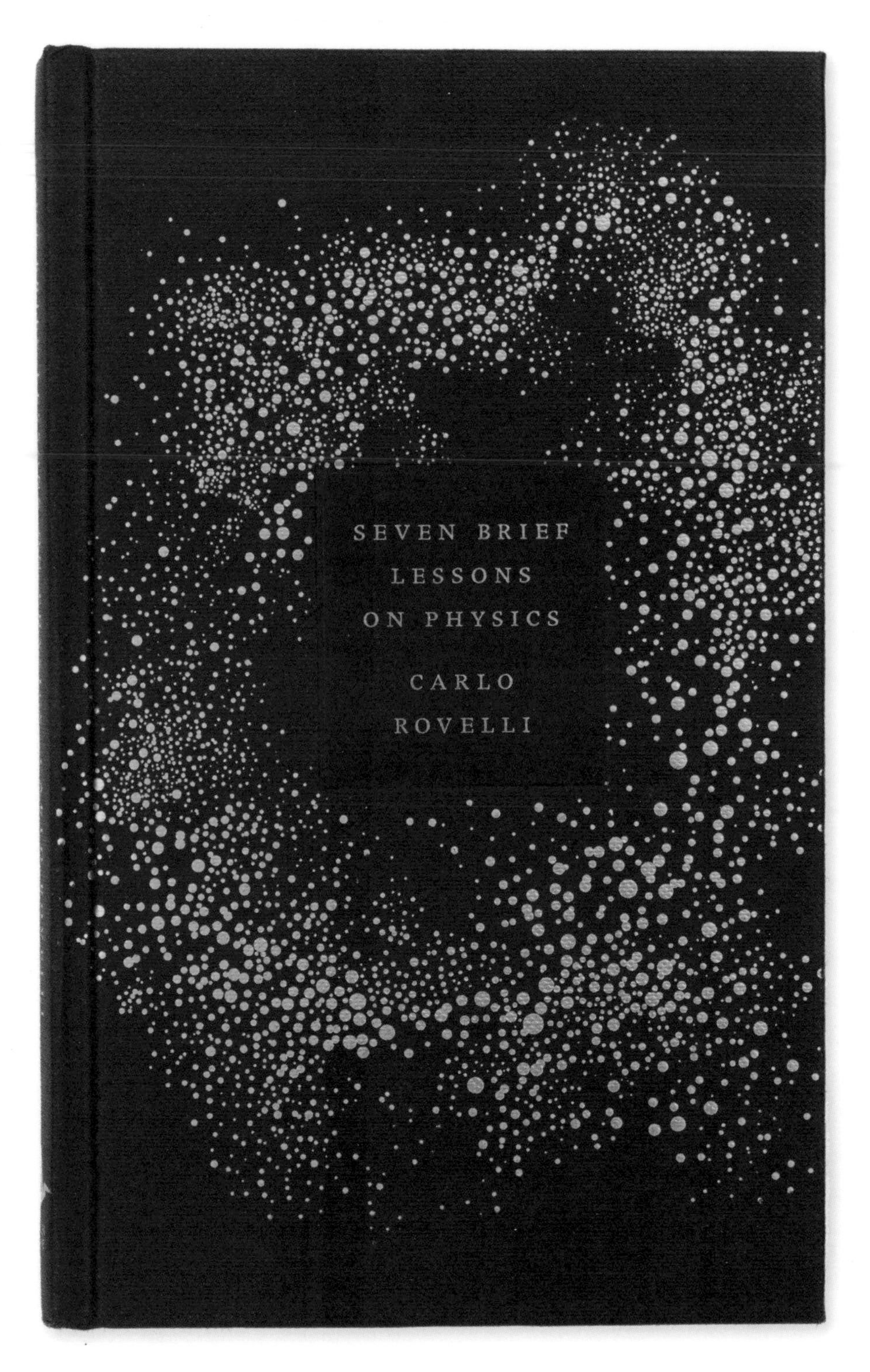

personality. Secondly, the shortness of the book itself can be seen as a benefit for those who find science titles hard going. And finally, the presentation of the book is expensive and sophisticated.

Like Carroll and Smolin, Rovelli knows his topic well. An active physicist, he works in one of the most dramatic areas of the field, the attempt to develop a theory of quantum gravity. Since Einstein, the two main aspects of physics have been quantum physics (the physics of the very small, which determines how almost everything we directly experience works) and the general theory of relativity (the physics behind gravity). Although gravity influences everything, it is such a relatively weak force (billions of billions times weaker than electromagnetism) that its main effect is on big things, such as stars and galaxies.

Although both theories are very successful, they are not compatible. They cannot be brought together. This is because gravity is not quantised – it doesn't deal with the universe in terms of small discrete chunks the way quantum theory does. So ever since the 1930s, attempts have been made to come up with a theory of gravity that is quantised and can be unified with the rest of quantum physics. The leading contender for this is string theory which, as we have seen, is still worked on by many physicists. But as Smolin showed, string theory may never provide a useful scientific structure. However, there are alternative theories.

The leading competitor is currently loop quantum gravity – and it is this field that Rovelli works in. In *Sette brevi lezioni*, Rovelli gives a series of sketches of key developments in, for example, cosmology and quantum gravity, but he gives a distorted view of the importance of loop quantum gravity. This theory may indeed come to supplant string theory, but as yet it isn't as well supported. The benefit of this kind of book is likely to be that some readers will be inspired to get hold of more in-depth titles and begin to appreciate science more.

Getting the topic right

Making more people aware of science and technology, given the impact of both on our lives, is crucial. This is why it's heartening that in Germany the numbers two and three slots for the most successful non-fiction books published between 2005 and 2015 went to titles by Eckart von Hirschhausen, a German medical doctor and television presenter. Most notable of the two was *Die Leber wächst mit ihren Aufgaben* (The Liver Grows with its Tasks), which addresses quirky scientific questions from the influence of acupuncture on cars (explaining the placebo effect) to the way that holes in cheese make you fat.

As we have seen, books like Hirschhausen's focusing on human science, along with those that cover physics and cosmology, have dominated the field of science book writing. Other areas of science have been relatively neglected. It's interesting to speculate why, for example, chemistry has had very little science writing dedicated to it.

A senior editor at one of the UK's leading science publishers, Oxford University Press, suggested that it can be hard for chemistry titles to appeal to the public. Jeremy Lewis commented in an interview with the author: 'I do think chemistry is under-represented in popular science. When you go to the bookstore, the small science section usually has an even smaller chemistry shelf. I think chemistry tends to just fly more under the radar when there are "sexier" breakthroughs in disciplines like physics and biology – topics such as quantum theory and gene editing.' According to research undertaken in 2015 for the Royal Society of Chemistry report *Public Attitudes to Chemistry*, 'People struggle to imagine how chemistry affects their everyday lives and regard chemists as lacking in agency: they do not recognize how chemists are involved in the end product of their own work.'

It seems that unless a popular science book is particularly well written, it needs to either appeal to personal interest (for example, health or psychology) or to be about something fundamental and dramatic, as is usually the case with physics and cosmology. It might seem that maths doesn't fit this general rule, but even this relatively inaccessible topic can be popular when it mixes stories of the peculiarities of mathematicians with the fundamental oddities of mathematics. The personal element is obviously there, for example, in *Fermat's Last Theorem*, in its description of Andrew Wiles's work. It was also important in the success of my own bestselling maths book from 2003, *A Brief History of Infinity*, which has a topic – infinity – that naturally engages curiosity, but also focuses on the mathematicians involved in developing our ideas of infinity.

Immortal lives

However, it is biology (mostly human) and physics that continue to dominate more recent influential titles. A significant example is Rebecca Skloot's 2010 *The Immortal Life of Henrietta Lacks*. Skloot is an American science writer whose medical-themed titles make use of strong storytelling to bring science to the public. The intriguing-sounding title refers to a cervical cancer patient, Henrietta Lacks, who died in 1951. Cells taken from Lacks' tumour were used to set up an important medical research tool known as an immortal cell line.

Usually, our cells cannot split and form new cells indefinitely, as they have built-in controls that track the number of splits that have occurred using a system called telomeres, which operate rather like a reel of tickets: the telomeres shorten as the cells divide and eventually kill their ability to do so. But in some cancer cells this restriction is permanently removed. Lacks' cells were the first such immortal cell line to be created, making them hugely important in the history of cellular medicine. The cells, known as HeLa, have been used to research both cancer and AIDS and are still going strong. Over 20 tonnes of Lacks' cells have now been grown.

What made Skloot's book (which has sold well over a million copies) so popular was the human interest. She covered Lacks' life, and the shock to her family, who did not discover the existence of the HeLa cell line until the 1970s. This did not prevent Skloot from also exploring the life and work of the scientists involved and the importance of the work for medical science, but the book emphasises once more the benefit of bringing humanity into the kind of science book that has dominated the field in the last 100 years – a book that helps the general public discover and engage with science.

Rebecca Skloot

THE IMMORTAL LIFE OF HENRIETTA LACKS

In 1951, a young black woman died of cancer.
Her death changed the history of medicine.

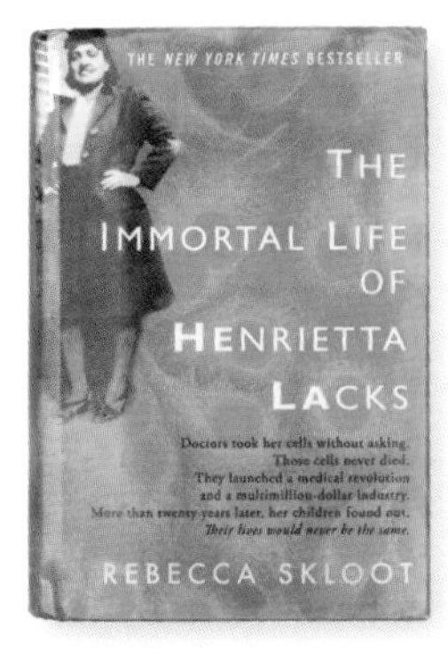

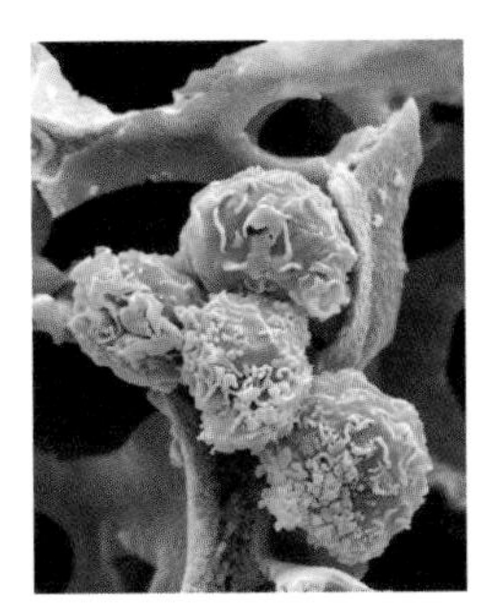

Rebecca Skloot
THE IMMORTAL LIFE OF HENRIETTA LACKS, 2010

An Australian first edition published by Picador (left) and an American edition by Crown Publishers above a scanning electron micrograph of HeLa cancer cells derived from the sample taken from Lacks in 1951.

A small example of the significance of the HeLa line can be seen in Professor Mark Pallen's engaging 2018 book, *The Last Days of Smallpox*, which tells the story of the eradication of the smallpox disease and the circumstances of an outbreak in Birmingham, England, after its eradication. The book notes that the virus samples taken from the carrier in this outbreak showed 'unusual behaviour when grown on layers of cultured human cells known as HeLa cells'. Henrietta Lacks' story continued to have resonance in a disease outbreak on a different continent.

Origin stories

As we have seen, science writing has tended to be dominated by English-language titles, which end up translated into other languages – English is now the international language of science, with the vast majority of important scientific papers published in English. Only occasionally does a science book published first in another language make it into English. Perhaps the most unusual example of this is *Ḳitsur Toldot Ha-enoshut* (A Brief History of Mankind, titled *Sapiens* in English), published in Hebrew in 2011 by Yuval Noah Harari.

Harari is anything but a typical popular science author – as a historian he has no science background. His field comes through more strongly in *Sapiens* than in his follow-up books, which stray even further from his area of expertise. In *Sapiens*, he effectively gives a history of the entirety of human existence, but with a genetic context that makes it clear this is intended to be popular science. Harari's follow-up title, *The History of Tomorrow* (published in English as *Homo Deus*) attempts futurology, the always difficult and rarely consistently scientific attempt to predict the future. These books have been international successes, but in the history of science writing they are more significant for their unusual progression from a second language into English than for their content, as the science in these books is widely considered weak by professionals in the field.

A useful contrast, delivering high-quality science, is English biologist Nick Lane's impressive title on the origins of life, *The Vital Question*, published in 2015. What Lane does so well here is to make it clear just how complex the cellular machinery is inside organisms with complex cells (like us). Perhaps even more impressively, Lane takes on the biggest biological question of them all: how life started.

Until recently it had been thought that life started in a primeval soup of organic material, perhaps spurred on by the energy of lightning strikes. However, Lane shows how this approach is incompatible with what we now know of the conditions on the early Earth and puts forward an alternative theory, based primarily on water and carbon dioxide. He also explores the way in which getting from simple cells to complex cells is as much a leap as getting started in the first place. Both steps appear to have happened only once, perhaps making life a much rarer phenomenon in the universe than is often thought.

If Lane gives us our best modern answer to where life came from, a 2018 title from English chemist Peter Atkins took things even further by trying to explain how the

THE MILLION COPY BESTSELLER
Yuval Noah Harari
Sapiens
A Brief
History of
Humankind
'I would recommend *Sapiens* to anyone who's interested
in the history and future of our species'
BILL GATES
VINTAGE

Noah Harari
KITSUR TOLDOT HA-ENOSHUT (A BRIEF HISTORY OF HUMANKIND)

First pubished in 2011 by DVIR, this is the 2013 updated edition alongside the 2014 English translation published by Vintage.

universe as a whole came into being. *Conjuring the Universe* is a slim volume, but takes on a mind-bending aspect of science as Atkins explores how it is possible to create a whole universe from scratch.

If we imagine the universe starting from nothing much at all, Atkins shows how many of the basic laws of physics emerge from what he describes as indolence and anarchy. The first of these is shorthand for the principle of least action, which says that the universe takes the easiest route in terms of use of energy and so on. Throw in the mathematics of symmetry devised by one of the world's greatest female mathematicians, Emmy Noether, and Atkins is able to deduce many of the basic principles of physics with a mellifluous writing style that makes it easy to go along with his arguments.

Atkins also goes further to suggest that some of the familiar constants of nature don't really exist. For example, the speed of light is one of the most important constants,

Nick Lane
THE VITAL QUESTION, PROFILE BOOKS, 2016

The paperback edition from Lane's book on the origins of life, first published in hardback in 2015.

Peter Atkins
CONJURING THE UNIVERSE, OXFORD UNIVERSITY PRESS, 2018

A surprisingly concise exploration of the origins of the universe.

with an exact value of 299,792,458 metres per second. (It's exact because the metre is defined from this value.) But Atkins points out that the apparent constant is simply a matter of our choice of units – if, as it's perfectly possible to do, we render such a constant in its most 'pure' unitless state, it effectively vanishes away. All in all, this book stirs up the reader's ideas of what the fundamentals of the universe involve.

Driving science forward

Although the majority of the authors included in this chapter are male, since the 1990s far more science books have been written by female authors, reflecting a lessening (though not eradication) of the gender bias we have seen throughout this book, both in terms of scientists and science writers. This whole topic became itself a subject of an important science book published in 2017, Angela Saini's *Inferior*.

A British science journalist who has largely worked in radio, Saini really manages to dig into the illogic and bias that has dogged science. *Inferior* explores a gender bias in science that far exceeds imbalances in many other professions and that is sometimes even defended by (older male) scientists as being the natural order. As we have seen, the idea that women's brains were different and less capable of scientific work was common before the twentieth century and was even shared by Charles Darwin (see page 190).

Saini demonstrates how social scientists in the twentieth century perpetuated ideas that mistakenly underlined a gender difference that for some traits is not there at all, and for others is so small that it has no significant impact on our abilities. It might seem that in the twenty-first century it should no longer be necessary to make these points. You only have to look at the society portrayed in a drama set in the 1960s, such as *Mad Men*, to see how much we've moved on. Yet, there are still unnecessary gender distinctions

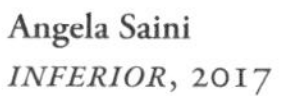

Angela Saini
INFERIOR, 2017

The intentionally controversial pink of Saini's UK first edition published by Fourth Estate, alongside the US first edition published by Beacon Press.

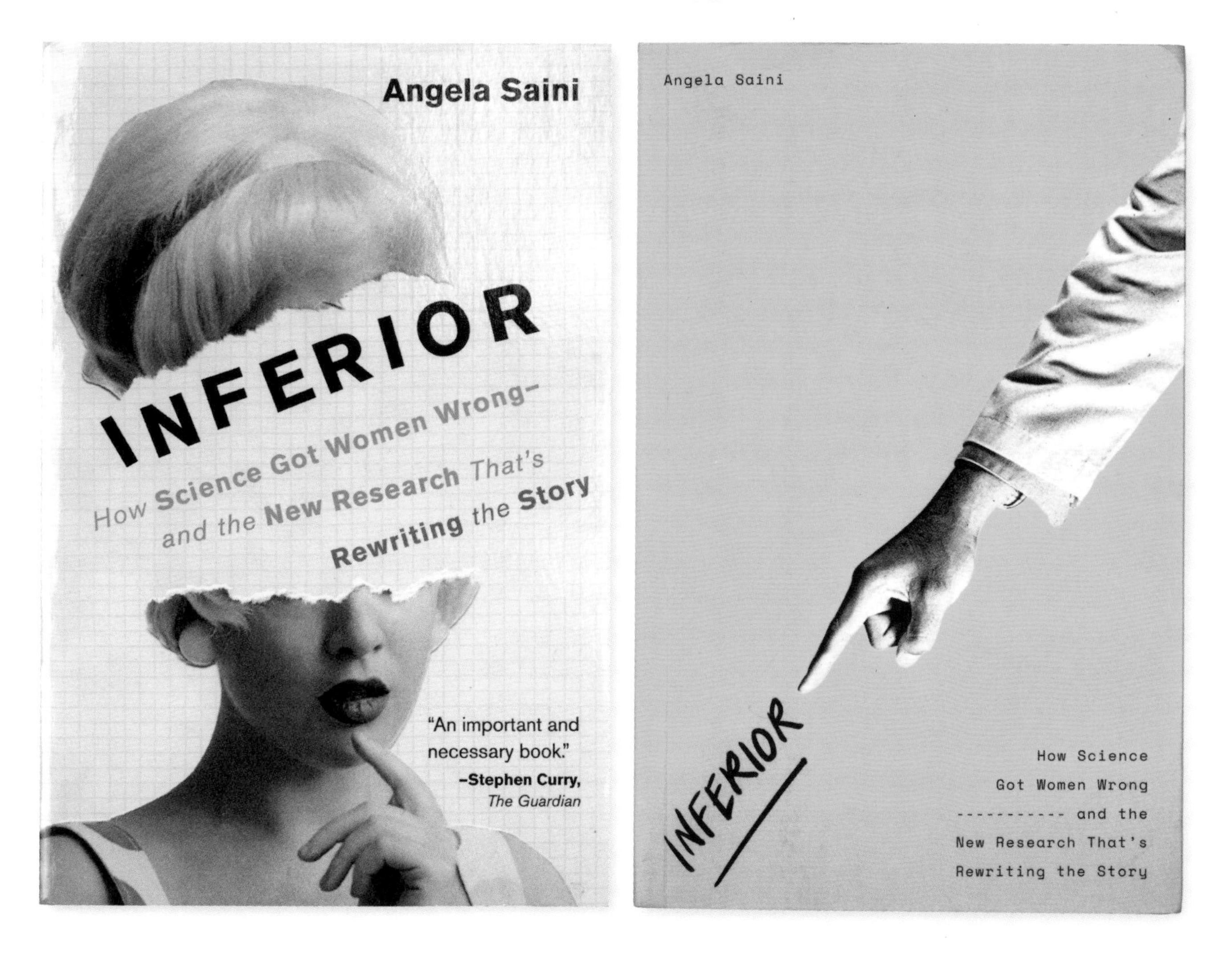

Stuart Firestein
IGNORANCE, OXFORD UNIVERSITY PRESS, 2012

A serious but approachable exploration of how science is undertaken and how we mistake its role.

being made, and in some areas of science, there remain strong advocates for theories that should have been left with the Victorians.

As well as challenging our gender stereotypes, modern science writing has also given us important titles that question the nature of science itself and how we undertake it, in a way that is far more realistic than some of the earlier titles. One very significant book in this respect was Stuart Firestein's *Ignorance*, published in 2012.

Subtitled 'How It Drives Science', *Ignorance* turns the common public understanding of science on its head, pointing out that it is not the facts that we know that are important, but rather the gaps in our knowledge, because it is these that drive science forward. Firestein comments: 'Working scientists don't get bogged down in the factual swamp because they don't care all that much for facts. It's not that they discount or ignore them, but rather that they don't see them as an end in themselves. They don't stop at the facts; they begin there, right beyond the facts, where the facts run out.'

This distinction between our imagined ideas of the work of scientists and what actually happens in the lab and the theory office is brought ably into the spotlight by German physicist Sabine Hossenfelder in her 2018 title *Lost in Math* (the book was first written in English and subsequently translated into her native German as *Das hässliche Universum: Warum unsere Suche nach Schönheit die Physik in die Sackgasse führt*).

Lost in Math is a powerful analysis of problems in the modern approach to physics. Until the twentieth century, experimenters made observations and undertook experiments, while theoreticians then looked for theories to explain these observations, which could then be tested against further experiments. Now, particularly in particle physics, it's more the case that physicists dream up whole rafts of theory supported only by mathematics, much of which can never be experimentally confirmed. What can be checked is often so expensive to work on that only a very small number of possibilities can be examined.

It's the maths that is in the driving seat, which surely is wrong. As Hossenfelder points out, string theory works best if the cosmological constant, a value that numerically reflects the rate of expansion or contraction of the universe, is negative. Unfortunately, the constant is actually positive, but most string theorists spend their time working with a negative cosmological constant. It makes for more beautiful mathematics – but has nothing to do with our universe.

Hossenfelder repeatedly comes back to two measures used to test theories: beauty, which is a subjective phenomenon, and naturalness, which appears more scientific as it involves numbers, but relies on a bizarre confidence that values in nature that are dimensionless (for example, ratios of masses) should be Goldilocks-like in not being too big or too small, but should be located around the value of 1. The physicists she interviews for the book (nearly all of them male), often cling onto these measures without being able to justify them, other than saying that everyone else likes them too.

Like Lee Smolin (see page 247), Hossenfelder shows that clinging to theories past their sell-by date is not surprising, because physicists are people too. If you've spent most of your career on a theory, you don't give it up easily. And if hundreds of other people are working on that theory, surely it must have some substance behind it? Hossenfelder points out that in a period of about a year after the Large Hadron Collider at CERN

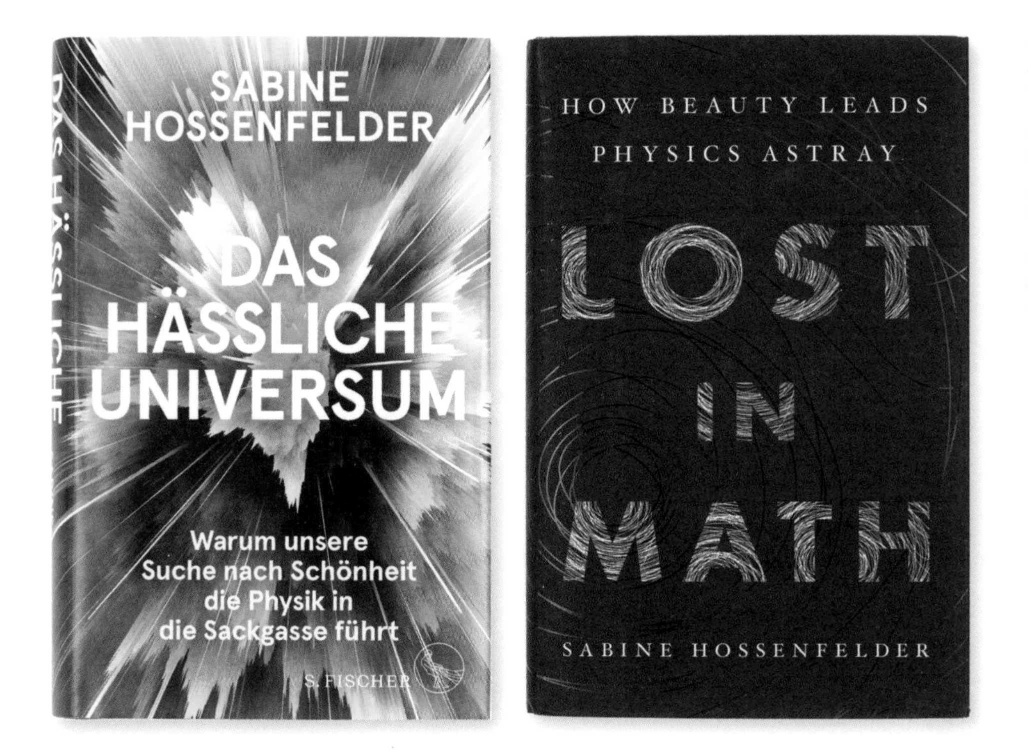

Sabine Hossenfelder
LOST IN MATH (DAS HÄSSLICHE UNIVERSUM), 2018

This title from German physicist Hossenfelder was first published in English by Basic Books before the author's German translation later the same year, published by S. Fischer.

produced data that looked interesting but turned out to be a statistical fluctuation, 500 papers were published exploring this non-event theoretically, many published in top journals.

Lost in Math is a science book for a new age where readers are less bound by authority. This is an approach that has always benefited science. The motto of the UK's Royal Society is *Nullius in verba*, roughly translating as 'Take no one's word for it'. In practice, the approach taken in science is closer to the Russian proverb quoted by US President Ronald Reagan in the 1980s, 'Trust but verify'. For a long time, the general public was spoon-fed the latest theories without any attempt to qualify uncertainty. But now, we are getting more science writing that helps the reader to question theories to gain a deeper understanding.

This is not a matter of being anti-science. The best such questioning comes from scientists themselves. But it gives the reader a better, more sophisticated picture of what science is all about.

There has never been more emphasis on the importance of public engagement. We need both to encourage a deeper interest in science and to counter anti-scientific views that can go hand in hand with some types of politics. Getting the public interested in science both helps to recruit new scientists of the future and spreads an understanding of why an area of scientific research deserves funding. The best science books continue to do this, but are now able to give us a deeper, more realistic understanding of science. And surely that is a good thing.

150 GREAT SCIENCE BOOKS

Works discussed in this book ordered by date of first publication:

1. Edwin Smith papyrus (1600 BCE)
2. Hippocrates *Hippocratic Corpus* (fifth/fourth century BCE)
3. Aristotle *History of Animals* (fourth century BCE)
4. Aristotle *Physics* (fourth century BCE)
5. Archimedes *The Sand-Reckoner* (third century BCE)
6. Euclid *Elements* (ca. 290 BCE)
7. Anon *Nine Chapters on the Mathematical Art* (ca. 200 BCE)
8. Titus Lucretius Carus *De Rerum Natura* (first century BCE)
9. Ptolemy *The Almagest* (ca. 150 CE)
10. Diophantus *Arithmetica* (third century CE)
11. Brahmagupta *Brāhmasphuṭasiddhānta* (628 CE)
12. Abū Ja'far Muḥammad ibn Mūsā al-Khwārizmī *Al-kitāb al-mukhtaṣar fī ḥisāb al-ğabr wa'l-muqābala* (ca. 820 CE)
13. Ḥunayn ibn Isḥāq *Al-Ashr Makalat Fi'l'ayn* (ninth century)
14. Abū 'Alī al-Ḥasan ibn al-Haytham *Kitāb al-Manāẓir* (tenth/eleventh century)
15. Abū 'Alī al-Ḥasan ibn al-Haytham *Mizan al-Hikmah* (tenth/eleventh century)
16. ibn Sīnā *al-Qānūn fī al-Ṭibb* (eleventh century)
17. Bhaskara *Siddhānta Śiromaṇī* (twelfth century)
18. Fibonacci *Liber Abaci* (1202)
19. Roger Bacon *De Mirabile Protestate Artis et Naturae* (1250)
20. Roger Bacon *Opus Majus* (1266–1267)
21. Leonardo da Vinci *Notebooks* (1452–1519)
22. Peter Bienewitz *Astronomicum Caesareum* (1540)
23. Nicolaus Copernicus *De Revolutionibus Orbium Coelestium* (1543)
24. Andreas Vesalius *De Humani Corporis Fabrica* (1543)
25. Sebastian Münster *Cosmographia* (1544)
26. Gerolamo Cardano *Ars Magna* (1545)
27. Georgius Agricola *De Re Metallica* (1556)
28. Rafael Bombelli *Algebra* (1572)
29. William Gilbert *De Magnete* (1600)
30. Johannes Kepler *De Stella Nova* (1606)
31. Johannes Kepler *Astronomia Nova* (1609)
32. Galileo Galilei *Sidereus Nuncius* (1610)
33. Johannes Kepler *Harmonices Mundi* (1619)

34. Francis Bacon *Novum Organum Scientiarum* (1620)
35. Johannes Kepler *Tabulae Rudolphinae* (1627)
36. William Harvey *Exercitatio Anatomica de Motu Cordis* (1628)
37. Galileo Galilei *Dialogo Sopra i due Massimi Sistemi del Mondo* (1632)
38. René Descartes *Discours de la Méthode (La Géométrie)* (1637)
39. Galileo Galilei *Discorsi e Dimostrazioni Matematiche Intorno a Due Nuove Scienze* (1638)
40. Nicolas Culpeper *The English Physitian (Complete Herbal)* (1652)
41. Robert Boyle *New Experiments Physico-Mechanical* (1660)
42. Robert Boyle *The Sceptical Chymist* (1661)
43. Gerolamo Cardano *Liber de Ludo Aleae* (1663)
44. Robert Hooke *Micrographia* (1665)
45. Otto von Guericke *Experimenta Nova* (1672)
46. Alain Manesson Mallet *Description de L'Univers* (1683)
47. Isaac Newton *Philosophiae Naturalis Principia Mathematica* (1687)
48. Isaac Newton *Opticks* (1704)
49. Carl von Linné *Systema Naturae* (1735)
50. Émilie du Châtelet *Institutions de Physique* (1740)
51. Comte de Buffon *Histoire Naturelle* (1749–1804)
52. Carl von Linné *Species plantarum* (1753)
53. Leonhard Euler *Lettres à une Princesse d'Allemagne* (1768)
54. Antoine Lavoisier *Traité Élémentaire de Chimie* (1789)
55. Erasmus Darwin *The Botanic Garden* (1791)
56. Erasmus Darwin *Zoonomia; or the Laws of Organic Life* (1794)
57. Thomas Malthus *An Essay on the Principle of Population* (1798)
58. Jöns Jacob Berzelius *Läroboken i Kemien* (1808)
59. John Dalton *A New System of Chemical Philosophy* (1808)
60. Jean-Baptiste Lamarck *Philosophie Zoologique* (1809)
61. Georges Cuvier *Le Règne Animal* (1817)
62. Sadi Carnot *Réflexions sur la Puissance Motrice du Feu* (1824)
63. John James Audubon *Birds of America* (1827)
64. Charles Lyell *Principles of Geology* (1830–1833)
65. John Herschel *A Preliminary Discourse on the Study of Natural Philosophy* (1831)
66. Charles Babbage *On the Economy of Machinery and Manufacture* (1832)
67. Robert Chambers *Vestiges of the Natural History of Creation* (1844)
68. Alexander von Humboldt *Kosmos* (1845–1862)
69. George Boole *An Investigation of the Laws of Thought on Which are Founded the Mathematical Theories of Logic and Probabilities* (1854)
70. Henry Gray *Anatomy* (1858)
71. Charles Darwin *On the Origin of Species* (1859)
72. Michael Faraday *The Chemical History of the Candle* (1861)
73. Ignaz Semmelweis *Die Ätiologie, der Begriff und die Prophylaxis des Kindbettfiebers* (1861)

74. Gregor Mendel *Versuche über Pflanzenhybriden* (1866)
75. John Venn *The Logic of Chance* (1866)
76. John Tyndall *Sound: delivered in eight lectures* (1867)
77. John Tyndall *Heat: a mode of motion* (1868)
78. Antoinette Brown Blackwell *Studies in General Science* (1869)
79. James Clerk Maxwell *Theory of Heat* (1871)
80. Charles Darwin *The Descent of Man* (1871)
81. James Clerk Maxwell *Treatise on Electricity and Magnetism* (1873)
82. John Tyndall *Six Lectures on Light* (1873)
83. Antoinette Brown Blackwell *The Sexes throughout nature* (1875)
84. Jean-Henri Fabre *Souvenirs Entomologiques* (1879)
85. John Venn *Symbolic Logic* (1881)
86. Edwin Abbott *Flatland* (1884)
87. Charles Darwin *The Autobiography of Charles Darwin* (1887)
88. David Hilbert *Grundlagen der Geometrie* (1899)
89. Eadweard Muybridge *Animals in Motion* (1899)
90. Eadweard Muybridge *The Human Figure in Motion* (1901)
91. Ernst Haeckel *Kunstformen der Natur* (1904)
92. Marie Curie *Traité de Radioactivité* (1910)
93. Alfred North Whitehead and Bertrand Russell *Principia Mathematica* (1910–1913)
94. Alfred Wegener *Die Entstehung der Kontinente und Ozeane* (1915)
95. Albert Einstein *Über die Spezielle und die Allgemeine Relativitätstheorie, Gemeinverständlich* (1917)
96. Arthur Eddington *The Nature of the Physical World* (1928)
97. Karl Popper *Logik der Forschung* (1934)
98. Hans Zinsser *Rats, Lice and History* (1935)
99. Lancelot Hogben *Mathematics for the Million* (1937)
100. Linus Pauling *The Nature of the Chemical Bond* (1939)
101. Erwin Schrödinger *What is Life?* (1944)
102. Donald Hebb *The Organization of Behavior* (1949)
103. Konrad Lorenz *Er Redete mit dem Viehden Vogeln und den Fischen* (1949)
104. Thomas Kuhn *The Structure of Scientific Revolutions* (1962)
105. Rachel Carson *Silent Spring* (1962)
106. Richard Feynman *The Feynman Lectures on Physics* (1963)
107. Desmond Morris *The Naked Ape* (1967)
108. James Watson *The Double Helix* (1968)
109. Jacques Monod *Le Hasard et la Nécessité* (1970)
110. Alvin Toffler *Future Shock* (1970)
111. Gustav Eckstein *The Body Has a Head* (1970)
112. Jacob Bronowski *The Ascent of Man* (1973)
113. Anne Sayre *Rosalind Franklin and DNA* (1975)
114. Richard Dawkins *The Selfish Gene* (1976)
115. Desmond Morris *Manwatching: A Field Guide to Human Behaviour* (1978)

116. Douglas R. Hofstadter *Gödel, Escher, Bach* (1979)
117. James Lovelock *Gaia* (1979)
118. David Bohm *Wholeness and the Implicate Order* (1980)
119. Carl Sagan *Cosmos* (1980)
120. John Gribbin *In Search of Schrödinger's Cat* (1984)
121. Richard Feynman *QED: The Strange Theory of Light and Matter* (1985)
122. Ralph Leighton *Surely You're Joking, Mr. Feynman!* (1985)
123. Oliver Sacks *The Man Who Mistook His Wife for a Hat* (1985)
124. James Gleick *Chaos: Making a New Science* (1987)
125. Stephen Hawking *A Brief History of Time* (1988)
126. Werner Heisenberg *Encounters with Einstein* (1989)
127. David Attenborough *Life on Earth* (1992)
128. Dava Sobel *Longitude* (1995)
129. Simon Singh *Fermat's Last Theorem* (1997)
130. Richard Dawkins *Unweaving the Rainbow* (1998)
131. Bill Bryson *A Short History of Nearly Everything* (2003)
132. Brian Clegg *A Brief History of Infinity* (2003)
133. Maurice Wilkins *The Third Man of the Double Helix* (2003)
134. Armand Leroi *Mutants* (2004)
135. Lee Smolin *The Trouble with Physics* (2006)
136. Eckart von Hirschhausen *Die Leber wächst mit ihren Aufgaben* (2008)
137. Brian Cox and Jeff Forshaw *Why Does $E=mc^2$?* (2010)
138. Rebecca Skloot *The Immortal Life of Henrietta Lacks* (2010)
139. Sean Carroll *From Eternity to Here* (2011)
140. Yuval Noah Harari *Ḳitsur Toldot Ha-enoshut* ('Sapiens' in English) (2011)
141. Brian Cox and Jeff Forshaw *The Quantum Universe* (2012)
142. Stuart Firestein *Ignorance* (2012)
143. Carlo Rovelli *Sette Brevi Lezioni di Fisica* (2014)
144. Nick Lane *The Vital Question* (2015)
145. David Wootton *The Invention of Science* (2015)
146. Angela Saini *Inferior* (2017)
147. Neil deGrasse Tyson *Astrophysics for People in a Hurry* (2017)
148. Peter Atkins *Conjuring the Universe* (2018)
149. Sabine Hossenfelder *Lost in Math* (2018)
150. Mark Pallen *The Last Days of Smallpox* (2018)

150 GREAT SCIENCE BOOKS

Works discussed in this book ordered by author name:

1. Edwin Abbott *Flatland* (1884)
2. Georgius Agricola *De Re Metallica* (1556)
3. Abū ʿAlī al-Ḥasan ibn al-Haytham *Kitāb al-Manāẓir* (tenth/eleventh century CE)
4. Abū ʿAlī al-Ḥasan ibn al-Haytham *Mizan al-Hikmah* (tenth/eleventh century CE)
5. Abū Ja'far Muḥammad ibn Mūsā al-Khwārizmī *Al-kitāb al-mukhtaṣar fī ḥisāb al-ğabr wa'l-muqābala* (ca. 820 CE)
6. Anon *Nine Chapters on the Mathematical Art* (ca. 200 BCE)
7. Archimedes *The Sand-Reckoner* (third century BCE)
8. Aristotle *History of Animals* (fourth century BCE)
9. Aristotle *Physics* (fourth century BCE)
10. Peter Atkins *Conjuring the Universe* (2018)
11. David Attenborough *Life on Earth* (1992)
12. John James Audubon *Birds of America* (1827)
13. Charles Babbage *On the Economy of Machinery and Manufacture* (1832)
14. Roger Bacon *De Mirabile Protestate Artis et Naturae* (1250)
15. Roger Bacon *Opus Majus* (1266–1267)
16. Francis Bacon *Novum Organum Scientiarum* (1620)
17. Jöns Jacob Berzelius *Läroboken i Kemien* (1808)
18. Peter Bienewitz *Astronomicum Caesareum* (1540)
19. Bhaskara *Siddhānta Śiromaṇī* (twelfth century CE)
20. Antoinette Brown Blackwell *Studies in General Science* (1869)
21. Antoinette Brown Blackwell *The Sexes throughout nature* (1875)
22. David Bohm *Wholeness and the Implicate Order* (1980)
23. Rafael Bombelli *Algebra* (1572)
24. George Boole *An Investigation of the Laws of Thought on Which are Founded the Mathematical Theories of Logic and Probabilities* (1854)
25. Robert Boyle *New Experiments Physico-Mechanical* (1660)
26. Robert Boyle *The Sceptical Chymist* (1661)
27. Brahmagupta *Brāhmasphuṭasiddhānta* (628 CE)
28. Jacob Bronowski *The Ascent of Man* (1973)
29. Bill Bryson *A Short History of Nearly Everything* (2003)
30. Comte de Buffon *Histoire Naturelle* (1749–1804)
31. Gerolamo Cardano *Ars Magna* (1545)
32. Gerolamo Cardano *Liber de Ludo Aleae* (1663)

33. Sadi Carnot *Réflexions sur la Puissance Motrice du Feu* (1824)
34. Sean Carroll *From Eternity to Here* (2011)
35. Rachel Carson *Silent Spring* (1962)
36. Robert Chambers *Vestiges of the Natural History of Creation* (1844)
37. Émilie du Châtelet *Institutions de Physique* (1740)
38. Brian Clegg *A Brief History of Infinity* (2003)
39. Nicolaus Copernicus *De Revolutionibus Orbium Coelestium* (1543)
40. Brian Cox and Jeff Forshaw *Why Does $E=mc^2$?* (2010)
41. Brian Cox and Jeff Forshaw *The Quantum Universe* (2012)
42. Nicolas Culpeper *The English Physitian (Complete Herbal)* (1652)
43. Marie Curie *Traité de Radioactivité* (1910)
44. Georges Cuvier *Le Règne Animal* (1817)
45. John Dalton *A New System of Chemical Philosophy* (1808)
46. Charles Darwin *On the Origin of Species* (1859)
47. Charles Darwin *The Descent of Man* (1871)
48. Charles Darwin *The Autobiography of Charles Darwin* (1887)
49. Erasmus Darwin *The Botanic Garden* (1791)
50. Erasmus Darwin *Zoonomia; or the Laws of Organic Life* (1794)
51. Richard Dawkins *The Selfish Gene* (1976)
52. Richard Dawkins *Unweaving the Rainbow* (1998)
53. René Descartes *Discours de la Méthode (La Géométrie)* (1637)
54. Diophantus *Arithmetica* (third century CE)
55. Gustav Eckstein *The Body Has a Head* (1970)
56. Arthur Eddington *The Nature of the Physical World* (1928)
57. Albert Einstein *Über die Spezielle und die Allgemeine Relativitätstheorie, Gemeinverständlich* (1917)
58. Euclid *Elements* (ca. 290 BCE)
59. Leonhard Euler *Lettres à une Princesse d'Allemagne* (1768)
60. Jean-Henri Fabre *Souvenirs Entomologiques* (1879)
61. Michael Faraday *The Chemical History of the Candle* (1861)
62. Richard Feynman *The Feynman Lectures on Physics* (1963)
63. Richard Feynman *QED: The Strange Theory of Light and Matter* (1985)
64. Fibonacci *Liber Abaci* (1202)
65. Stuart Firestein *Ignorance* (2012)
66. Galileo Galilei *Sidereus Nuncius* (1610)
67. Galileo Galilei *Dialogo Sopra i due Massimi Sistemi del Mondo* (1632)
68. Galileo Galilei *Discorsi e Dimostrazioni Matematiche Intorno a Due Nuove Scienze* (1638)
69. William Gilbert *De Magnete* (1600)
70. James Gleick *Chaos: Making a New Science* (1987)
71. Henry Gray *Anatomy* (1858)
72. John Gribbin *In Search of Schrödinger's Cat* (1984)
73. Otto von Guericke *Experimenta Nova* (1672)
74. Ernst Haeckel *Kunstformen der Natur* (1904)

75. Yuval Noah Harari *Ḳitsur Toldot Ha-enoshut* ('Sapiens' in English) (2011)
76. William Harvey *Exercitatio Anatomica de Motu Cordis* (1628)
77. Stephen Hawking *A Brief History of Time* (1988)
78. Donald Hebb *The Organization of Behavior* (1949)
79. Werner Heisenberg *Encounters with Einstein* (1989)
80. John Herschel *A Preliminary Discourse on the Study of Natural Philosophy* (1831)
81. David Hilbert *Grundlagen der Geometrie* (1899)
82. Hippocrates *Hippocratic Corpus* (fifth/fourth century BCE)
83. Eckart von Hirschhausen *Die Leber wächst mit ihren Aufgaben* (2008)
84. Douglas R. Hofstadter *Gödel, Escher, Bach* (1979)
85. Lancelot Hogben *Mathematics for the Million* (1937)
86. Robert Hooke *Micrographia* (1665)
87. Sabine Hossenfelder *Lost in Math* (2018)
88. Alexander von Humboldt *Kosmos* (1845–1862)
89. Ḥunayn ibn Isḥāq *Al-Ashr Makalat Fi'l'ayn* (ninth century CE)
90. ibn Sīnā *al-Qānūn fī al-Ṭibb* (eleventh century CE)
91. Johannes Kepler *De Stella Nova* (1606)
92. Johannes Kepler *Astronomia Nova* (1609)
93. Johannes Kepler *Harmonices Mundi* (1619)
94. Johannes Kepler *Tabulae Rudolphinae* (1627)
95. Thomas Kuhn *The Structure of Scientific Revolutions* (1962)
96. Jean-Baptiste Lamarck *Philosophie Zoologique* (1809)
97. Nick Lane *The Vital Question* (2015)
98. Antoine Lavoisier *Traité Élémentaire de Chimie* (1789)
99. Ralph Leighton *Surely You're Joking, Mr. Feynman!* (1985)
100. Leonardo da Vinci *Notebooks* (1452–1519)
101. Armand Leroi *Mutants* (2004)
102. Carl von Linné *Systema Naturae* (1735)
103. Carl von Linné *Species plantarum* (1753)
104. Konrad Lorenz *Er Redete mit dem Viehden Vogeln und den Fischen* (1949)
105. James Lovelock *Gaia* (1979)
106. Titus Lucretius Carus *De Rerum Natura* (first century BCE)
107. Charles Lyell *Principles of Geology* (1830–1833)
108. Alain Manesson Mallet *Description de L'Univers* (1683)
109. Thomas Malthus *An Essay on the Principle of Population* (1798)
110. James Clerk Maxwell *Theory of Heat* (1871)
111. James Clerk Maxwell *Treatise on Electricity and Magnetism* (1873)
112. Gregor Mendel *Versuche über Pflanzenhybriden* (1866)
113. Jacques Monod *Le Hasard et la Nécessité* (1970)
114. Desmond Morris *The Naked Ape* (1967)
115. Desmond Morris *Manwatching: A Field Guide to Human Behaviour* (1978)
116. Sebastian Münster *Cosmographia* (1544)
117. Eadweard Muybridge *Animals in Motion* (1899)

118. Eadweard Muybridge *The Human Figure in Motion* (1901)
119. Isaac Newton *Philosophiae Naturalis Principia Mathematica* (1687)
120. Isaac Newton *Opticks* (1704)
121. Mark Pallen *The Last Days of Smallpox* (2018)
122. Linus Pauling *The Nature of the Chemical Bond* (1939)
123. Karl Popper *Logik der Forschung* (1934)
124. Ptolemy *The Almagest* (ca. 150 CE)
125. Carlo Rovelli *Sette Brevi Lezioni di Fisica* (2014)
126. Oliver Sacks *The Man Who Mistook His Wife for a Hat* (1985)
127. Carl Sagan *Cosmos* (1980)
128. Angela Saini *Inferior* (2017)
129. Anne Sayre *Rosalind Franklin and DNA* (1975)
130. Erwin Schrödinger *What is Life?* (1944)
131. Ignaz Semmelweis *Die Ätiologie, der Begriff und die Prophylaxis des Kindbettfiebers* (1861)
132. Rebecca Skloot *The Immortal Life of Henrietta Lacks* (2010)
133. Simon Singh *Fermat's Last Theorem* (1997)
134. Edwin Smith papyrus (1600 BCE)
135. Lee Smolin *The Trouble with Physics* (2006)
136. Dava Sobel *Longitude* (1995)
137. Alvin Toffler *Future Shock* (1970)
138. John Tyndall *Sound: delivered in eight lectures* (1867)
139. John Tyndall *Heat: a mode of motion* (1868)
140. John Tyndall *Six Lectures on Light* (1873)
141. Neil deGrasse Tyson *Astrophysics for People in a Hurry* (2017)
142. John Venn *The Logic of Chance* (1866)
143. John Venn *Symbolic Logic* (1881)
144. Andreas Vesalius *De Humani Corporis Fabrica* (1543)
145. James Watson *The Double Helix* (1968)
146. Alfred Wegener *Die Entstehung der Kontinente und Ozeane* (1915)
147. Alfred North Whitehead and Bertrand Russell *Principia Mathematica* (1910–1913)
148. Maurice Wilkins *The Third Man of the Double Helix* (2003)
149. David Wootton *The Invention of Science* (2015)
150. Hans Zinsser *Rats, Lice and History* (1935)

INDEX

PICTURE CREDITS

NB Picture caption headings list publication date of the edition photographed.

Alamy/ Artokoloro Quint Lox Limited: 15L; Chronicle: 150; The Granger Collection: 139R; The Natural History Museum: 148, 149T, 149B; Photo12: 216T; Science History Images: 30R; sjbooks: 209; The World History Archive: 102T

Bibliothèque nationale de France: 153, 163BR

The Bodleian Library, University of Oxford /Digby 235, f.270: 71; /Huntington 214, f.004-005: 8

Bridgeman Images/ Archives Charmet: 9L, 38L; Archives Larousse, Paris, France: 182; © British Library Board: 34, 35L, 60, 77, 85; British Library, London, UK: 12–13, 55B, 72; © Christie's Images: 33, 109; Costa: 7, 82, 83; De Agostini Picture Library: 84, /G. Dagli Orti: 11T; © Devonshire Collection, Chatsworth: 121L; Fisher Collection, Pittsburgh, PA, USA: 121R; Musee Conde, Chantilly, France: 51R; Natural History Museum, London, UK: 132; Orlicka Galerie, Rychnov nad Kneznou, Czech Republic: 100L; Pictures from History: 9R, 14; Private Collection: 14, 16, 54L; © S. Beaucourt/Novapix: 54R, 55T; © S. Bianchetti: 116; Universal History Archive/UIG: 64; With kind permission of the University of Edinburgh: 43TR; The University of St. Andrews, Scotland, UK: 102B; The Worshipful Company of Clockmakers' Collection, UK: 244

The British Library, Harley MS 3487: 40

CERN: 214T

Getty/ Alfred Eisenstaedt: 211; Apic: 70; Bettmann: 19T, 37B, 199, 239R; DEA/ A. DAGLI ORTI: 29, /ARCHIVIO J. LANGE: 28, /G. NIMATALLAH: 11B, /S. VANNINI: 27; Dominique BERRETTY: 222L; Florence Vandamm: 186R; Fratelli Alinari IDEA S.p.A.: 47, 81; Heritage Images: endpapers, 46, 56, 191; Historical Picture Archive: 57; Hulton Archive: 186M; Popperfoto:19B; Photo12: 143, 152R; Royal Geographical Society: 166R; Science & Society Picture Library: 63, 120, 142, 146, 151, 169R, 174; 198L; Susan Wood: 223R; Ted Spiegel: 38R; ullstein picture Dtl.:161; Universal History Archive: 128T, 201T; Universal Images Group: 37T; Wallace Kirkland: 220L

Reprinted by permission of HarperCollins Publishers ltd. © David Attenborough, 1991; Angela Saini, 2017; Simon Singh, 1997: Dava Sobel, 1995; Gustav Eckstein, 1969

Heritage Auctions, HA.com: 195T

Internet Archive/ Biblioteca de la Universidad de Sevilla: 163BL; Boston Public Library: 139L; Brandeis University Library: 137T; The Computer Museum Archive: 152L; Cornell University Archive: 171T, 186L; Duke University Libraries: 76, 171T; e rara: 90; Fisher – University of Toronto: 43BL, 43BR, 100R, 101, 196, 197; Francis A. Countway Library of

Medicine: 172, 173; John Carter Brown Library: 2, 92, 93, 94-95; Library of Congress: 190; Missouri Botanical Garden: 129, 157, 158, 159, 160; Mugar Memorial Library, Boston University: 175TL, 175TR; Osmania Library: 171BL, 171BR; Project Gutenberg: 187, 188, 189; Smithsonian Libraries: 49, 98, 123, 130–31, 144, 145, 162, 163T, 164R, 165, 167, 168; University of California Libraries: 177; University of Connecticut Libraries: 61; Wellcome Library: 133, 135, 136, 137BL, 137BR 175BL, 175BM, 175BR; West Virginia University Libraries: 112, 113 Yale University, Cushing/Whitney Medical Library: 164L

Jerusalem – The National Library of Israel, Ms. Yah. Ar. 384: 65T

Library of Congress: 86, 87, 104–105, 106, 107T, 114, 115B, 126, 127, 134, 179, 180, 181; /Geography and Map Division: 78–79, 88–89; /Rare Book and Special Collections Division: 17, 107B, 111, 117, 118, 119, 154, 155, 156

The Metropolitan Museum of Art/ Gift of Dr. Alfred E. Cohn, in honor of William M. Ivins Jr., 1953: 110; /The Elisha Whittelsey Collection, The Elisha Whittelsey Fund, 1951 by exchange: 115T; /Harris Brisbane Dick Fund, 1934: 108; /Rogers Fund, 1913: 58, 59L

With permission of the Ministry for cultural assets and activities / Biblioteca Nazionale Centrale Firenze: 68–69

Nature Picture Library/ John Sparks: 240 OSU Libraries Special Collections & Archives Research Center: 202; /Cornell University Archives: 201B

Peter Harrington, Rare Books: 205R

Rare Book & Manuscript Library, Columbia University in the City of New York: [Euclid's Elements, manuscript from ca. 1294], [Plimpton MS 165], 44; [Jiuzhang Suanshu, Liu Hui, third century], 59R; [Lilavati of Bhaskara, 1650], 66–67

Royal Belgian Institute of Natural Sciences: 25

Science Photo Library/ A. Barrington Brown, © Gonville & Caius College: 219R; Anthony Haworth: 226L; Bodleian Museum, Oxford University Images: 42R; Jean-Loup Charmet: 35R; King's College London: 169L; Paul D Stewart: cover, 91; Royal Astronomical Society: 42L, 43TL; Steve Gschmeissner: 251BR

Shutterstock/ David Brimm: 26

TBCL Modern First Editions: 198R

Wellcome Collection: 15R, 32, 50, 51L, 65BL, 65BR, 65BM, 96, 97, 99, 124, 125, 128B, 166L, 183, 184, 195B

Wikimedia Commons: 52–53; /Rama: 10

World Digital Library: 48

ABOUT THE AUTHOR

BRIAN CLEGG With MAs in Natural Sciences from Cambridge University and Operational Research from Lancaster University, Brian Clegg is a full-time science writer with over 30 titles published from *A Brief History of Infinity* (2003) to *The Quantum Age* (2015), and most recently *Professor Maxwell's Duplicitous Demon* (2019). He has also written for a range of publications from the *Wall Street Journal* to *BBC Science Focus* and *Playboy* magazines. He lives in Wiltshire, England where he edits the book review site Popular Science.

ACKNOWLEDGEMENTS

For Gillian, Rebecca and Chelsea

With thanks to Elizabeth Clinton, Tom Kitch, Claire Saunders, Kate Shanahan and all at Quarto involved in this book. Particularly thanks to Simon Singh and John Gribbin amongst the many science writers whose work has fascinated me.

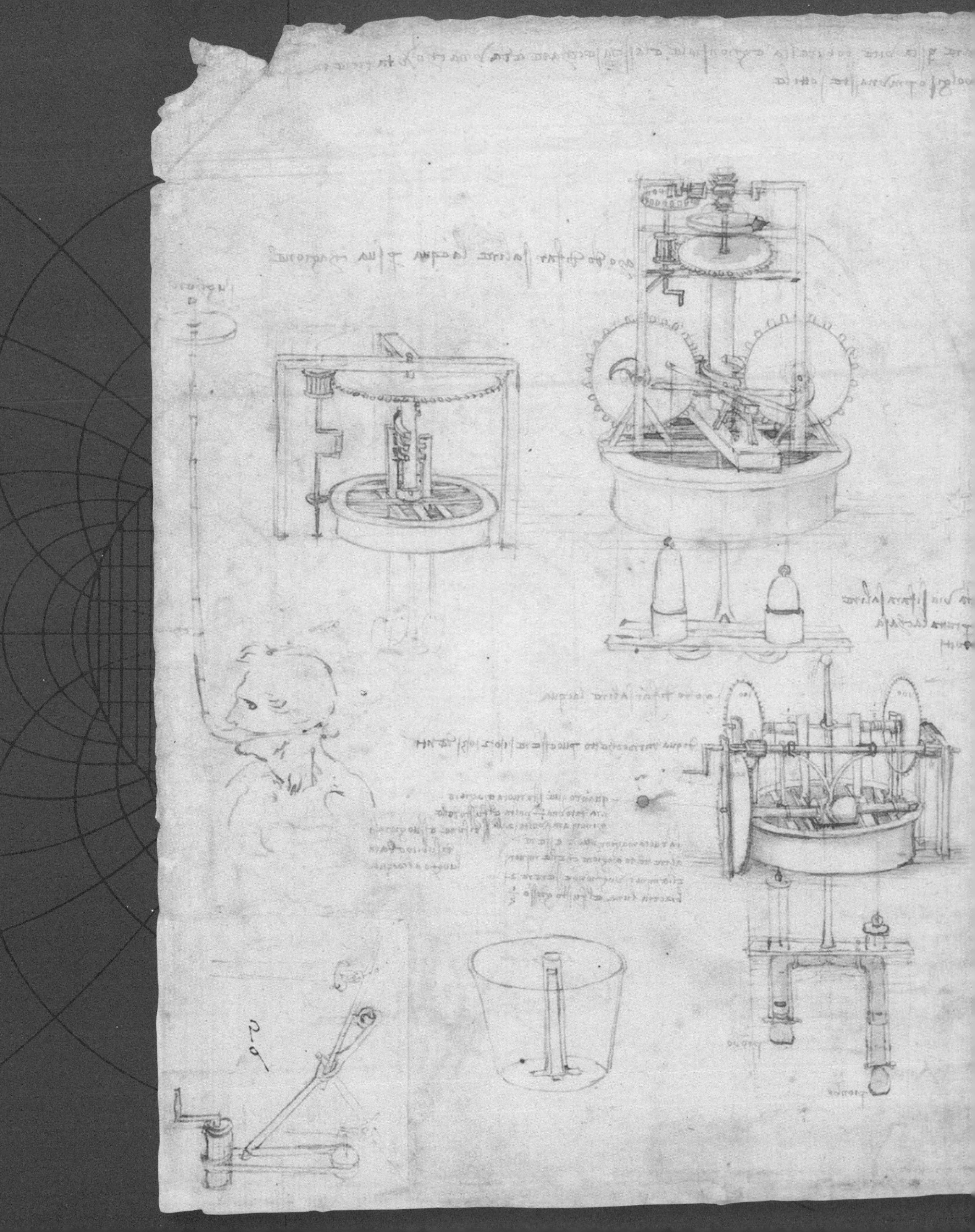